LINEAR PROGRAMMING
WITH MATLAB

MPS-SIAM Series on Optimization

This series is published jointly by the Mathematical Programming Society and the Society for Industrial and Applied Mathematics. It includes research monographs, books on applications, textbooks at all levels, and tutorials. Besides being of high scientific quality, books in the series must advance the understanding and practice of optimization. They must also be written clearly and at an appropriate level.

Series Volumes

Ferris, Michael C., Mangasarian, Olvi L., and Wright, Stephen J., *Linear Programming with MATLAB*

Attouch, Hedy, Buttazzo, Giuseppe, and Michaille, Gérard, *Variational Analysis in Sobolev and BV Spaces: Applications to PDEs and Optimization*

Wallace, Stein W. and Ziemba, William T., editors, *Applications of Stochastic Programming*

Grötschel, Martin, editor, *The Sharpest Cut: The Impact of Manfred Padberg and His Work*

Renegar, James, *A Mathematical View of Interior-Point Methods in Convex Optimization*

Ben-Tal, Aharon and Nemirovski, Arkadi, *Lectures on Modern Convex Optimization: Analysis, Algorithms, and Engineering Applications*

Conn, Andrew R., Gould, Nicholas I. M., and Toint, Phillippe L., *Trust-Region Methods*

LINEAR PROGRAMMING
WITH MATLAB

Michael C. Ferris
Olvi L. Mangasarian
Stephen J. Wright
University of Wisconsin–Madison
Madison, Wisconsin

Society for Industrial and Applied Mathematics
Philadelphia

Mathematical Programming Society
Philadelphia

Library of Congress Cataloging-in-Publication Data

Ferris, Michael C.
  Linear programming with MATLAB / Michael C. Ferris, Olvi L. Mangasarian, Stephen J. Wright.
     p. cm. — (MPS-SIAM series on optimization ; 7)
  Includes bibliographical references and index.
  ISBN 978-0-898716-43-6 (alk. paper)
  1. Linear programming—Data processing. 2. MATLAB. 3. Mathematical optimization. 4. Algebras, Linear. I. Mangasarian, Olvi L., 1934- II. Wright, Stephen J., 1960- III. Title.

QA402.5.F425 2007
519.7'2—dc22

2007061748

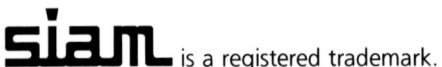 is a registered trademark.

 is a registered trademark.

To Jane, Claire, and Jill

Contents

Preface

This text has grown over many years from a set of class notes for an undergraduate linear programming course offered at the University of Wisconsin-Madison. Though targeted to Computer Science undergraduates, the course has attracted both undergraduates and beginning graduate students from many other departments, including Industrial and Systems Engineering, Statistics, and Mathematics. The course aims to provide a one-semester elementary introduction to linear programming formulations, algorithms, computations, and applications. Only basic knowledge of linear algebra and calculus is required.

One feature of our approach is the use of MATLAB codes to demonstrate the computational aspects of the course, from the elementary manipulations that form the building blocks of algorithms to full implementations of revised simplex and interior-point methods. (The latter are clearly *not* robust or efficient enough to solve larger practical problems, but they do illustrate the basic principles of the computational methods in question.) The MATLAB codes (and associated mex files) are distributed on the web site associated with the book: www.siam.org/books/mp07.

We have included a chapter on quadratic programs and complementarity problems, which are topics whose importance in a number of application areas appears to be growing by the day. The final chapter deals with approximation and classification problems, which are of interest to statisticians and others, showing how these problems can be formulated and solved as linear or quadratic programs. An earlier chapter deals with the topic of duality, which is of interest not only because of the insight it provides into the beautiful theory underlying linear programming but also because of its usefulness in formulating practical problems. (The dual of a problem may be easier to solve than the primal, or it might provide a bound on the optimal solution value or other useful information.)

A one-semester undergraduate class should include most of the chapters in the text. If time is pressing, some of the later chapters could be omitted in part or in their entirety. However, we believe that all topics covered are interesting and relevant to the intended student audience, and so we hope that most teachers can find a way to incorporate them into their curriculum.

We thank the students and colleagues who have given us feedback on the manuscript during its development, particularly our colleague Bob Meyer. We are also grateful to the referees of the manuscript who read it thoroughly and provided valuable suggestions for improvement, most of which we adopted. Finally, we thank our wives and families for their love and support over many years.

Madison, Wisconsin, USA
Spring 2007

Chapter 1

Introduction

Nothing happens in the universe that does not have a sense of either certain maximum or minimum.   L. Euler, Swiss Mathematician and Physicist, 1707–1783

Optimization is a fundamental tool for understanding nature, science, engineering, economics, and mathematics. Physical and chemical systems tend to a state that minimizes some measure of their energy. People try to operate man-made systems (for example, a chemical plant, a cancer treatment device, an investment portfolio, or a nation's economy) to optimize their performance in some sense. Consider the following examples:

1. Given a range of foods to choose from, what is the diet of lowest cost that meets an individual's nutritional requirements?

2. What is the most profitable schedule an airline can devise given a particular fleet of planes, a certain level of staffing, and expected demands on the various routes?

3. Where should a company locate its factories and warehouses so that the costs of transporting raw materials and finished products are minimized?

4. How should the equipment in an oil refinery be operated, so as to maximize rate of production while meeting given standards of quality?

5. What is the best treatment plan for a cancer patient, given the characteristics of the tumor and its proximity to vital organs?

Simple problems of this type can sometimes be solved by common sense, or by using tools from calculus. Others can be formulated as optimization problems, in which the goal is to select values that maximize or minimize a given *objective function*, subject to certain *constraints*. In the next section, we show how a practical problem can be formulated as a particular type of optimization problem known as a *linear program*.

1.1   An Example: The Professor's Dairy

1.1.1   The Setup

University professors sometimes engage in businesses to make a little extra cash. Professor Snape and his family run a business that produces and sells dairy products from the milk of the family cows, Daisy, Ermentrude, and Florence. Together, the three cows produce 22 gallons of milk each week, and Snape and his family turn the milk into ice cream and butter that they then sell at the Farmer's Market each Saturday morning.

The butter-making process requires 2 gallons of milk to produce one kilogram of butter, and 3 gallons of milk is required to make one gallon of ice cream. Professor Snape owns a huge refrigerator that can store practically unlimited amounts of butter, but his freezer can hold at most 6 gallons of ice cream.

Snape's family has at most 6 hours per week in total to spend on manufacturing their delicious products. One hour of work is needed to produce either 4 gallons of ice cream or one kilogram of butter. Any fraction of one hour is needed to produce the corresponding fraction of product.

Professor Snape's products have a great reputation, and he always sells everything he produces. He sets the prices to ensure a profit of $5 per gallon of ice cream and $4 per kilogram of butter. He would like to figure out how much ice cream and butter he should produce to maximize his profit.

1.1.2   Formulating the Problem and a Graphical Solution

The first step in formulating this problem is to identify the two variables, which are the quantities that we are able to vary. These are the number of gallons of ice cream, which we denote by x, and the number of kilograms of butter, which we denote by y. Next, we figure out how the objective function depends on these variables. We denote the objective (which in this case is the profit) by z, and note that it is simply $z = 5x + 4y$ dollars in this example.

Since we aim to maximize the production, it is generally in our interest to choose x and y as large as possible. However, the constraints on production mentioned above prevent us from making these variables *too* large. We now formulate the various constraints in the description above algebraically.

- The 6-gallon constraint on freezer capacity causes us to impose the constraint $x \leq 6$.

- The total amount of labor required to produce x gallons of ice cream and y kilograms of butter is $.25x + y$. Since the family can labor for a total of at most 6 hours during the week, we have the constraint $.25x + y \leq 6$.

- We look at the amount of milk needed by the production process. The total number of gallons of milk used is $3x + 2y$, and since there are 22 gallons of milk available, we have the constraint $3x + 2y \leq 22$.

- Finally, the problem must include the simple constraints $x \geq 0$, $y \geq 0$, because it does not make sense to produce negative amounts of ice cream or butter.

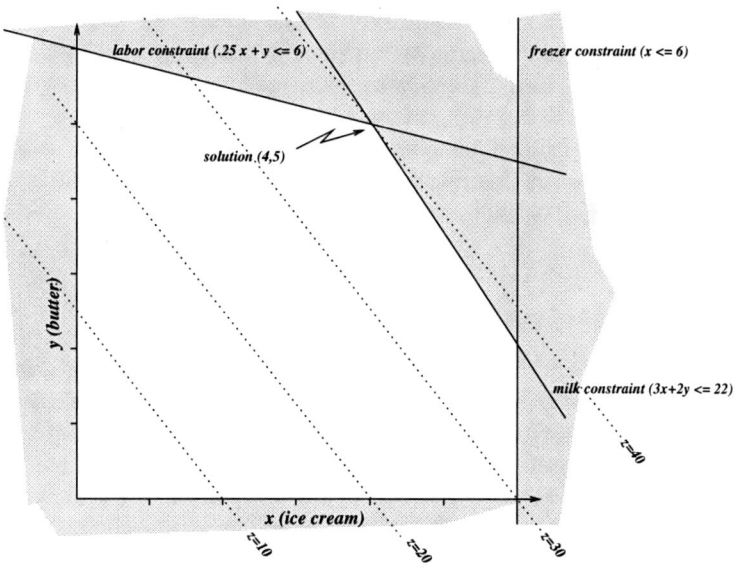

Figure 1.1. *The Professor's dairy: Constraints and objective.*

Summarizing, we can express the linear program mathematically as follows:

$$
\begin{array}{llrcrcl}
\max\limits_{x,y} & z = 5x + 4y \\
\text{subject to} & x & & & & \leq & 6, \\
& .25x & + & y & & \leq & 6, \\
& 3x & + & 2y & & \leq & 22, \\
& & & x, y & & \geq & 0.
\end{array}
\tag{1.1}
$$

Figure 1.1 illustrates this problem graphically, plotting the variable x along the horizontal axis and y along the vertical axis. Each constraint is represented by a line, the shaded side of the line representing the region of the (x, y) plane that fails to satisfy the constraint. For example, the constraint $3x + 2y \leq 22$ is represented by the line $3x + 2y = 22$ (obtained by replacing the inequality by an equality), with the "upper" side of the line shaded. In general, we can determine which side of the line satisfies the constraint and which does not by picking a point that does not lie on the line and determining whether or not the constraint is satisfied at this point. If so, then *all* points on this side of the line are *feasible*; if not, then all points on this side of the line are *infeasible*.

The set of points satisfying all five of the constraints is known as the *feasible region*. In this problem the feasible region is the five-sided polygonal region in the middle of the figure.

The linear programming problem is to the find a point in this feasible region that maximizes the objective $z = 5x + 4y$. As a step towards this goal, we plot in Figure 1.1 a dotted line representing the set of points at which $z = 20$. This line indicates feasible points such as $(x, y) = (0, 5)$ and $(x, y) = (2, 2.5)$ that yield a profit of \$20. Similarly, we plot the line $z = 5x + 4y = 30$—the set of points that achieves a profit of \$30. Note that this line (and all other lines of constant z) is parallel to the line $z = 20$. In fact, we can maximize profit over the feasible region by moving this line as far as possible to the right while keeping some overlap with the feasible region and keeping it parallel to the $z = 20$ line. It is not difficult to see that this process will lead us to a profit of $z = 40$ and that this line intersects the feasible region at the single point $(x, y) = (4, 5)$. Note that this point is a "corner point" of the feasible region, corresponding to the point at which two of the constraints—the limit of milk supply and the limit on labor supply—are satisfied as equalities.

1.1.3   Changing the Problem

The graphical representation of Figure 1.1 can be used to see how the solution changes when the data is changed in certain ways. An investigation of this type is known as *sensitivity analysis* and will be discussed in Chapter 6. We discuss two possible changes to the example problem here. A first time reader may skip this section without loss of continuity since it is meant primarily as an intuitive graphical introduction to duality and sensitivity.

First, we look at what happens if Professor Snape decides to increase the price of ice cream, while leaving the price of butter (and all the other problem data) the same. We ask the question, How much can we increase the price of ice cream without changing the solution $(4, 5)$? It is intuitively clear that if the profit on ice cream is *much* greater than on butter, it would make sense to make as much ice cream as possible subject to meeting the constraints, that is, 6 gallons. Hence, if the price of ice cream increases by more than a certain amount, the solution will move away from the point $(4, 5)$.

Suppose for instance that we increase the profit on ice cream to \$5.50, so that the objective function becomes $z = 5.5x + 4y$. If we plot the contours of this new objective (see Figure 1.2), we find that they are rotated slightly in the clockwise direction from the contours in Figure 1.1. It is clear that for a \$42 profit, $(4, 5)$ is still the optimum. However, if the profit on ice cream is increased further, the contours will eventually have exactly the same slope as the "milk" constraint, at which point every point on the line joining $(4, 5)$ to $(6, 2)$ will be a solution. What ice cream profit p will make the contours of the objective $z = px + 4y$ parallel to the line $3x + 2y = 22$? By matching the slopes of these two lines, we find that the operative value is $p = 6$. If the price of ice cream is slightly higher than 6, the point $(6, 2)$ will be the unique optimum.

Exercise 1-1-1. Plot a figure like Figures 1.1 and 1.2 for the case in which the objective is $z = 8x + 4y$, while the constraints remain the same. Verify from your figure that $(6, 2)$ is the optimum.

Returning to the original problem, we could ask a slightly different question. Suppose that Professor Snape's neighbor, Professor Crouch, has some excess milk and is offering to sell it to Snape for \$1 per gallon. Given that Snape still wants to maximize his profits, and given that his other constraints are still in place (labor and freezer capacity), should he buy any milk from Crouch and, if so, how much?

To answer this question, we note first that if Snape purchases c gallons, the milk constraint becomes $3x + 2y \le 22 + c$. Graphically, the boundary of this constraint shifts upward and to the right, as we see in Figure 1.3. Provided c is not too large, the contours of the objective will not be greatly affected by this change to the problem, and so the solution

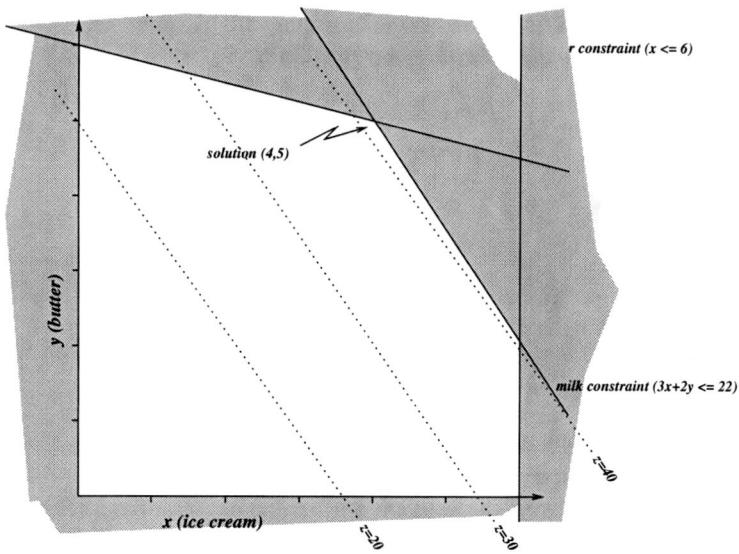

Figure 1.2. *The Professor's dairy: After increasing the profit on ice cream to $5.50, the objective contours rotate slightly clockwise, but the optimum is still (4, 5).*

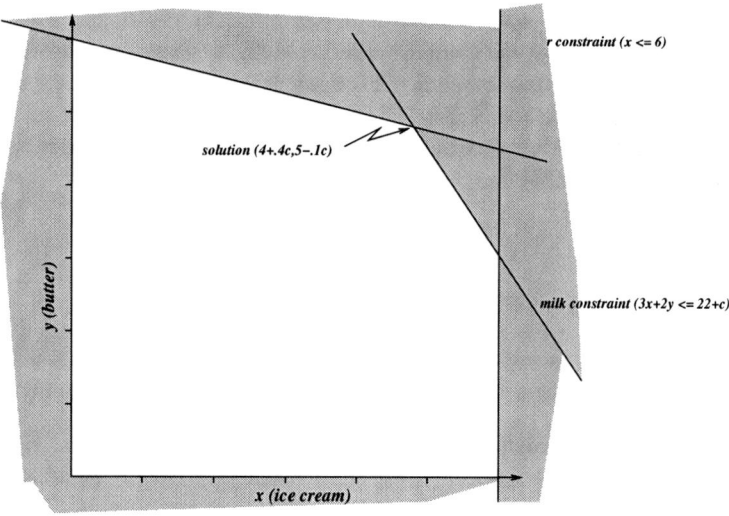

Figure 1.3. *The Professor's dairy: If the professor purchases c gallons from his neighbor, the milk constraint shifts upward and to the right.*

will still occur at the intersection of the labor constraint with the milk constraint, that is, at the point (x, y) that satisfies the following two equalities:

$$.25x + y = 6,$$
$$3x + 2y = 22 + c.$$

The solution is

$$(x, y) = (4 + .4c, 5 - .1c),$$

and the objective function value at this point (allowing for the \$1 per gallon purchase price of milk from Crouch) is

$$z = 5x + 4y - c = 5(4 + .4c) + 4(5 - .1c) - c = 40 + .6c.$$

It follows that it is definitely to Snape's advantage to buy *some* milk from Crouch, as he earns an extra 60 cents in profit for each gallon purchased.

However, if c is too large, the solution will no longer be at the intersection of the labor and milk constraints, and there is no further advantage to be gained. This happens when the milk constraint is shifted so far that it intersects with both the labor limit and the freezer limit at the point $(6, 4.5)$, which is true when $c = 5$. As c increases above this value, the solution stays at $(6, 4.5)$ while the profit actually starts to decline, as Snape is buying surplus milk unnecessarily without producing any more of either butter or ice cream.

Analysis of this type will be discussed further when we cover the subject of duality in Chapter 4.

The graphical analysis used in this section is sufficient for understanding problems with two variables. However, when extra variables are added (for example, if the professor decides to make cottage cheese and gourmet yogurt as well), it is hard to solve or analyze the problem using graphical techniques alone. This book describes computational techniques, motivated by the graphical analysis above, that can be used to solve problems with many variables and constraints. Solution of this problem using an algebraic approach, namely the simplex method, is given in Section 3.

1.1.4   Discussion

The example of this section has three important properties.

- Its variables (the amounts of ice cream and butter to produce) are *continuous* variables. They can take on any real value, subject to satisfying the bounds and constraints.

- All constraints and bounds involve *linear functions* of the variables. That is, each term of the sum is either a constant or else a constant multiple of one of the variables.

- The objective function—profit, in this case—is also a linear function of the variables.

Problems with these three essential properties are known as *linear programming* problems or *linear programs*. Most of our book is devoted to algorithms for solving this class of problems. Linear programming can be extended in various ways to give broader classes of optimization problems. For instance, if we allow the objective function to be a quadratic function of the variables (but still require the constraint to be linear and the variables to be continuous), we obtain *quadratic programming* problems, which we study in Chapter 7. If we allow both constraints and objective to be nonlinear functions (but still require continuous variables), the problem becomes a *nonlinear program*. If we restrict some of the variables to take on integer values, the problem becomes an *integer program*. We give several references for nonlinear and integer programming in the Notes and References at the end of this chapter.

Since 1947, when George B. Dantzig proposed his now classic simplex method for solving linear programs, the utilization of linear programming as a tool for modeling and computation has grown tremendously. Besides becoming a powerful tool in the area for which it was originally designed (economic planning), it has found a myriad of applications in such diverse areas as numerical analysis, approximation theory, pattern recognition, and machine learning. It has become a key tool in the important disciplines of operations research and management science.

1.2 Formulations

Throughout this book, we will refer to the following form of the linear program as the *standard form*:

$$
\begin{aligned}
\min_{x_1, x_2, \ldots, x_n} \quad & z = p_1 x_1 + \cdots + p_n x_n \\
\text{subject to} \quad & A_{11} x_1 + \cdots + A_{1n} x_n \geq b_1, \\
& \quad\vdots \qquad\quad \ddots \qquad \vdots \qquad\quad \vdots \\
& A_{m1} x_1 + \cdots + A_{mn} x_n \geq b_m, \\
& \qquad\qquad\qquad x_1, x_2, \ldots, x_n \geq 0.
\end{aligned}
\tag{1.2}
$$

By grouping the variables $x_1, x_2, \ldots, x_n$ into a vector x and constructing the following matrix and vectors from the problem data,

$$
A = \begin{bmatrix} A_{11} & \cdots & A_{1n} \\ \vdots & \ddots & \vdots \\ A_{m1} & \cdots & A_{mn} \end{bmatrix}, \qquad b = \begin{bmatrix} b_1 \\ \vdots \\ b_m \end{bmatrix}, \qquad p = \begin{bmatrix} p_1 \\ \vdots \\ p_n \end{bmatrix},
$$

we can restate the standard form compactly as follows:

$$
\begin{aligned}
\min_{x} \quad & z = p'x \\
\text{subject to} \quad & Ax \geq b, \ x \geq 0,
\end{aligned}
$$

where p' denotes the transpose of the column vector p, which is known as the *cost vector*.

Every linear program can be put into this standard form. We show in Chapter 3 how problems with equality constraints, free variables, and so on can be reformulated as standard-form problems. Problem (1.1) of the previous section can be expressed in standard form by setting x to be the vector made up of the two scalars x and y, while

$$
A = -\begin{bmatrix} 1 & 0 \\ .25 & 1 \\ 3 & 2 \end{bmatrix}, \qquad b = -\begin{bmatrix} 6 \\ 6 \\ 22 \end{bmatrix}, \qquad p = -\begin{bmatrix} 5 \\ 4 \end{bmatrix}.
$$

To perform this conversion, we changed "$\leq$" inequality constraints into "$\geq$" inequalities by simply multiplying both sides by -1. We also noted that maximization of a function (which we do in (1.1)) is equivalent to minimization of the negation of this function, which is why we have negative entries in p above.

In Chapter 5 we introduce another formulation in which all the general constraints are assumed to be equality constraints. This is known as the *canonical form* and is written as follows:

$$\begin{array}{ll} \min & z = p'x \\ \text{subject to} & \mathcal{A}x = b, \quad x \geq 0. \end{array}$$

As with the standard form, any linear program can be put into this form by appropriate transformations of the constraints and variables. We could express our example (1.1) in canonical form by first replacing (x, y) by (x_1, x_2) in (1.1) and then introducing three slack variables x_3, x_4, and x_5 to represent the amount by which the right-hand sides exceed the left-hand sides of the three constraints. We then obtain the following formulation:

$$\begin{array}{llllllll}
\min\limits_{x} & z = -5x_1 - 4x_2 \\
\text{subject to} & x_1 & & & + & x_3 & & & & = & 6, \\
& .25x_1 & + & x_2 & & & + & x_4 & & = & 6, \\
& 3x_1 & + & 2x_2 & & & & & + & x_5 & = & 22, \\
& & & & & x_1, x_2, x_3, x_4, x_5 & & & \geq & 0.
\end{array}$$

We can verify that the problem is in canonical form by setting

$$\mathcal{A} = \begin{bmatrix} 1 & 0 & 1 & 0 & 0 \\ .25 & 1 & 0 & 1 & 0 \\ 3 & 2 & 0 & 0 & 1 \end{bmatrix}, \quad b = \begin{bmatrix} 6 \\ 6 \\ 22 \end{bmatrix}, \quad p = -\begin{bmatrix} 5 \\ 4 \\ 0 \\ 0 \\ 0 \end{bmatrix}, \quad x = \begin{bmatrix} x_1 \\ x_2 \\ x_3 \\ x_4 \\ x_5 \end{bmatrix}.$$

1.3 Applications

In this section, we discuss several other practical problems that can be formulated as linear programs.

1.3.1 The Diet Problem

In an early application, linear programming was used to determine the daily diet for a person. From among a large number of possible foods, a diet was determined that achieved all the nutritional requirements of the individual while minimizing total cost.

To formulate as a linear program, we suppose that the n possible foods are indexed by $j = 1, 2, \ldots, n$ and that the m nutritional categories are indexed by $i = 1, 2, \ldots, m$. We let x_j be the amount of food j to be included in the diet (measured in number of servings), and denote by p_j the cost of one serving of food j. We let b_i denote the minimum daily requirement of nutrient i and A_{ij} be the amount of nutrient i contained in one serving of food j. By assembling this data into matrices and vectors in the usual way, we find that the linear program to determine the optimal diet can be formulated as follows:

$$\min_{x} \quad z = p'x$$
$$\text{subject to} \quad Ax \geq b, \quad x \geq 0.$$

The bounds $x \geq 0$ indicate that only nonnegative amounts of each food will be considered, while the "$\geq$" inequality constraints require the diet to meet or exceed the nutritional requirements in each category $i = 1, 2, \ldots, m$. If we wish to place an upper limit of d_j on the number of servings of food j to be included in the diet (to ensure that the diet does not become too heavy on any one particular food), we could add the constraints $x_j \leq d_j$, $j = 1, 2, \ldots, n$, to the model.

1.3.2 Linear Surface Fitting

Suppose that we have a set of observations $(A_i., b_i)$, $i = 1, 2, \ldots, m$, where each $A_i.$ is a (row) vector with n real elements, and each b_i is a single real number. We would like to find a vector $x \in \mathbf{R}^n$ and a constant γ such that

$$A_i.x + \gamma \approx b_i \qquad \text{for each } i = 1, 2, \ldots, m.$$

The elements of the vector x can be thought of as "weights" that are applied to the components of $A_i.$ to yield a prediction of each scalar b_i. For example, m could be the number of people in a population under study, and the components of each $A_i.$ could represent the income of person i, the number of years they completed in school, the value of their house, their number of dependent children, and so on. Each b_i could represent the amount of federal income tax they pay.

To find the "best" pair (x, γ), we need to measure the misfit between $A_i.x + \gamma$ and b_i over all the i. One possible technique is to sum the absolute values of all the mismatches, that is,

$$\sum_{i=1}^{m} |A_i.x + \gamma - b_i|.$$

We can formulate a linear program to find the (x, γ) that minimizes this measure. First, define the matrix A and the vector b by

$$A = \begin{bmatrix} A_1. \\ A_2. \\ \vdots \\ A_m. \end{bmatrix}, \qquad b = \begin{bmatrix} b_1 \\ b_2 \\ \vdots \\ b_m \end{bmatrix}.$$

Next, write the linear program as follows:

$$\min_{x, \gamma, y} \quad z = e'y$$
$$\text{subject to} \quad -y \leq Ax + \gamma e - b \leq y.$$

In this formulation, $e = (1, 1, \ldots, 1) \in \mathbf{R}^m$, so that the objective is the sum of the elements of y. The constraints ensure that each y_i is no smaller than the absolute value $|A_i.x + \gamma - b_i|$, while the fact that we are *minimizing* the sum of y_i's ensures that each y_i is chosen no larger than it really needs to be. Hence, the minimization process chooses each y_i to be *equal* to $|A_i.x + \gamma - b_i|$.

When $n = 1$ (that is, each $A_i.$ has just a single element), this problem has a simple geometric interpretation. Plotting $A_i.$ on the horizontal axis and b_i on the vertical axis, this formulation finds the line in the two-dimensional $(A_i., b_i)$ space such that the sum of vertical distances from the line to the data points b_i is minimized.

1.3.3   Load Balancing Problem

Consider the task of balancing computational work among n processors, some of which may already be loaded with other work. We wish to distribute the new work in such a way that the lightest-loaded processor has as heavy a load as possible. We define the data for the problem as follows:

$$p_i = \quad \text{current load of processor } i = 1, 2, \ldots, n \text{ (nonnegative),}$$
$$L = \quad \text{additional total load to be distributed,}$$
$$x_i = \quad \text{fraction of additional load } L \text{ distributed to processor } i, \text{ with } x_i \geq 0 \text{ and } \textstyle\sum_{i=1}^{n} x_i = 1,$$
$$\gamma = \quad \text{minimum of final loads after distribution of workload } L.$$

Assuming that the new work can be distributed among multiple processors without incurring any overhead, we can formulate the problem as follows:

$$\max_{x,\gamma} \quad \gamma$$
$$\text{subject to} \quad \gamma e \leq p + xL, \quad e'x = 1, \quad x \geq 0,$$

where $e = (1, 1, \ldots, 1)'$ is the vector of 1's with n components.

Interestingly, this is one of the few linear programs that can be solved in closed form. When $p_i \leq L/n$ for all $i = 1, 2, \ldots, n$, the optimal γ is $(e'p + L)/n$, and all processors have the same workload γ. Otherwise, the processors that had the heaviest loads to begin with do not receive any new work; the solution is slightly more complicated in this case but can be determined by sorting the p_i's. Similar solutions are obtained for the continuous knapsack problem that we mention later.

1.3.4   Resource Allocation

Consider a company that needs to decide how to allocate its resources (for example, raw materials, labor, or time on rented equipment) in a certain period to produce a variety of finished products. Suppose the company is able to to produce m types of finished products (indexed $i = 1, 2, \ldots, m$) and that it uses n resources (indexed by $j = 1, 2, \ldots, n$). Each unit of finished product i yields c_i dollars in revenue, whereas each unit of resource j costs d_j dollars. Suppose too that one unit of product i requires A_{ij} units of resource j to manufacture and that a maximum of b_j units of resource j are available in this period. The manufacturer aims to maximize their profit (defined as total revenue minus total cost) subject to using no more resources than are available.

The variables in this problem are y_i, $i = 1, 2, \ldots, m$, which is the number of units of product i, and x_j, $j = 1, 2, \ldots, n$, the number of units of resource j consumed. The linear programming formulation is as follows:

$$\max_{x, y} \quad z = c'y - d'x$$
$$\text{subject to} \quad x = A'y, \quad x \leq b, \quad x, y \geq 0.$$

To further explain the constraint $x = A'y$ better, we consider the jth equation of this system, which is

$$x_j = A_{1j}y_1 + A_{2j}y_2 + \cdots + A_{mj}y_m.$$

Each term $A_{ij}y_i$ indicates the amount of resource j used to manufacture the desired amount of product i, and so the summation represents the total amount of resource j required to make the specified amounts of the products. The bound $x \leq b$ ensures that we do not exceed the available resources, and the nonnegativity constraint $y \geq 0$ constrains us to produce a nonnegative amount of each product. (The constraint $x \geq 0$ is actually redundant and can be omitted from the formulation; since all the elements of y and A are nonnegative, all elements of $x = A'y$ must also be nonnegative.)

1.3.5 Classification

In classification problems, we are given two sets of points in the space of n dimensions $\mathbf{R}^n$. Our aim is to find a hyperplane in the space $\mathbf{R}^n$ that separates these two sets as accurately as possible. We use this hyperplane to classify any new points that arise; if the new point lies on one side of the hyperplane, we classify it as an element of the first set, while if it lies on the other side, we place it in the second set.

Linear programming can be used to find the separating hyperplane, which is defined by a vector $w \in \mathbf{R}^n$ and a scalar γ. Ideally, we would like each point t in the first set to satisfy $w't \geq \gamma$, while each point t in the second set satisfies $w't \leq \gamma$. To guard against a trivial answer (note that the conditions just specified are trivially satisfied by $w = 0$ and $\gamma = 0$!), we seek to enforce the stronger conditions $w't \geq \gamma + 1$ for points in the first set and $w't \leq \gamma - 1$ for points in the second set. Moreover, because the two sets may be intermingled, it may not be able to enforce a clean separation. We define the objective function in the linear program to be the sum of the average violations of the classification conditions over each set.

We set up the linear program by constructing an $m \times n$ matrix M whose ith row contains the n components of the ith points in the first set. Similarly, we construct a $k \times n$ matrix B from the points in the second set. The violations of the condition $w't \geq \gamma + 1$ for points in the first set are measured by a vector y, which is defined by the inequalities $y \geq -(Mw - \gamma e) + e$, $y \geq 0$, where $e = (1, 1, \ldots, 1)' \in \mathbf{R}^m$. Similarly, violations of the condition $w't \leq \gamma - 1$ for points in the second set are measured by the vector z defined by $z \geq (Bw - \gamma e) + e$, $z \geq 0$, where $e \in \mathbf{R}^k$. In general, e will be a vector of ones of appropriate dimension. The average violation on the first set is $e'y/m$ and on the second

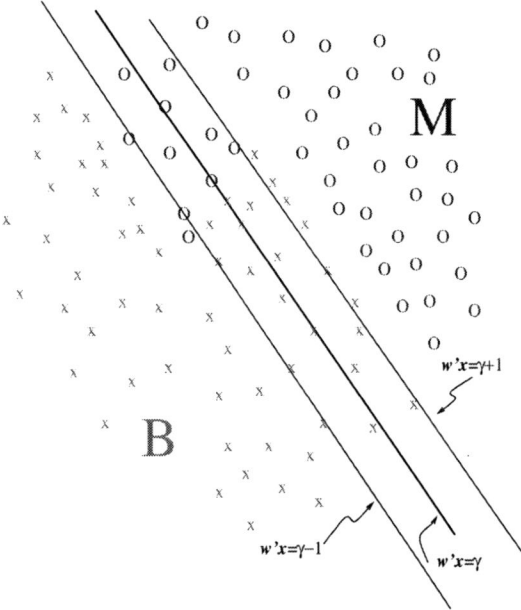

Figure 1.4. *Classification using the plane $w'x = \gamma$.*

set is $e'z/k$, and so we can write the linear program as follows:

$$
\begin{aligned}
\min_{w,\gamma,y,z} \quad & \tfrac{1}{m}e'y + \tfrac{1}{k}e'z \\
\text{subject to} \quad y \;&\geq\; -(Mw \;-\; \gamma e) \;+\; e, \\
z \;&\geq\; (Bw \;-\; \gamma e) \;+\; e, \\
(y, z) &\geq 0.
\end{aligned}
$$

Figure 1.4 shows the separation in a particular example arising in breast cancer diagnosis (Mangasarian, Street & Wolberg (1995)). The first set M (indicated by circles in the diagram) consists of fine needle aspirates (samples) taken from malignant tumors. Their location in the two-dimensional space is defined by the measures of two properties of each tumor, for example, the average cell size and the average deviation from "roundness" of the cells in the sample. The second set B (indicated by crosses) consists of fine needle aspirates taken from benign tumors. Note that the hyperplane $w'x = \gamma$ (which in two dimensions is simply a line) separates most of the benign points from most of the malignant points.

Another interesting application of the linear programming approach to classification is described by Bosch & Smith (1998), who use a separating plane in three dimensions that count the frequencies of certain words to determine that 12 disputed Federalist Papers were probably authored by James Madison rather than Alexander Hamilton.

1.3.6 Minimum-Cost Network Flow

Network problems, which involve the optimization of a flow pattern in a network of nodes and arcs, are important because of their applicability to many diverse practical problems.

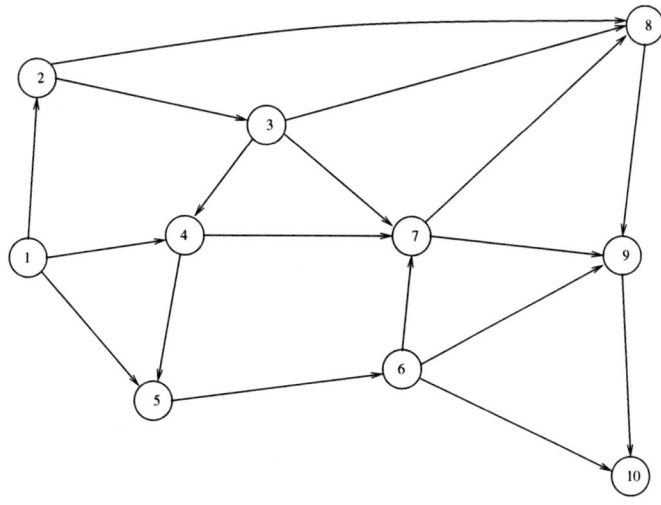

Figure 1.5. *Nodes and arcs in a network.*

We consider here a particular kind of network problem known as *minimum-cost network flow*, where the "flow" consists of the movement of a certain commodity along the arcs of a network, from the nodes at which the commodity is produced to the nodes where it is consumed. If the cost of transporting the commodity along an arc is a fixed multiple of the amount of commodity, then the problem of minimizing the total cost can be formulated as a linear program.

Networks, such as that depicted in Figure 1.5, consist of nodes $\mathcal{N}$ and arcs $\mathcal{A}$, where the arc (i, j) connects an origin node i to a destination node j. Associated with each node i is a *divergence* b_i, which represents the amount of product produced or consumed at node i. When $b_i > 0$, node i is a *supply node*, while if $b_i < 0$, it is a *demand node*. Associated with each arc (i, j) are a lower bound l_{ij} and an upper bound u_{ij} of the amount of the commodity that can be moved along that arc. Each variable x_{ij} in the problem represents the amount of commodity moved along the arc (i, j). The cost of moving one unit of flow along arc (i, j) is c_{ij}. We aim to minimize the total cost of moving the commodity from the supply nodes to the demand nodes.

Using this notation, we can formulate the minimum-cost network flow problem as follows:

$$\min_{x} \quad z = \sum_{(i,j)\in\mathcal{A}} c_{ij}x_{ij}$$

$$\text{subject to} \quad \sum_{j:(i,j)\in\mathcal{A}} x_{ij} - \sum_{j:(j,i)\in\mathcal{A}} x_{ji} = b_i \qquad \text{for all nodes } i \in \mathcal{N},$$

$$l_{ij} \le x_{ij} \le u_{ij} \qquad \text{for all arcs } (i, j) \in \mathcal{A}.$$

The first constraint states that the net flow through each node should match its divergence. The first summation represents the total flow *out of* node i, summed over all the arcs that have node i as their origin. The second summation represents total flow *into* node i, summed over all the arcs having node i as their destination. The difference between inflow and outflow is constrained to be the divergence b_i.

By relabeling the flow variables as $x_1, x_2, \ldots, x_n$, where n is the total number of arcs, we can put the problem into a more general programming form. However, the special notation used above reveals the structure of this application, which can be used in designing especially efficient versions of the simplex method. Note, in particular, that the coefficient matrix arising from the flow constraints contains only the numbers 0, 1, and -1. If all the problem data is integral, it can be shown that the solution x also contains only integer components.

1.4   Algorithms and Complexity

Though easy to state, linear programs can be quite challenging to solve computationally. The essential difficulty lies in determining which of the inequality constraints and bounds are *active* (that is, satisfied as equalities) at the solution and which are satisfied but *inactive*. (For example, the constraint $2x_1 + x_2 \le 8$ is active at the point $(x_1, x_2) = (1, 6)$; it is satisfied but inactive at the point $(2, 2)$; it is violated at the point $(4, 1)$.) To determine which constraints are active at the solution would seem to be a combinatorial problem: If there are l inequality constraints and bounds, and each of them can be either active or inactive, we may have a total of 2^l active/inactive combinations. The situation hardly improves if we make use of the fact that a solution occurs at one of the vertices of the feasible region, defined as a point at which at least n of the constraints are active. A problem in $\mathbf{R}^n$ with a total of l inequality constraints and bounds (and no equality constraints) may have as many as

$$\binom{l}{n} = \frac{l!}{(l-n)!n!}$$

vertices. Even for a small problem with $n = 10$ variables and $l = 20$ inequality constraints and bounds, there may be $184{,}756$ vertices, and possibly $1{,}048{,}576$ active/inactive combinations. A "brute force" algorithm that examines all these possibilities will be much too slow for practical purposes.

1.4.1   The Simplex Method

From a geometrical point of view, the simplex method is easy to understand. It starts by determining whether the feasible region is empty. If so, it declares the problem to be *infeasible* and terminates. Otherwise, it finds a vertex of the feasible region to use as a starting point. It then moves from this vertex to an adjacent vertex for which the value of the objective z is lower—in effect, sliding along an edge of the feasible region until it can proceed no further without violating one of the constraints. This process is repeated; the algorithm moves from vertex to (adjacent) vertex, decreasing z each time. The algorithm can terminate in one of two ways. First, it may encounter a vertex whose value of z is less than or equal to all adjacent vertices. In this case, it declares this vertex to be a solution of the linear program. Second, it may detect that the problem is unbounded. That is, it may find a direction leading away from the current vertex that remains feasible (no matter how long a step is taken along it) such that the objective z decreases to $-\infty$ along this direction. In this case, it declares the problem to be *unbounded*.

Suppose in our two-variable example of Figure 1.1 that the simplex algorithm starts at the origin $(0, 0)$. It could find the optimum $(4, 5)$ by moving along one of two paths where, due to conversion to a minimization problem, z is the negative of that depicted in Figure 1.1.

	Path 1		Path 2
$(0,0)$	$z = 0$		
$(6,0)$	$z = -30$	$(0,0)$	$z = 0$
$(6,2)$	$z = -38$	$(0,6)$	$z = -24$
$(4,5)$	$z = -40$	$(4,5)$	$z = -40$

Note that *both* adjacent vertices of the initial point $(0, 0)$ have lower objective values, and hence each one is a valid choice for the next iterate. The simplex method uses a *pivot selection rule* to select from among these possibilities; different variants of the simplex method use different pivot rules, as we see in Chapters 3 and 5.

1.4.2   Interior-Point Methods

Although the simplex method performs well on most practical problems, there are pathological examples (Klee & Minty (1972)) in which the number of iterations required is exponential in the number of variables. On such examples, linear programming seems to reveal a combinatorial nature. A surprising development occurred in 1979, when a (theoretically) more efficient method was discovered by Khachiyan (1979). For problems in which the data A, b, c were integer or rational numbers, Khachiyan's *ellipsoid method* can solve the problem in a time that is bounded by a polynomial function of the number of bits L needed to store the data and the number of variables n. However, the ellipsoid method proved to be difficult to implement and disappointingly slow in practice. Karmarkar (1984) proposed a new algorithm with a similar polynomial bound. He made the additional claim that a computational implementation of his algorithm solved large problems faster than existing simplex codes. Though this claim was never fully borne out, Karmarkar's announcement started a surge of new research into *interior-point methods*, so named because their iterates move through the interior of the feasible region toward a solution, rather than traveling from vertex to vertex around the boundary. Software based on interior-point methods is often significantly faster than simplex codes on large practical problems. We discuss these methods further in Chapter 8.

Notes and References

The use of the word "programming" in connection with linear programming is somewhat anachronistic. It refers to the step-by-step mathematical procedure used to solve this optimization problem, not specifically to its implementation in a computer program. The term "linear programming" was coined in the 1940s, well before the word "programming" became strongly associated with computers.

The definition of the term *standard form* is itself not "standard"; other authors use a definition different from the one we provide in (1.2). The term *canonical form* is not widely used and is also not standard terminology, but we use it here as a convenient way to distinguish between the two formulations, both of which appear throughout the book.

The classic text on the simplex method is by the inventor of this method, George B. Dantzig (1963). In 1939, the Russian Nobel Laureate Leonid V. Kantorovich had also proposed a method for solving linear programs; see Kantorovich (1960).

More advanced treatments of linear programming than ours include the books of Chvátal (1983) and Vanderbei (1997). Wright (1997) focuses on interior-point methods. Several advanced chapters on linear programming (both simplex and interior-point) also appear in the text of Nocedal & Wright (2006). The latter text also contains material on more general optimization problems, especially nonlinear optimization problems with and without constraints. The text of Wolsey (1998) provides an excellent introduction to integer programming.

Chapter 2

Linear Algebra: A Constructive Approach

In Section 1.4 we sketched a geometric interpretation of the simplex method. In this chapter, we describe the basis of an algebraic interpretation that allows it to be implemented on a computer. The fundamental building block for the simplex method from linear algebra is the *Jordan exchange*. In this chapter, we describe the Jordan exchange and its implementation in MATLAB. We use it in a constructive derivation of several key results in linear algebra concerning linear independence and the solution of linear systems of equations. In the latter part of the chapter, we discuss the LU factorization, another linear algebra tool that is important in implementations of the simplex method.

In this chapter and the rest of the book, we assume basic familiarity with MATLAB. There are many books and web sites that will get you started in MATLAB; we recommend the MATLAB Primer by Sigmon & Davis (2004).

We first describe the Jordan exchange, a fundamental building block of linear algebra and the simplex algorithm for linear programming.

2.1 Jordan Exchange

Consider the following simple linear equation in the one-dimensional variables x and y:

$$y = ax.$$

The form of the equation indicates that x is the *independent* variable and y is the *dependent* variable: Given a value of x, the equation tells us how to determine the corresponding value of y. Thus we can think of the dependent variable as a function of the independent variable; that is, $y(x) := ax$. If we assume that $a \neq 0$, we can reverse the roles of y and x as follows:

$$x = \tilde{a}y, \qquad \text{where } \tilde{a} = \frac{1}{a}.$$

Note that now we have a function $x(y)$ in which x is determined as a function of y. This exchange in roles between dependent and independent variables gives a very simple procedure for solving either the *equation $ax - b = 0$* or the *inequality $ax - b \geq 0$*, using the following simple equivalences:

$$ax - b = 0 \iff ax = y, \quad y = b \iff x = \tilde{a}y, \quad y = b,$$
$$ax - b \geq 0 \iff ax = y, \quad y \geq b \iff x = \tilde{a}y, \quad y \geq b. \tag{2.1}$$

In particular, the second formula gives an explicit characterization of the values of x that satisfy the inequality $ax - b \geq 0$, in terms of an independent variable y for which $y \geq b$.

The *Jordan exchange* is a generalization of the process above. It deals with the case in which $x \in \mathbf{R}^n$ is a *vector* of independent variables and $y \in \mathbf{R}^m$ is a *vector* of dependent variables, and we wish to exchange one of the independent variables with one of the dependent variables. The Jordan exchange plays a crucial role in solving systems of equations, linear inequalities, and linear programs. In addition, it can be used to derive fundamental results of linear algebra and linear programming.

We now demonstrate a Jordan exchange on the system $y = Ax$ by exchanging the roles of a component y_r of y and a component x_s of x. First, we write this system equation-wise as

$$y_i = A_{i1}x_1 + A_{i2}x_2 + \cdots + A_{in}x_n, \qquad i = 1, 2, \ldots, m, \tag{2.2}$$

where the independent variables are $x_1, x_2, \ldots, x_n$ and the dependent variables are $y_1, y_2, \ldots, y_m$, and the A_{ij}'s are the coefficients. We can think of the y_i's as (linear) functions of x_j's, that is,

$$y_i(x) := A_{i1}x_1 + A_{i2}x_2 + \cdots + A_{in}x_n, \qquad i = 1, 2, \ldots, m, \tag{2.3}$$

or, more succinctly, $y(x) := Ax$. This system can also be represented in the following *tableau* form:

$$
\begin{array}{rc|ccccc|}
 & & x_1 & \cdots & x_s & \cdots & x_n \\
\cline{3-7}
y_1 & = & A_{11} & \cdots & A_{1s} & \cdots & A_{1n} \\
\vdots & & \vdots & & \vdots & & \vdots \\
y_r & = & A_{r1} & \cdots & A_{rs} & \cdots & A_{rn} \\
\vdots & & \vdots & & \vdots & & \vdots \\
y_m & = & A_{m1} & \cdots & A_{ms} & \cdots & A_{mn} \\
\cline{3-7}
\end{array}
$$

Note that the tableau is nothing more than a compact representation of the system of equations (2.2) or the functions determining the dependent variables from the independent variables (2.3). All the operations that we perform on the tableau are just simple algebraic operations on the system of equations, rewritten to conform with the tableau representation.

We now describe the *Jordan exchange* or *pivot operation* with regard to the tableau representation. The dependent variable y_r will become independent, while x_s changes from being independent to being dependent. The process is carried out by the following three steps.

(a) Solve the rth equation

$$y_r = A_{r1}x_1 + \cdots + A_{rs}x_s + \cdots + A_{rn}x_n$$

for x_s in terms of $x_1, x_2, \ldots, x_{s-1}, y_r, x_{s+1}, \ldots, x_n$. Note that this is possible if and only if $A_{rs} \neq 0$. (A_{rs} is known as the *pivot element*, or simply the *pivot*.)

(b) Substitute for x_s in all the remaining equations.

(c) Write the new system in a new tableau form as follows:

		x_1	$\cdots$	x_{s-1}	y_r	x_{s+1}	$\cdots$	x_n
y_1	=	B_{11}	$\cdots$		B_{1s}		$\cdots$	B_{1n}
$\vdots$		$\vdots$			$\vdots$			$\vdots$
y_{r-1}	=							
x_s	=	B_{r1}	$\cdots$		B_{rs}		$\cdots$	B_{rn}
y_{r+1}	=							
$\vdots$		$\vdots$			$\vdots$			$\vdots$
y_m	=	B_{m1}	$\cdots$		B_{ms}		$\cdots$	B_{mn}

To determine the elements B_{ij} in terms of the elements A_{ij}, let us carry out the algebra specified by the Jordan exchange. As will be our custom in this book, we will describe and produce corresponding MATLAB m-files for the important algebraic operations that we perform. Solution of the rth equation

$$y_r = \sum_{\substack{j=1 \\ j \neq s}}^n A_{rj}x_j + A_{rs}x_s$$

for x_s gives

$$x_s = \frac{1}{A_{rs}}y_r + \sum_{\substack{j=1 \\ j \neq s}}^n \frac{-A_{rj}}{A_{rs}}x_j = B_{rs}y_r + \sum_{\substack{j=1 \\ j \neq s}}^n B_{rj}x_j, \tag{2.4}$$

where

$$B_{rs} = \frac{1}{A_{rs}}, \qquad B_{rj} = \frac{-A_{rj}}{A_{rs}} \qquad \forall j \neq s. \tag{2.5}$$

These formulae define the rth row $B_{r.}$ of the transformed tableau. We can express them in MATLAB by first defining J to represent the columns of the tableau excluding the sth column, that is,

```
≫ J = [1:s-1,s+1:n];
```

and then writing

```
≫ B(r,s) = 1.0/A(r,s); B(r,J) = -A(r,J)/A(r,s);
```

This "vector index" facility is an important feature of MATLAB which enables terse coding of expressions such as those given in (2.5).

We can now proceed with part (b) of the Jordan exchange. Substituting of the expression for x_s from (2.4) in the ith equation of the tableau ($i \neq r$), we have

$$y_i = \sum_{\substack{j=1 \\ j \neq s}}^{n} A_{ij}x_j + A_{is} \left(\frac{1}{A_{rs}}y_r + \sum_{\substack{j=1 \\ j \neq s}}^{n} \frac{-A_{rj}}{A_{rs}}x_j \right)$$

$$= \sum_{\substack{j=1 \\ j \neq s}}^{n} B_{ij}x_j + B_{is}y_r,$$

where

$$B_{is} = \frac{A_{is}}{A_{rs}}, \qquad B_{ij} = \left(A_{ij} - \frac{A_{is}}{A_{rs}}A_{rj} \right) = \left(A_{ij} - B_{is}A_{rj} \right) \qquad \forall i \neq r, j \neq s. \quad (2.6)$$

These formulae define rows $B_i.$, $i = 1, 2, \ldots, m, i \neq r$, of the transformed tableau. We can also write these equations succinctly in MATLAB notation by defining J as above and defining I to represent all the rows of the tableau except the rth row, that is,

```
≫ I = [1:r-1,r+1:m];
```

and writing

```
≫ B(I,s)  = A(I,s)/A(r,s);
≫ B(I,J)  = A(I,J)  - B(I,s)*A(r,J);
```

The complete description of one step of the Jordan exchange with pivot A_{rs} is coded in jx.m—the function jx in MATLAB.

Note that we have introduced the "function" facility of MATLAB. Any function can be defined in a file of the same name as the function, but with suffix .m, just as the function jx is stored in jx.m. It can then be invoked from within MATLAB—either from the command window or from within other functions—by simply typing the function name together with its arguments. The following example shows a call to the function jx.

Example 2-1-1. Solve the following system of equations for x_1, x_2 in terms of y_1, y_2:

$$y_1 = 2x_1 + x_2,$$
$$y_2 = 3x_1 + x_2.$$

Working from the MATLAB command window, one first loads the data file containing the matrix A of Example 2-1-1 and then invokes jx twice to perform the two required Jordan exchanges:

```
≫ load ex2-1-1
≫ B = jx(A,1,1)
≫ B = jx(B,2,2)
```

MATLAB file jx.m: Jordan exchange

```
function B = jx(A,r,s)
% syntax: B = jx(A,r,s)
% input: matrix A, integers r,s
% perform a Jordan exchange with pivot A(r,s)

[m,n] = size(A); B = zeros(m,n);
I = [1:r-1,r+1:m]; J = [1:s-1,s+1:n];

% update pivot row
B(r,s) = 1.0/A(r,s); B(r,J) = -A(r,J)/A(r,s);

% update pivot column
B(I,s) = A(I,s)/A(r,s);

% update remainder of tableau
B(I,J) = A(I,J)-B(I,s)*A(r,J);
return;
```

Note that we overwrite B at each step to hold the cumulative effects of the sequence of exchanges.

We now introduce a MATLAB structure that stores a complete tableau—the row and column labels corresponding to the dependent and independent variables, along with the matrix of coefficients that defines the relationship between these quantities. The totbl command can be used to construct a tableau as follows:

```
>> load ex2-1-1
>> T = totbl(A);
```

$$
\begin{array}{c c}
 & \begin{array}{c c} x_1 & x_2 \end{array} \\
\begin{array}{c} y_1 = \\ y_2 = \end{array} &
\boxed{\begin{array}{c c} 2 & 1 \\ 3 & 1 \end{array}}
\end{array}
$$

The row labels (dependent variables) are assigned the default values y_1 and y_2 and the column labels (independent variables) the default values x_1 and x_2; other forms of the totbl command, discussed later, will allow the user to define their own labels. The command tbl can be used to print out the tableau along with its associated labels.

To perform Jordan exchanges on the tableau representation (rather than on just the matrix), we use the labeled Jordan exchange function ljx in place of jx as follows:

```
>> T = ljx(T,1,1);
```

$$
\begin{array}{c c}
 & \begin{array}{c c} y_1 & x_2 \end{array} \\
\begin{array}{c} x_1 = \\ y_2 = \end{array} &
\boxed{\begin{array}{c c} 0.5 & -0.5 \\ 1.5 & -0.5 \end{array}}
\end{array}
$$

```
>> T = ljx(T,2,2);
```

$$
\begin{array}{c}
 \quad \begin{array}{cc} y_1 & y_2 \end{array} \\
\begin{array}{cc}
x_1 & = \\
x_2 & =
\end{array}
\begin{array}{|cc|}
\hline
-1 & 1 \\
3 & -2 \\
\hline
\end{array}
\end{array}
$$

In addition to making the algebraic changes to the matrix, `ljx` swaps the row and column labels as required by the exchange and prints the modified tableau using `tbl`. The trailing semicolon should not be omitted after the call to `ljx` since it results in the printing of additional unnecessary information about the structure `T`. ∎

The following simple theorem provides a formal justification of the Jordan exchange formulae (2.5) and (2.6) as well as their extension to multiple pivots in succession. The result will enable us to use the Jordan exchange to give some constructive derivations of key results in linear algebra and linear programming.

Theorem 2.1.1. *Consider the linear function y defined by $y(x) := Ax$, where $A \in \mathbf{R}^{m \times n}$. After k pivots (with appropriate reordering of rows and columns) denote the initial and kth tableaus as follows:*

$$
\begin{array}{cc}
& \begin{array}{cc} x_{J_1} & x_{J_2} \end{array} \\
\begin{array}{cc}
y_{I_1} & = \\
y_{I_2} & =
\end{array}
\begin{array}{|cc|}
\hline
A_{I_1 J_1} & A_{I_1 J_2} \\
A_{I_2 J_1} & A_{I_2 J_2} \\
\hline
\end{array}
\end{array}
\qquad
\begin{array}{cc}
& \begin{array}{cc} y_{I_1} & x_{J_2} \end{array} \\
\begin{array}{cc}
x_{J_1} & = \\
y_{I_2} & =
\end{array}
\begin{array}{|cc|}
\hline
B_{I_1 J_1} & B_{I_1 J_2} \\
B_{I_2 J_1} & B_{I_2 J_2} \\
\hline
\end{array}
\end{array}
$$

Here I_1, I_2 is a partition of $\{1, 2, \ldots, m\}$ and J_1, J_2 is a partition of $\{1, 2, \ldots, n\}$, with I_1 and J_1 containing the same number of elements. Then for all values of $x \in \mathbf{R}^n$

$$
x_{J_1} = B_{I_1 J_1} y_{I_1}(x) + B_{I_1 J_2} x_{J_2},
$$
$$
y_{I_2}(x) = B_{I_2 J_1} y_{I_1}(x) + B_{I_2 J_2} x_{J_2}.
$$

That is, the original linear functions y satisfy the new linear relationships given by the kth tableau.

Proof. We show the result for one pivot. The result for k pivots follows by induction.

For a pivot on the (r, s) element, we have $I_1 = \{r\}$, $I_2 = \{1, \ldots, r-1, r+1, \ldots, m\}$, $J_1 = \{s\}$, and $J_2 = \{1, \ldots, s-1, s+1, \ldots, n\}$. Then for all x, we have

$$
x_{J_1} - B_{I_1 J_1} y_{I_1}(x) - B_{I_1 J_2} x_{J_2} = x_s - B_{rs} y_r(x) - \sum_{\substack{j=1 \\ j \neq s}}^{n} B_{rj} x_j
$$

$$
= x_s - \frac{1}{A_{rs}} \left(\sum_{j=1}^{n} A_{rj} x_j \right) - \sum_{\substack{j=1 \\ j \neq s}}^{n} \frac{-A_{rj}}{A_{rs}} x_j
$$

$$
= 0
$$

and

$$y_{l_2}(x) - B_{l_2 j_1} y_{l_1}(x) - B_{l_2 j_2} x_{j_2}$$

$$= \left[y_i(x) - B_{is} y_r(x) - \sum_{\substack{j=1 \\ j \neq s}}^{n} B_{ij} x_j \right], \quad \begin{array}{l} i = 1, 2, \ldots, m, \\ i \neq r \end{array}$$

$$= \left[\sum_{j=1}^{n} A_{ij} x_j - \frac{A_{is}}{A_{rs}} \sum_{j=1}^{n} A_{rj} x_j - \sum_{\substack{j=1 \\ j \neq s}}^{n} \left(A_{ij} - \frac{A_{is} A_{rj}}{A_{rs}} \right) x_j \right], \quad \begin{array}{l} i = 1, 2, \ldots, m, \\ i \neq r \end{array}$$

$$= 0,$$

verifying the claims.   □

The following result shows that if two tableaus have identical dependent variables for all possible values of the independent variables, then the tableau entries are also identical.

Proposition 2.1.2. *If the linear function y is defined by $y(x) = Ax$ and also by $y(x) = Bx$, then $A = B$.*

Proof. Since $(A - B)x = 0$ for all $x \in \mathbf{R}^n$, we can choose $x = I_{\cdot i}$, where $I_{\cdot i}$ is the ith column of the identity matrix. We deduce that the ith columns of A and B are identical. Since this fact is true for all $i = 1, 2, \ldots, n$, we conclude that $A = B$.   □

2.2 Linear Independence

A simple geometric way to solve a system of two equations in two unknowns is to plot the corresponding lines and determine the point where they intersect. Of course, this technique fails when the lines are parallel to one another. A key idea in linear algebra is that of *linear dependence*, which is a generalization of the idea of parallel lines. Given a matrix $A \in \mathbf{R}^{m \times n}$, we may ask if any of its rows are redundant. In other words, is there a row $A_{k \cdot}$ that can be expressed as a linear combination of the other rows? That is,

$$A_{k \cdot} = \sum_{\substack{i=1 \\ i \neq k}}^{m} \lambda_i A_{i \cdot}. \tag{2.7}$$

If so, then the rows of A are said to be *linearly dependent*.

As a concrete illustration, consider the matrix

$$A = \begin{bmatrix} 1 & 2 & 4 \\ 3 & 4 & 8 \\ 5 & 6 & 12 \end{bmatrix}.$$

The third row of this matrix is redundant, because it can be expressed as a linear combination of the first two rows as follows:

$$A_{3.} = 2A_{2.} - A_{1.}.$$

If we rearrange this equation, we see that

$$\begin{bmatrix} -1 & 2 & -1 \end{bmatrix} A = 0,$$

that is, for $z' = \begin{bmatrix} -1 & 2 & -1 \end{bmatrix}$, $z'A = 0$ with $z \neq 0$.

We define linear dependence of the rows of a matrix A formally as follows:

$$z'A = 0 \qquad \text{for some nonzero } z \in \mathbf{R}^m.$$

(We see that (2.7) can be expressed in this form by taking $z_i = \lambda_i$, $i \neq k$, $z_k = -1$.) The negation of linear dependence is *linear independence* of the rows of A, which is defined by the implication

$$z'A = 0 \implies z = 0.$$

The idea of linear independence extends also to functions, including the linear functions y defined by $y(x) := Ax$ that we have been considering above. The functions $y_i(x)$, $i = 1, 2, \ldots, m$, defined by $y(x) := Ax$ are said to be linearly dependent if

$$z'y(x) = 0 \qquad \forall x \in \mathbf{R}^n \text{ for some nonzero } z \in \mathbf{R}^m$$

and linearly independent if

$$z'y(x) = 0 \qquad \forall x \in \mathbf{R}^n \implies z = 0. \tag{2.8}$$

The equivalence of the linear independence definitions for matrices and functions is clear when we note that

$$z'Ax = 0 \qquad \forall x \in \mathbf{R}^n \text{ for some nonzero } z \in \mathbf{R}^m$$
$$\Longleftrightarrow z'A = 0 \qquad \text{for some nonzero } z \in \mathbf{R}^m.$$

Thus the functions $y(x)$ are linearly independent if and only if the rows of the matrix A are linearly independent.

Proposition 2.2.1. *If the m linear functions y_i are linearly independent, then any p of them are also linearly independent, where $p \leq m$.*

Proof. The proof is obvious from contrapositive statement: y_i, $i = 1, 2, \ldots, p$, are linearly dependent implies y_i, $i = 1, 2, \ldots, m$, are linearly dependent. □

Proposition 2.2.2. *If the linear functions y defined by $y(x) = Ax$, $A \in \mathbf{R}^{m \times n}$, are linearly independent, then $m \leq n$. Furthermore, in the tableau representation, all m dependent y_i's can be made independent; that is, they can be exchanged with m independent x_j's.*

Proof. Suppose that the linear functions $y(x) = Ax$ are linearly independent. Exchange y's and x's in the tableau until no further pivots are possible, at which point we are blocked by a tableau of the following form (after a possible rearrangement of rows and columns):

$$
\begin{array}{c|cc}
 & y_{I_1} & x_{J_2} \\
\hline
x_{J_1} = & B_{I_1 J_1} & B_{I_1 J_2} \\
y_{I_2} = & B_{I_2 J_1} & 0
\end{array}
$$

If $I_2 \neq \emptyset$, we have by Theorem 2.1.1 that

$$ y_{I_2}(x) = B_{I_2 J_1} y_{I_1}(x) \qquad \forall x \in \mathbf{R}^n, $$

which we can rewrite as follows:

$$ \begin{bmatrix} -B_{I_2 J_1} & I \end{bmatrix} \begin{bmatrix} y_{I_1}(x) \\ y_{I_2}(x) \end{bmatrix} = 0. $$

By taking z to be any row of the matrix $\begin{bmatrix} -B_{I_2 J_1} & I \end{bmatrix}$, note that z is nonzero and that $z'y(x) = 0$. According to definition (2.8), the existence of z implies that the functions $y(x) = Ax$ are linearly dependent. Hence, we must have $I_2 = \emptyset$, and therefore $m \leq n$ and all the y_i's have been pivoted to the top of the tableau, as required. $\square$

Example 2-2-1. For illustration, consider the matrix defined earlier:

$$ A = \begin{bmatrix} 1 & 2 & 4 \\ 3 & 4 & 8 \\ 5 & 6 & 12 \end{bmatrix}. $$

Then

```
>> load ex2-2-1
>> T = totbl(A);
```

$$
\begin{array}{c|ccc}
 & x_1 & x_2 & x_3 \\
\hline
y_1 = & 1 & 2 & 4 \\
y_2 = & 3 & 4 & 8 \\
y_3 = & 5 & 6 & 12
\end{array}
$$

```
>> T = ljx(T,1,1);
```

$$
\begin{array}{c|ccc}
 & y_1 & x_2 & x_3 \\
\hline
x_1 = & 1 & -2 & -4 \\
y_2 = & 3 & -2 & -4 \\
y_3 = & 5 & -4 & -8
\end{array}
$$

```
>> T = ljx(T,2,2);
```

$$
\begin{array}{c|ccc}
 & y_1 & y_2 & x_3 \\
\hline
x_1 = & -2 & 1 & 0 \\
x_2 = & 1.5 & -0.5 & -2 \\
y_3 = & -1 & 2 & 0
\end{array}
$$

Note that we cannot pivot any more y's to the top and that $y_3 = -y_1 + 2y_2$. This final relationship indicates the linear dependence relationship amongst the rows of A, namely that $A_{3.} = -A_{1.} + 2A_{2.}$. ∎

The above result can be strengthened to the following fundamental theorem, which can be taken as an alternative and constructive definition of linear independence.

Theorem 2.2.3 (Steinitz). *For a given matrix $A \in \mathbf{R}^{m \times n}$, the linear functions y, defined by $y(x) := Ax$, are linearly independent if and only if for the corresponding tableau all the y_i's can be exchanged with some m independent x_j's.*

Proof. The "only if" part follows from Proposition 2.2.2, and so we need to prove just the "if" part. If all the y_i's can be exchanged to the top of the tableau, then we have (by rearranging rows and columns if necessary) that

$$
y = \begin{array}{cc} x_{J_1} & x_{J_2} \\ \boxed{\begin{array}{cc} A_{.J_1} & A_{.J_2} \end{array}} \end{array} \longrightarrow \quad x_{J_1} = \begin{array}{cc} y & x_{J_2} \\ \boxed{\begin{array}{cc} B_{.J_1} & B_{.J_2} \end{array}} \end{array}
$$

Suppose now that there is some z such that $z'A = 0$. We therefore have that $z'Ax = 0$ for all $x \in \mathbf{R}^n$. In the right-hand tableau above, we may set the independent variables $y = z$, $x_{J_2} = 0$, whereupon $x_{J_1} = B_{.J_1}z$. For this particular choice of x and y, we have $y = Ax$ from Theorem 2.1.1, and so it follows that

$$0 = z'Ax = z'y = z'z,$$

implying that $z = 0$. We have shown that the only z for which $z'A = 0$ is the zero vector $z = 0$, verifying that the rows of A and hence the functions y are linearly independent. ☐

A consequence of this result is that given a matrix A, the number of linearly independent rows in A is the maximum number of components of y that can be exchanged to the top of the tableau for the functions $y(x) := Ax$.

When not all the rows of A are linearly independent, we reach a tableau in which one or more of the y_i's are expressed in terms of other components of y. These relationships show the linear dependencies between the functions $y(x)$ and, therefore, between the rows of the matrix A.

Example 2-2-2. Let the matrix A be defined by

$$
A = \begin{bmatrix} -1 & 0 & 3 \\ 2 & -2 & 4 \\ 0 & -2 & 10 \end{bmatrix}.
$$

By using `ljx.m`, find out how many linearly independent rows A has. If there are linear dependencies, write them out explicitly.

After entering the matrix A into MATLAB, we construct a tableau and perform two Jordan exchanges to make y_1 and y_2 independent variables:

```
>> T=totbl(A);
```

		x_1	x_2	x_3
y_1	=	-1	0	3
y_2	=	2	-2	4
y_3	=	0	-2	10

```
>> T=ljx(T,2,1);
>> T=ljx(T,1,2);
```

$$
\begin{array}{cc}
 & \begin{array}{ccc} y_2 & y_1 & x_3 \end{array} \\
\begin{array}{c} x_2 = \\ x_1 = \\ y_3 = \end{array} &
\left[\begin{array}{ccc}
-0.5 & -1 & 5 \\
0 & -1 & 3 \\
1 & 2 & 0
\end{array}\right]
\end{array}
$$

We cannot exchange y_3 with x_3 because there is a zero in the pivot position. We conclude that this matrix has two linearly independent rows. By reading across the final row of the tableau, we see that the components of y are related as follows:

$$y_3 = y_2 + 2y_1.$$

In this example, we could have done the Jordan exchanges in some other way; for example, `T=ljx(T,1,1)` followed by `T=ljx(T,3,3)`. However, the relationship that we derive between the components of y will be equivalent. ∎

Exercise 2-2-3. Let

$$
A = \begin{bmatrix}
1 & 2 & 3 & 4 \\
3 & 1 & 3 & 0 \\
1 & 3 & -3 & -8
\end{bmatrix}.
$$

Using `ljx.m`, find out how many linearly independent rows A has. By working with A', find out how many linearly independent columns it has. If there are linear dependencies, write them out explicitly.

2.3   Matrix Inversion

An $n \times n$ matrix is nonsingular if the rows of A are linearly independent; otherwise the matrix is singular.

Theorem 2.3.1. *The system* $y = Ax$ *with* $A \in \mathbf{R}^{n \times n}$ *can be inverted to* $x = By$ *if and only if* A *is nonsingular. In this case, the matrix* B *is unique and is called the inverse of* A *and is denoted by* A^{-1}. *It satisfies* $AA^{-1} = A^{-1}A = I.$

Proof. Apply Steinitz's theorem (Theorem 2.2.3) to A to get B such that $x = By$. B is unique by Proposition 2.1.2. Finally, $y = Ax = ABy$ for all y shows that $I - AB = 0$, and $x = By = BAx$ for all x shows that $I - BA = 0$. □

Example 2-3-1. Invert the matrix

$$
A = \begin{bmatrix}
2 & -1 & -1 \\
0 & 2 & -1 \\
0 & -1 & 1
\end{bmatrix}
$$

using MATLAB and the `ljx` function.

```
>> load ex2-3-1
>> T = totbl(A);
```

$$
\begin{array}{cc}
 & \begin{array}{ccc} x_1 & x_2 & x_3 \end{array} \\
\begin{array}{c} y_1 = \\ y_2 = \\ y_3 = \end{array} &
\left[\begin{array}{ccc}
2 & -1 & -1 \\
0 & 2 & -1 \\
0 & -1 & 1
\end{array}\right]
\end{array}
$$

```
≫ T = ljx(T,1,1);
```

		y_1	x_2	x_3
x_1	=	0.5	0.5	0.5
y_2	=	0	2	−1
y_3	=	0	−1	1

```
≫ T = ljx(T,2,2);
```

		y_1	y_2	x_3
x_1	=	0.5	0.25	0.75
x_2	=	0	0.5	0.5
y_3	=	0	−0.5	0.5

```
≫ T = ljx(T,3,3);
```

		y_1	y_2	y_3
x_1	=	0.5	1	1.5
x_2	=	0	1	1
x_3	=	0	1	2

Note that the inverse of A is now found in this tableau, that is,

```
≫ invA = T.val
```

$$A^{-1} = \begin{bmatrix} 0.5 & 1 & 1.5 \\ 0 & 1 & 1 \\ 0 & 1 & 2 \end{bmatrix}$$

The command T.val extracts the matrix from the tableau T (that is, it strips off the labels).

We can check that the computed matrix is indeed A^{-1} by evaluating $AA^{-1} - I$ using the following code:

```
≫ A*invA-eye(3)
```

(where eye(3) is MATLAB notation for a 3×3 identity matrix). This code should result in a 3×3 matrix whose elements are zero or very small numbers (which may not be exactly zero because of numerical roundoff error).

As an alternative method for solving this problem, we show that it is possible not to pivot along the diagonal. For the matrix B of Theorem 2.3.1 to be A^{-1}, the subscript indices of x and y must be both arranged in ascending order. If this is not the case, reordering of rows and/or columns is necessary.

```
≫ load ex2-3-1
≫ T = totbl(A);
≫ T = ljx(T,1,3);
```

		x_1	x_2	y_1
x_3	=	2	−1	−1
y_2	=	−2	3	1
y_3	=	2	−2	−1

≫ T = ljx(T,3,1);

		y_3	x_2	y_1
x_3	=	1	1	0
y_2	=	-1	1	0
x_1	=	0.5	1	0.5

≫ T = ljx(T,2,2);

		y_3	y_2	y_1
x_3	=	2	1	0
x_2	=	1	1	0
x_1	=	1.5	1	0.5

Notice that the numbers in this final tableau are identical to those obtained after the previous sequence of pivots but that the rows and columns have been reordered according to the labels. To restore the correct ordering of the rows and columns and recover the inverse A^{-1}, we use the command

≫ Atemp = T.val;

to extract the matrix from the tableau and then use standard MATLAB commands to reorder the rows and columns of the 3×3 matrix Atemp to obtain A^{-1}. In this case, we note from the row labels that rows 1, 2, and 3 of T must appear as rows 3, 2, and 1 of A^{-1}, respectively; while from the column labels we see that columns 1, 2, and 3 of T must appear as rows 3, 2, and 1 of A^{-1}, respectively. We can define permutation vectors I and J to effect the reordering and then define A^{-1} as follows:

≫ I=[3 2 1]; J=[3 2 1];

≫ invA(I,J)=Atemp;

Alternatively, we can avoid generating Atemp and instead use

≫ invA(I,J)=T.val;

Note that the permutations correspond to the row and column label orderings after the final Jordan exchange. That is, the vector I shows the final ordering of the row labels (x_3, x_2, x_1), and the vector J shows the final ordering of the column labels (y_3, y_2, y_1). This scheme for choosing the reordering vectors will work in general, provided we put the reordering on the left-hand side of the assignment of T.val to invA, as above. ∎

Of course, for the example above, one can avoid the final reordering step by pivoting down the diagonal in order. However, there may be problems for which such a pivot sequence is not possible. A simple example is given by the matrix

$$A = \begin{bmatrix} 0 & 1 \\ 1 & 1 \end{bmatrix},$$

for which the (1, 1) element cannot be used as the first pivot. Another example follows.

Example 2-3-2. Calculate A^{-1} using ljx, where

$$A = \begin{bmatrix} 0 & 1 & 3 \\ 4 & 3 & 2 \\ 1 & 6 & 6 \end{bmatrix}.$$

≫ load ex2-3-2

≫ T = totbl(A);

		x_1	x_2	x_3
y_1	=	0	1	3
y_2	=	4	3	2
y_3	=	1	6	6

≫ T = ljx(T,3,1);

		y_3	x_2	x_3
y_1	=	0	1	3
y_2	=	4	-21	-22
x_1	=	1	-6	-6

≫ T = ljx(T,1,2);

		y_3	y_1	x_3
x_2	=	0	1	-3
y_2	=	4	-21	41
x_1	=	1	-6	12

≫ T = ljx(T,2,3);

		y_3	y_1	y_2
x_2	=	0.2927	-0.5366	-0.0732
x_3	=	-0.0976	0.5122	0.0244
x_1	=	-0.1707	0.1463	0.2927

We now extract the numerical values from the tableau, define the permutation vectors, and perform the reordering as follows:

≫ I=[2 3 1]; J=[3 1 2];

≫ invA(I,J) = T.val;

$$A^{-1} = \begin{bmatrix} 0.1463 & 0.2927 & -0.1707 \\ -0.5366 & -0.0732 & 0.2927 \\ 0.5122 & 0.0244 & -0.0976 \end{bmatrix}$$ ■

Exercise 2-3-3. Calculate A^{-1} using ljx.m, where

$$A = \begin{bmatrix} 1 & 2 & 3 \\ 2 & 4 & 2 \\ 1 & 1 & 1 \end{bmatrix}.$$

Exercise 2-3-4. Use $\mathtt{ljx.m}$ to find the inverses of the following matrices in MATLAB:

$$A = \begin{bmatrix} 1 & 0 & 1 \\ 2 & 1 & 1 \\ 3 & 0 & 0 \end{bmatrix}, \qquad B = \begin{bmatrix} 1 & 0 & 1 \\ 2 & 1 & 1 \\ 1 & 1 & 0 \end{bmatrix}, \qquad C = \begin{bmatrix} 2 & 0 & 1 \\ 1 & 2 & 0.5 \\ 0 & 1 & 1 \end{bmatrix}.$$

If a matrix is singular, show the linear dependence between the rows of the matrix. (Use $\mathtt{T.val}$ and perform any reordering needed on the resulting MATLAB matrix to obtain the final result.)

At this point, we note that the pivot rules described in Section 2.1 in terms of matrix elements have simple matrix block analogues. That is, instead of pivoting on A_{rs}, we could instead pivot on the submatrix A_{RS}, where $R \subseteq \{1, 2, \ldots, m\}$ and $S \subseteq \{1, 2, \ldots, n\}$ are two index sets with the same number of elements. The only real difference is that the inverse of A_{RS} is used in place of $1/A_{rs}$. MATLAB code for the block Jordan exchange is given in $\mathtt{bjx.m}$. The code incorporates changes to the labels, and so it is an extension of $\mathtt{ljx}$ rather than of $\mathtt{jx}$. In addition, the code updates the matrix A instead of creating the new matrix B, and so it needs to perform the operations in a slightly different order from the analogous operations in $\mathtt{ljx}$.

Two successive Jordan exchanges can be effected as one block pivot of order two. Thus by induction, any sequence of Jordan exchanges can be effected by a single block pivot. What are the algebraic consequences of this observation? Consider a system that leads to the following tableau:

$$(2.9)$$

$$\begin{array}{|cc|} \hline A & B \\ C & D \\ \hline \end{array}$$

where A is square and invertible. Applying a block pivot to the matrix A, the transformed tableau is

$$(2.10)$$

$$\begin{array}{|cc|} \hline A^{-1} & -A^{-1}B \\ CA^{-1} & D - CA^{-1}B \\ \hline \end{array}$$

The matrix $D - CA^{-1}B$ is called the Schur complement of A in $\begin{bmatrix} A & B \\ C & D \end{bmatrix}$. The algebraic formula for the block pivot operation will prove to be very useful in what follows. For example, if A^{-1} exists, then the original matrix (2.9) is invertible if and only if the Schur complement $D - CA^{-1}B$ is invertible.

Exercise 2-3-5. The following tableau expresses how the dependent variables y_{I_1} and y_{I_2} can be expressed in terms of the independent variables x_{J_1} and x_{J_2}:

$$\begin{array}{c} \\ y_{I_1} = \\ y_{I_2} = \end{array} \begin{array}{cc} x_{J_1} & x_{J_2} \\ \hline A & B \\ C & D \\ \hline \end{array}$$

By performing some simple manipulations, derive the following formulae:

$$x_{J_1} = A^{-1}y_{I_1} - A^{-1}Bx_{J_2} \quad \text{and} \quad y_{I_2} = CA^{-1}y_{I_1} + (D - CA^{-1}B)x_{J_2},$$

thus justifying the block Jordan exchange formula (2.10).

MATLAB file bjx.m: Labeled block Jordan exchange

```
function A = bjx(A,R,S)
% syntax: B = bjx(A,R,S)
% input: tableau A, integer vectors R,S
% perform a block Jordan exchange with pivot A(R,S)

R = R(:); S = S(:);
[m,n] = size(A.val);

% setdiff(1:m,R) := {1,...,m}\R
I = setdiff(1:m,R); J = setdiff(1:n,S);

% note that values are updated in place
% update pivot column
A.val(R,S) = inv(A.val(R,S));
A.val(I,S) = A.val(I,S)*A.val(R,S);

% update remainder of tableau
A.val(I,J) = A.val(I,J)-A.val(I,S)*A.val(R,J);

% update pivot row
A.val(R,J) = -A.val(R,S)*A.val(R,J);

% now update the labels
swap = A.bas(R);
A.bas(R) = A.nonbas(S);
A.nonbas(S) = swap;

if isfield(A,'dualbas')
  swap = A.dualbas(S);
  A.dualbas(S) = A.dualnonbas(R);
  A.dualnonbas(S) = swap;
end

tbl(A);

return;
```

2.4   Exact Solution of *m* Equations in *n* Unknowns

At this stage, we have considered only square systems where the number of variables is the same as the number of equations. In optimization applications, it is more likely that the systems under consideration have different numbers of variables than equations. Such systems pose additional concerns. For example, the system

$$x_1 + x_2 = 1$$

having one equation and two variables has infinitely many solutions, whereas the system

$$x_1 = 1, \qquad x_1 = 2$$

clearly has no solution.

Suppose we wish to solve $Ax = b$ (or determine that no solution exists), where $A \in \mathbf{R}^{m \times n}$ and $b \in \mathbf{R}^m$, with m and n not necessarily equal. The Jordan exchange gives a simple method for solving this problem under no assumption whatsoever on A.

1. Write the system in the following tableau form:

$$
y \quad = \quad \begin{array}{c} \\ \boxed{\begin{array}{c|c} A & -b \end{array}} \end{array}
$$

$$
\begin{array}{cc} x & 1 \end{array}
$$

Our aim is to seek x and y related by this tableau *such that $y = 0$*.

2. Pivot as many of the y_i's to the top of the tableau, say y_{I_1}, until no more can be pivoted, in which case we are blocked by a tableau as follows (with row and column reordering):

$$
\begin{array}{ccc}
 & y_{I_1} & x_{J_2} & 1 \\
x_{J_1} = & \boxed{\begin{array}{cc|c} B_{I_1 J_1} & B_{I_1 J_2} & d_{I_1} \end{array}} \\
y_{I_2} = & \boxed{\begin{array}{cc|c} B_{I_2 J_1} & 0 & d_{I_2} \end{array}}
\end{array}
$$

We now ask the question: Is it possible to find x and y related by this tableau such that $y = 0$?

3. The system is solvable if and only if $d_{I_2} = 0$, since *we require $y_{I_1} = 0$ and $y_{I_2} = 0$*. When $d_{I_2} = 0$, we obtain by writing out the relationships in the tableau explicitly that

$$
\begin{aligned}
y_{I_1} &= 0, \\
y_{I_2} &= B_{I_2 J_1} y_{I_1} = 0, \\
x_{J_2} &\text{ is arbitrary}, \\
x_{J_1} &= B_{I_1 J_2} x_{J_2} + d_{I_1}.
\end{aligned}
$$

Note that m could be less than or greater than n.

Example 2-4-1. Solve the following system:

$$
\begin{array}{rcrcrcr}
x_1 & - & x_2 & + & x_3 & = & 2, \\
-x_1 & + & 2x_2 & + & x_3 & = & 3, \\
x_1 & - & x_2 & - & x_3 & = & -2.
\end{array}
$$

Remember that our technique is to pivot as many of the y_i's to the top of the tableau as possible and then check that the resulting tableau is still valid. In particular, the elements in the last column corresponding to the y_i's that we cannot pivot to the top—the subvector d_{I_2} in the notation above—should be zero.

Rewrite the problem as follows:

$$
\begin{array}{rcrcrcrcr}
y_1 & = & x_1 & - & x_2 & + & x_3 & - & 2, \\
y_2 & = & -x_1 & + & 2x_2 & + & x_3 & - & 3, \\
y_3 & = & x_1 & - & x_2 & - & x_3 & + & 2,
\end{array}
$$

and carry out the following pivot operations on its tableau representation:

```
>> load ex2-4-1
>> T = totbl(A,b);
```

		x_1	x_2	x_3	1
y_1	=	1	−1	1	−2
y_2	=	−1	2	1	−3
y_3	=	1	−1	−1	2

This gives rise to the following sequence of tableaus:

```
>> T = ljx(T,1,1);
```

		y_1	x_2	x_3	1
x_1	=	1	1	−1	2
y_2	=	−1	1	2	−5
y_3	=	1	0	−2	4

```
>> T = ljx(T,2,2);
```

		y_1	y_2	x_3	1
x_1	=	2	1	−3	7
x_2	=	1	1	−2	5
y_3	=	1	0	−2	4

```
>> T = ljx(T,3,3);
```

		y_1	y_2	y_3	1
x_1	=	0.5	1	1.5	1
x_2	=	0	1	1	1
x_3	=	0.5	0	−0.5	2

The final solution can be read off the tableau by setting $y_1 = y_2 = y_3 = 0$. We get $x_1 = 1$, $x_2 = 1$, and $x_3 = 2$. Note that if the calculations are being carried out by hand, the columns of the tableau that are labeled with a y_i can be suppressed since their values are never needed. This can result in a significant saving in computation time, particularly if you are performing the steps by hand.

```
>> load ex2-4-1
>> T = ljx(T,1,1);
>> T = delcol(T,'y1');
```

		x_2	x_3	1
x_1	=	1	−1	2
y_2	=	1	2	−5
y_3	=	0	−2	4

```
>> T = ljx(T,2,1);
>> T = delcol(T,'y2');
```

$$
\begin{array}{rcl}
 & & x_3 \quad 1 \\
x_1 & = & \boxed{\begin{array}{c|c} -3 & 7 \\ -2 & 5 \\ -2 & 4 \end{array}} \\
x_2 & = & \\
y_3 & = &
\end{array}
$$

```
>> T = ljx(T,3,1);
>> T = delcol(T,'y3');
```

$$
\begin{array}{rcl}
 & & 1 \\
x_1 & = & \boxed{\begin{array}{c} 1 \\ 1 \\ 2 \end{array}} \\
x_2 & = & \\
x_3 & = &
\end{array}
$$

Of course, by deleting the y_i columns, we lose access to the linear dependence relationships between these variables. ∎

Exercise 2-4-2. Test yourself on the following example:

$$
\begin{aligned}
x_1 + x_2 + x_3 &= 1, \\
x_1 - x_2 - x_3 &= 1, \\
x_1 - x_2 + x_3 &= 3,
\end{aligned}
$$

which has solution $x = (1, -1, 1)$ using

```
>> load ex2-4-2
>> T = totbl(A,b);
>> ...
```

Exercise 2-4-3. Solve the following systems of equations:

1.

$$
\begin{array}{rcrcrcr}
2u & + & 3v & + & 3w & = & 2, \\
   &   & 5v & + & 7w & = & 2, \\
6u & + & 9v & + & 8w & = & 5.
\end{array}
$$

2.

$$
\begin{array}{rcrcrcr}
 u & + & 4v & + & 2w & = & -2, \\
-2u & - & 8v & + & 3w & = & 32, \\
   &   & v & + & w & = & 1.
\end{array}
$$

Example 2-4-4. Solve the following system of 3 equations in 4 unknowns:

$$
\begin{array}{rcrcrcrcr}
x_1 & - & x_2 &   &   & + & x_4 & = & 1, \\
x_1 &   &   & + & x_3 &   &   & = & 1, \\
x_1 & + & x_2 & + & 2x_3 & - & x_4 & = & 0.
\end{array}
$$

The data file `ex2-4-4.mat` can be loaded into MATLAB enabling the following sequence of tableaus to be constructed:

```
>> load ex2-4-4
>> T = totbl(A,b);
```

		x_1	x_2	x_3	x_4	1
y_1	=	1	-1	0	1	-1
y_2	=	1	0	1	0	-1
y_3	=	1	1	2	-1	0

```
>> T =ljx(T,2,1);
```

		y_2	x_2	x_3	x_4	1
y_1	=	1	-1	-1	1	0
x_1	=	1	0	-1	0	1
y_3	=	1	1	1	-1	1

```
>> T =ljx(T,3,2);
```

		y_2	y_3	x_3	x_4	1
y_1	=	2	-1	0	0	1
x_1	=	1	0	-1	0	1
x_2	=	-1	1	-1	1	-1

At this point we are blocked—we cannot pivot y_1 to the top of the tableau because the pivot elements corresponding to x_3 and x_4 are both zero. (It makes no sense to exchange y_1 with either y_2 or y_3 since such a move would not increase the number of y_i's at the top of the tableau.) Since the element in the final column of the row labeled y_1 is nonzero, the system has *no solution* because, whenever $y_2 = 0$ and $y_3 = 0$, it follows that $y_1 = 1$. We are unable to set $y_1 = y_2 = y_3 = 0$ in the final tableau. In fact, we have

$$y_1(x) = 2y_2(x) - y_3(x) + 1.$$

Note that the tableau also shows that the rows of the coefficient matrix A of the original linear system are related by $A_1 = 2A_2 - A_3$. and thus are linearly dependent. ∎

Example 2-4-5. We now consider a modification of Example 2-4-4 in which the right-hand side of the first equation is changed. Our system is now as follows:

$$\begin{aligned}
x_1 &- x_2 & &+ x_4 &= 2, \\
x_1 & &+ x_3 & &= 1, \\
x_1 &+ x_2 &+ 2x_3 &- x_4 &= 0.
\end{aligned}$$

The data file ex2-4-5.mat can be loaded into MATLAB enabling the following sequence of tableaus to be constructed:

```
>> load ex2-4-5
>> T = totbl(A,b);
```

		x_1	x_2	x_3	x_4	1
y_1	=	1	-1	0	1	-2
y_2	=	1	0	1	0	-1
y_3	=	1	1	2	-1	0

≫ T =ljx(T,2,1);

		y_2	x_2	x_3	x_4	1
y_1	=	1	-1	-1	1	-1
x_1	=	1	0	-1	0	1
y_3	=	1	1	1	-1	1

≫ T =ljx(T,3,2);

		y_2	y_3	x_3	x_4	1
y_1	=	2	-1	0	0	0
x_1	=	1	0	-1	0	1
x_2	=	-1	1	-1	1	-1

This system has *infinitely many solutions* because the final column of the tableau contains a zero in the location corresponding to the y_i column label *and* some of the x_j's are still independent variables, and therefore their values can be chosen arbitrarily. Following the procedure above, we characterize the solution set by allowing x_3 and x_4 to be arbitrary and defining

$$\begin{aligned} x_1 &= -x_3 &&+ 1, \\ x_2 &= -x_3 + x_4 &&- 1. \end{aligned}$$

Another way to express this result is to introduce arbitrary variables λ_1 and λ_2 (representing the arbitrary values x_3 and x_4, respectively) and writing the solution as

$$\begin{bmatrix} x_1 \\ x_2 \\ x_3 \\ x_4 \end{bmatrix} = \begin{bmatrix} 1 \\ -1 \\ 0 \\ 0 \end{bmatrix} + \begin{bmatrix} -1 \\ -1 \\ 1 \\ 0 \end{bmatrix} \lambda_1 + \begin{bmatrix} 0 \\ 1 \\ 0 \\ 1 \end{bmatrix} \lambda_2. \quad \blacksquare$$

Exercise 2-4-6. Using the MATLAB function `ljx`, solve the following systems of equations by carrying out all pivot operations. If a system has no solution, give a reason. If a system has infinitely many solutions, describe the solution set as in Example 2-4-5. If the rows of the matrix are linearly dependent, write down the actual linear dependence relations.

1. $Ax = a$, where

$$A = \begin{bmatrix} 2 & -1 & 1 & 1 \\ -1 & 2 & -1 & -2 \\ 4 & 1 & 1 & -1 \end{bmatrix}, \qquad a = \begin{bmatrix} 1 \\ 1 \\ 5 \end{bmatrix}.$$

2. $Bx = b$, where

$$B = \begin{bmatrix} 1 & -1 & 1 & 2 \\ 1 & 1 & 0 & -1 \\ 1 & -3 & 2 & 5 \end{bmatrix}, \qquad b = \begin{bmatrix} 2 \\ 1 \\ 1 \end{bmatrix}.$$

3. $Cx = c$, where

$$C = \begin{bmatrix} 1 & -1 & 1 \\ 2 & 1 & 1 \\ -1 & -1 & 2 \\ 1 & 1 & -1 \end{bmatrix}, \qquad c = \begin{bmatrix} 3 \\ 2 \\ 2 \\ -1 \end{bmatrix}.$$

Exercise 2-4-7. Find all solutions of $Ax = b$, where

$$A = \begin{bmatrix} 1 & 2 & 3 & 4 \\ 1 & 0 & 1 & 0 \\ 1 & -1 & 0 & -1 \end{bmatrix}, \qquad b = \begin{bmatrix} 1 \\ 2 \\ 3 \end{bmatrix}.$$

Describe all solutions to the following set of equations and inequalities: $Ax = b$, $x_3 \geq 0$.

We end this section by deriving a fundamental theorem of linear algebra by means of the Jordan exchange. In what follows we shall refer to the initial tableau as the $y = Ax$ tableau and (if it is possible) the $x = By$ tableau as a tableau where all the y's have been pivoted to the top of the tableau.

Theorem 2.4.1. *Let* $A \in \mathbf{R}^{n \times n}$. *The following are equivalent:*

(a) *A is nonsingular.*

(b) $Ax = 0 \implies x = 0$, *that is,* $\ker A := \{x \mid Ax = 0\} = \{0\}$.

(c) $Ax = b$ *has a unique solution for* each $b \in \mathbf{R}^n$.

(d) $Ax = b$ *has a unique solution for* some $b \in \mathbf{R}^n$.

Proof.

(a) $\implies$ (c) By Theorem 2.3.1, A^{-1} exists and $A^{-1}b$ solves $Ax = b$ for each b. Furthermore, if x^1 and x^2 both solve $Ax = b$, then $A(x^1 - x^2) = 0$. This implies that $x^1 - x^2 = A^{-1}0 = 0$, and so solution is unique.

(c) $\implies$ (b) If we take $b = 0$, then (c) implies that $Ax = 0$ has a unique solution. Clearly, $x = 0$ is that unique solution.

(b) $\implies$ (d) (d) is a special case of (b).

(d) $\implies$ (a) We actually prove the contrapositive $\sim$(a) $\implies$ $\sim$(d). Suppose that A is singular. Then, by definition, its rows are linearly dependent, and so by the Steinitz theorem (Theorem 2.2.3), it will not be possible to pivot all the y_i's to the top of the tableau. By carrying out Jordan exchanges until we are blocked, we obtain a tableau of the form

$$\begin{array}{rc|cc|c}
 & & y_{I_1} & x_{J_2} & 1 \\
\hline
x_{J_1} & = & B_{I_1 J_1} & B_{I_1 J_2} & d_{I_1} \\
y_{I_2} & = & B_{I_2 J_1} & 0 & d_{I_2}
\end{array}$$

(after a possible rearrangement). Since A is square, some x_j's will remain as column labels in this final tableau; that is, J_2 is not empty. If $d_{I_2} = 0$, this final tableau indicates infinitely many solutions, since x_{J_2} is arbitrary. Alternatively, if $d_{I_2} \neq 0$, there are no solutions. In no case is it possible to have a *unique* solution, and so (d) cannot hold. $\square$

Exercise 2-4-8. Find matrices A, not necessarily square, for which the number of solutions to $Ax = b$ has the following properties. Explain your answers.

  1. one solution, regardless of b;

2. zero or infinitely many solutions, depending on b;

3. zero or one solution, depending on b;

4. infinitely many solutions, independent of b.

Note that using a block pivot on an $n \times n$ system of equations as above is equivalent to the following tableau manipulation:

$$
y = \begin{array}{cc} x & 1 \\ \hline A & -b \\ \hline \end{array} \qquad x = \begin{array}{cc} y & 1 \\ \hline A^{-1} & A^{-1}b \\ \hline \end{array}
$$

2.5 Solving Linear Equations Efficiently

We turn our attention now to the computational cost of solving systems of equations of the form $Ax = b$, where A is a square $n \times n$, nonsingular matrix. We measure the cost by counting the floating-point operations (flops), $+$, $-$, $*$, and $/$ (addition, subtraction, multiplication, and division), required to obtain the solution x. We consider the Jordan-exchange-based approach of Sections 2.3 and 2.4, along with a more efficient approach based on the LU factorization.

The flops needed to perform one Jordan exchange are shown in the following table.

Element	Transform	Flops
Pivot	$B(r, s) = 1/A(r, s)$	1
Pivot row	$B(r, J) = -A(r, J)/A(r, s)$	$n - 1$
Pivot column	$B(I, s) = A(I, s)/A(r, s)$	$m - 1$
Other elements	$B(I, J) = A(I, J) - B(I, s) * A(r, J)$	$2(m - 1)(n - 1)$

Thus there are $2mn - m - n + 1$ flops per Jordan exchange.

It requires n Jordan exchanges to invert an $n \times n$ matrix, each costing $2n^2 - 2n + 1$ flops, giving a total of $2n^3 - 2n^2 + n$ flops. For large n, we can approximate this count by $2n^3$. Having obtained A^{-1}, we can multiply it by b to obtain $x = A^{-1}b$. This matrix-vector multiplication requires an additional $2n^2$ flops, but for large n, this cost is small relative to the cost of calculating A^{-1}.

We now describe an alternative and faster method for solving square systems of equations called *Gaussian elimination*. This technique avoids calculating the inverse A^{-1} explicitly, and it will be used in Chapter 5 as a tool for implementing the simplex method efficiently. Gaussian elimination requires only about 1/3 as many flops to solve $Ax = b$, as does the method above, based on Jordan exchanges.

Gaussian elimination can be described in terms of the LU factorization of the matrix A. This factorization computes a unit lower triangular matrix L, an upper triangular matrix U, and a permutation matrix P such that

$$PA = LU. \tag{2.11}$$

By *unit lower triangular*, we mean that $L_{ij} = 0$ for $j > i$, and $L_{ii} = 1$ for all i; that is, L has the following form:

$$L = \begin{bmatrix} 1 & & & & & \\ * & 1 & & & & \\ * & * & 1 & & & \\ \vdots & \vdots & \vdots & \ddots & & \\ * & * & * & \cdots & 1 \end{bmatrix}.$$

The upper triangular matrix U has $U_{ij} = 0$ for $i > j$, that is,

$$U = \begin{bmatrix} * & * & \cdots & * \\ & * & \cdots & * \\ & & \ddots & \vdots \\ & & & * \end{bmatrix},$$

while the permutation matrix P is simply a rearrangement of the rows of the identity which captures the reordering of the rows of the system $Ax = b$ that takes place during the factorization process.

Details of computing the LU factorization are discussed in the next section. We describe here the recovery of the solution x once the factors L and U and the permutation matrix P are known. We note that

$$Pb = PAx = LUx = Lw, \qquad \text{where } w := Ux.$$

We first calculate w by solving $Lw = Pb$, using a process known as forward substitution, because it sweeps in a forward direction through the components of the solution w. The following MATLAB code obtains the solution:

```
>> w = P*b; w(1) = w(1)/L(1,1);
   for i=2:n
   J=1:i-1;
   w(i) = (w(i) - L(i,J)*w(J))/L(i,i);
   end
```

Having obtained w, we can now solve for x in $Ux = w$ by applying the back substitution procedure, so called because it sweeps through the components of x in reverse order.

```
>> x = w; x(n) = x(n)/U(n,n);
   for i=n-1:-1:1
   J=i+1:n;
   x(i) = (x(i) - U(i,J)*x(J))/U(i,i);
   end
```

(Forward and back substitution are known collectively as *triangular substitution*.) It is not difficult to see that the flop count for each triangular substitution is approximately n^2. This fact will be used in the next section to compute the flop count for solutions of equations using LU decomposition.

Exercise 2-5-1.   1. Prove that the product of two lower triangular matrices is lower triangular and that the inverse of a nonsingular lower triangular matrix is lower triangular. (Hint: see Appendix A.1.)

   2. Prove that the product AB of two square matrices is nonsingular if and only if both A and B are nonsingular. (Hint: use Theorem 2.3.1.)

MATLAB has built-in utilities that perform forward and backward substitution in a very efficient manner. The MATLAB code to solve a system $Ax = b$ can be carried out by the following 3 lines:

```
≫ [L,U,P] = lu(A);
≫ w = L\(P*b);
≫ x = U\w;
```

If the permutation matrix is omitted from the `lu` command, then MATLAB stores $P^{-1}L$ as the output L, removing the need to explicitly form the matrix P. Thus

```
≫ [L,U] = lu(A);
≫ w = L\b;
≫ x = U\w;
```

generates exactly the same solution, and the second command performs forward substitution implicitly handling the permutation. In fact, MATLAB provides a shorthand even for these three operations, namely

```
≫ x = A\b;
```

However, in Chapter 5, we see that explicit calculation and updates of L and U improve the efficiency of the revised simplex method, and so we do not discuss this simplified form further.

   For practical efficiency, we will always use the LU decomposition when solving a system of linear equations. Moreover, modern linear algebra implementations exploit the sparsity of A when forming L and U and are typically much more time and space efficient than explicit inversion. However, for exposition and proofs we prefer to use the Jordan exchange since it is very closely related to the proof techniques and the basic notions of linear algebra and linear programming that we exploit.

2.6   *LU* Decomposition

This section shows how the LU decomposition is carried out formally and can be omitted without any loss of continuity of the subject matter.

   Let $A \in \mathbf{R}^{n \times n}$, and suppose $A = LU$ for some matrices L and U. Then from Appendix A we know that

$$A = LU = \sum_{i=1}^{n} L_{\cdot i} U_{i \cdot};\tag{2.12}$$

that is, A is the sum of the outer products of the columns of L with the rows of U. Since we require L to be unit lower triangular and U to be upper triangular,

$$L_{ji} = 0, \qquad U_{ij} = 0, \qquad \text{if } j < i,$$

and $L_{ii} = 1$. If we compare the first rows of both sides of (2.12), then since the elements of the sum for $i > 1$ play no role, it follows that

$$U_{1.} = A_{1.}$$

Comparing the first columns of (2.12) gives (assuming $U_{11} \neq 0$)

$$L_{.1} = A_{.1}/U_{11}.$$

Now rearrange (2.12) to get

$$A - L_{.1}U_{1.} = \sum_{i=2}^{n} L_{.i}U_{i.}$$

The first row and column of both sides of this equation are zero. Comparing the second rows and columns gives

$$U_{2.} = A_{2.} - L_{21}U_{1.}$$

and

$$L_{.2} = \frac{A_{.2} - L_{.1}U_{12}}{U_{22}}.$$

Thus we have now calculated the first two columns of L and the first two rows of U. In the general case, the ith row of U and the ith column of L can be calculated using

$$U_{i.} = A_{i.} - \sum_{j=1}^{i-1} L_{ij}U_{j.}$$

and

$$L_{.i} = \frac{A_{.i} - \sum_{j=1}^{i-1} L_{.j}U_{ji}}{U_{ii}},$$

under the assumption that the pivot element $U_{ii} \neq 0$.

The only way that the above process can break down is if the pivot element $U_{ii} = 0$. In this case, either partial pivoting or complete pivoting is required to complete the process. In fact, these techniques are required for numerical stability even when the elements U_{ii} are small, as the following example demonstrates.

Example 2-6-1. Consider

$$A(\epsilon) = \begin{bmatrix} -\epsilon & 1 \\ 1 & 1 \end{bmatrix},$$

where ϵ is a small positive number. Using the scheme outlined above, with exact arithmetic we get

$$L_{.1} = \begin{bmatrix} 1 \\ -\epsilon^{-1} \end{bmatrix}, \qquad U_{1.} = \begin{bmatrix} -\epsilon & 1 \end{bmatrix}$$

and

$$L_{.2} = \frac{1}{1 + \epsilon^{-1}} \left(\begin{bmatrix} 1 \\ 1 \end{bmatrix} - \begin{bmatrix} 1 \\ -\epsilon^{-1} \end{bmatrix} \right) = \begin{bmatrix} 0 \\ 1 \end{bmatrix},$$

$$U_{2.} = \begin{bmatrix} 1 & 1 \end{bmatrix} - (-\epsilon^{-1}) \begin{bmatrix} -\epsilon & 1 \end{bmatrix} = \begin{bmatrix} 0 & 1 + \epsilon^{-1} \end{bmatrix}.$$

Thus

$$L = \begin{bmatrix} 1 & 0 \\ -\epsilon^{-1} & 1 \end{bmatrix}, \qquad U = \begin{bmatrix} -\epsilon & 1 \\ 0 & 1 + \epsilon^{-1} \end{bmatrix},$$

whose entries become increasingly large (and small), leading to numerical difficulties as $\epsilon \to 0$. In fact, as $\epsilon \to 0$ the matrix U becomes increasingly close to being singular, and on a finite-precision computer U becomes singular and the backward substitution process breaks down. At this stage it is becoming hard to use L as well!

For example, suppose that $\epsilon = 0.001$ and the machine can store only 3 significant digits. Then for

$$b = \begin{bmatrix} 1 \\ 2 \end{bmatrix},$$

the forward and backward substitution process gives

$$\begin{bmatrix} 1 & 0 \\ -1000 & 1 \end{bmatrix} \begin{bmatrix} w_1 \\ w_2 \end{bmatrix} = \begin{bmatrix} 1 \\ 2 \end{bmatrix} \implies \begin{bmatrix} w_1 \\ w_2 \end{bmatrix} = \begin{bmatrix} 1 \\ 1000 \end{bmatrix},$$

resulting in

$$\begin{bmatrix} -0.001 & 1 \\ 0 & 1000 \end{bmatrix} \begin{bmatrix} x_1 \\ x_2 \end{bmatrix} = \begin{bmatrix} 1 \\ 1000 \end{bmatrix} \implies \begin{bmatrix} x_1 \\ x_2 \end{bmatrix} = \begin{bmatrix} 0 \\ 1 \end{bmatrix},$$

where all values are rounded to 3 significant figures. However, it is easy to see that the solution of the problem (rounded to 3 significant figures) is $x_1 = 1, x_2 = 1$. $\blacksquare$

We conclude from the example that we should choose a large pivot element (U_{ii}) whenever possible, which was not the case in the very first pivot above. Two strategies are common: *partial pivoting* chooses the pivot element as the largest in a particular column under consideration; *complete pivoting* chooses the largest possible pivot element.

In complete pivoting the rows and columns are reordered so that the pivot element is the largest element in magnitude remaining to be processed, that is, in the submatrix of A with row and column index at least i. If the pivot element is zero, then the remaining submatrix must be identically zero, and hence the factorization process is trivially completed by setting the remaining columns of L as columns of the identity and the remaining rows of U to zero.

Thus in the example, the rows of $A(\epsilon)$ are interchanged, resulting in

$$P(\epsilon) = \begin{bmatrix} 0 & 1 \\ 1 & 0 \end{bmatrix}, \qquad L(\epsilon) = \begin{bmatrix} 1 & 0 \\ \epsilon & 1 \end{bmatrix}, \qquad U(\epsilon) = \begin{bmatrix} 1 & 1 \\ 0 & 1 - \epsilon \end{bmatrix}$$

when $\epsilon < 1$. Note that all these matrices are invertible when $\epsilon \to 0$.

Exercise 2-6-2. Show how the permutation helps the example, even when the machine can store only 3 significant digits.

In partial pivoting only the rows are reordered so that the pivot element is always the largest element in magnitude remaining on or below the diagonal in the pivot column. Thus, the only way the pivot element can be zero is if all the elements on or below the diagonal are zero, in which case the pivot column is a linear combination of the columns already processed. This is impossible if A is nonsingular. In our example, this is the same as complete pivoting.

The above discussion has therefore proved the following result.

Theorem 2.6.1. *Let A be any matrix in $\mathbf{R}^{n \times n}$. There exist permutation matrices P and Q, a lower triangular matrix L, and an upper triangular matrix U such that*

$$PAQ = LU.$$

If A is nonsingular, then Q may be taken as I and

$$PA = LU.$$

The flop count on iteration i can be calculated as follows:

Calculation	Flops
Column of L	$2(i-1)(n-i) + (n-i)$
Row of U	$2(i-1)(n-i+1) + (n-i+1)$

Thus the total number of flops is

$$\sum_{i=1}^{n} 4i(n-i) + O(n^2) = \frac{2n^3}{3} + O(n^2).$$

Notice that calculating the factors requires fewer flops than inversion by Jordan exchange. In fact, inversion by Jordan exchange requires $2n^3$ flops, that is, about three times as many flops. Solving linear systems requires essentially the same number of flops as calculating the factors; the amount of extra work required to solve a linear system using the factors within the forward and backward substitution routines is just $n^2 + O(n)$. Clearly, this does not impinge on the total flop count. In the unlikely event that the actual inverse is needed, this can be recovered by solving n linear systems, one for each column of the inverse, resulting in a total flop count of $O(n^3)$.

Exercise 2-6-3. Consider the matrix

$$A(\epsilon) = \begin{bmatrix} \epsilon & 1 \\ 1 & 2 \end{bmatrix}.$$

(a) Determine $L(\epsilon)$ and $U(\epsilon)$ without permuting the rows or columns of $A(\epsilon)$. Are $L(0)$ and $U(0)$ invertible?

(b) Use complete pivoting to factor $A(\epsilon)$. Write down $P(\epsilon)$, $Q(\epsilon)$, $L(\epsilon)$, and $U(\epsilon)$ explicitly.

Chapter 3

The Simplex Method

All linear programs can be reduced to the following *standard form*:

$$\min_{x} \quad z = p'x$$
$$\text{subject to} \quad Ax \geq b, \quad x \geq 0, \tag{3.1}$$

where $p \in \mathbf{R}^n, b \in \mathbf{R}^m$, and $A \in \mathbf{R}^{m \times n}$. To create the initial tableau for the simplex method, we rewrite the problem in the following *canonical form*:

$$\min_{x_B, x_N} \quad z = p'x_N + 0'x_B$$
$$\text{subject to} \quad x_B = Ax_N - b, \quad x_B, x_N \geq 0, \tag{3.2}$$

where the index sets N and B are defined initially as $\mathrm{N} = \{1, 2, \ldots, n\}$ and $\mathrm{B} = \{n + 1, \ldots, n + m\}$. The variables $x_{n+1}, \ldots, x_{n+m}$ are introduced to represent the slack in the inequalities $Ax \geq b$ (the difference between left- and right-hand sides of these inequalities) and are called *slack variables*. We shall represent this canonical linear program by the following tableau:

$$\tag{3.3}$$

		x_1	$\cdots$	x_n	1
x_{n+1}	=	A_{11}	$\cdots$	A_{1n}	$-b_1$
$\vdots$		$\vdots$	$\ddots$	$\vdots$	$\vdots$
x_{n+m}	=	A_{m1}	$\cdots$	A_{mn}	$-b_m$
z	=	p_1	$\cdots$	p_n	0

In this tableau, the slack variables $x_{n+1}, \ldots, x_{n+m}$ (the variables that make up x_B) are the dependent variables, while the original problem variables $x_1, \ldots, x_n$ (the variables that make up x_N) are independent variables. It is customary in the linear programming literature to call the dependent variables *basic* and the independent variables *nonbasic*, and we will adopt this terminology for the remainder of the book. A more succinct form of the initial tableau is known as the *condensed tableau*, which is written as follows:

$$\tag{3.4}$$

		x_N	1
x_B	=	A	$-b$
z	=	p'	0

We "read" a tableau by setting the nonbasic variables x_N to zero, thus assigning the basic variables x_B and the objective variable z the values in the last column of the tableau. The tableau above represents the point $x_N = 0$ and $x_B = -b$ (that is, $x_{n+i} = -b_i$ for $i = 1, 2, \ldots, m$), with an objective of $z = 0$. The tableau is said to be *feasible* if the values assigned to the basic variables by this procedure are nonnegative. In the above, the tableau will be feasible if $b \leq 0$.

At each iteration of the simplex method, we exchange one element between B and N, performing the corresponding Jordan exchange on the tableau representation, much as we did in Chapter 2 in solving systems of linear equations. We ensure that the tableau remains feasible at every iteration, and we try to choose the exchanged elements so that the objective function z decreases at every iteration. We continue in this fashion until either

1. a solution is found, or

2. we discover that the objective function is unbounded below on the feasible region, or

3. we determine that the feasible region is empty.

The simplex method can be examined from two viewpoints, which must be understood separately and jointly in order to fully comprehend the method:

1. an algebraic viewpoint represented by tableaus;

2. a geometric viewpoint obtained by plotting the constraints and the contours of the objective function in the space of original variables $\mathbf{R}^n$.

Later, we show that the points represented by each feasible tableau correspond to *vertices* of the feasible region.

3.1   A Simple Example

We now illustrate how the simplex method moves from a feasible tableau to an optimal tableau, one pivot at a time, by means of the following two-dimensional example.

Example 3-1-1.
$$
\begin{aligned}
\min_{x_1, x_2} \quad & 3x_1 - 6x_2 \\
\text{subject to} \quad & x_1 + 2x_2 \geq -1, \\
& 2x_1 + x_2 \geq 0, \\
& x_1 - x_2 \geq -1, \\
& x_1 - 4x_2 \geq -13, \\
& -4x_1 + x_2 \geq -23, \\
& x_1, x_2 \geq 0. \quad \blacksquare
\end{aligned}
$$

The first step is to add slack variables to convert the constraints into a set of general equalities combined with nonnegativity requirements on all the variables. The slacks are defined as follows:
$$
\begin{aligned}
x_3 &= x_1 + 2x_2 + 1, \\
x_4 &= 2x_1 + x_2, \\
x_5 &= x_1 - x_2 + 1, \\
x_6 &= x_1 - 4x_2 + 13, \\
x_7 &= -4x_1 + x_2 + 23.
\end{aligned}
$$

(When we use MATLAB to form the initial tableau, it adds the slacks automatically; there is no need to define them explicitly as above.) We formulate the initial tableau by assembling the data for the problem (that is, the matrix A and the vectors p and b) as indicated in the condensed tableau (3.4). The MATLAB command `totbl` performs this task:

```
>> load ex3-1-1
>> T = totbl(A,b,p);
```

		x_1	x_2	1
x_3	=	1	2	1
x_4	=	2	1	0
x_5	=	1	−1	1
x_6	=	1	−4	13
x_7	=	−4	1	23
z	=	3	−6	0

The labels associated with the original and slack variables are stored in the MATLAB structure T. The point represented by the tableau above can be deduced by setting the nonbasic variables x_1 and x_2 both to zero. The resulting point is feasible, since the corresponding values of the basic variables, which initially are the same as the slack variables $x_3, x_4, \ldots, x_7$, are all nonnegative. The value of the objective in this tableau, $z = 0$, is obtained from the bottom right element.

We now seek a pivot—a Jordan exchange of a basic variable with a nonbasic variable—that yields a decrease in the objective z. The first issue is to choose the nonbasic variable which is to become basic, that is, to choose a pivot column in the tableau. In allowing a nonbasic variable to become basic, we are allowing its value to possibly increase from 0 to some positive value. What effect will this increase have on z and on the dependent (basic) variables? In the given example, let us try increasing x_1 from 0. We assign x_1 the (nonnegative) value λ while holding the other nonbasic variable x_2 at zero; that is,

$$x_1 = \lambda, \qquad x_2 = 0.$$

The tableau tells us how the objective z depends on x_1 and x_2, and so for the values given above we have

$$z = 3(\lambda) - 6(0) = 3\lambda > 0 \qquad \text{for } \lambda > 0.$$

This expression tells us that z *increases* as λ increases—the opposite of what we want. Let us try instead choosing x_2 as the variable to increase, and set

$$x_1 = 0, \qquad x_2 = \lambda > 0. \tag{3.5}$$

For this choice, we have

$$z = 3(0) - 6\lambda = -6\lambda < 0 \qquad \text{for } \lambda > 0,$$

thus decreasing z, as we wished. The general rule is to choose the pivot column to have a *negative* value in the last row, as this indicates that z will decrease as the variable corresponding to that column increases away from 0. We use the term *pricing* to indicate selection of the pivot column. We call the label of the pivot column the *entering variable*, as this variable is the one that "enters" the basis at this step of the simplex method.

To determine which of the basic variables is to change places with the entering variable, we examine the effect of increasing the entering variable on each of the basic variables. Given (3.5), we have the following relationships:

$$
\begin{aligned}
x_3 &= 2\lambda + 1, \\
x_4 &= \lambda, \\
x_5 &= -\lambda + 1, \\
x_6 &= -4\lambda + 13, \\
x_7 &= \lambda + 23.
\end{aligned}
$$

Since $z = -6\lambda$, we clearly would like to make λ as large as possible to obtain the largest possible decrease in z. On the other hand, we cannot allow λ to become *too* large, as this would force some of the basic variables to become negative. By enforcing the nonnegativity restrictions on the variables above, we obtain the following restrictions on the value of λ:

$$
\begin{aligned}
x_3 &= 2\lambda + 1 \geq 0 &\implies& \quad \lambda \geq -1/2, \\
x_4 &= \lambda \geq 0 &\implies& \quad \lambda \geq 0, \\
x_5 &= -\lambda + 1 \geq 0 &\implies& \quad \lambda \leq 1, \\
x_6 &= -4\lambda + 13 \geq 0 &\implies& \quad \lambda \leq 13/4, \\
x_7 &= \lambda + 23 \geq 0 &\implies& \quad \lambda \geq -23.
\end{aligned}
$$

We see that the largest nonnegative value that λ can take without violating any of these constraints is $\lambda = 1$. Moreover, we observe that the *blocking variable*—the one that will become negative if we increase λ above its limit of 1—is x_5. We choose the row for which x_5 is the label as the pivot row and refer to x_5 as the *leaving variable*—the one that changes from being basic to being nonbasic. The pivot row selection process just outlined is called the *ratio test*.

By setting $\lambda = 1$, we have that x_1 and x_5 are zero, while the other variables remain nonnegative. We obtain the tableau corresponding to this point by performing the Jordan exchange of the row labeled x_5 (row 3) with the column labeled x_2 (column 2). The new tableau is as follows:

≫ T = ljx(T,3,2);

		x_1	x_5	1
x_3	=	3	-2	3
x_4	=	3	-1	1
x_2	=	1	-1	1
x_6	=	-3	4	9
x_7	=	-3	-1	24
z	=	-3	6	-6

Note that z has decreased from 0 to -6.

Before proceeding with this example, let us review the procedure above for a single step of the simplex method, indicating the general rules for selecting pivot columns and rows. Given the tableau   (3.6)

		x_N	1
x_B	=	H	h
z	=	c'	α

where B represents the current set of basic variables and N represents the current set of nonbasic variables, a pivot step of the simplex method is a Jordan exchange between a basic and nonbasic variable according to the following pivot selection rules:

1. *Pricing* (selection of pivot column s): The pivot column is a column s with a negative element in the bottom row. These elements are called *reduced costs*.

2. *Ratio Test* (selection of pivot row r): The pivot row is a row r such that

$$-h_r/H_{rs} = \min_i \{-h_i/H_{is} \mid H_{is} < 0\}.$$

Note that there is considerable flexibility in selection of the pivot column, as it is often the case that many of the reduced costs are negative. One simple rule is to choose the column with the most negative reduced cost. This gives the biggest decrease in z *per unit* increase in the entering variable. However, since we cannot tell *how much* we can increase the entering variable until we perform the ratio test, it is not generally true that this choice leads to the best decrease in z on this step, among all possible pivot columns.

Returning to the example, we see that column 1, the one labeled x_1, is the only possible choice for pivot column. The ratio test indicates that row 4, labeled by x_6, should be the pivot row. We thus obtain

$\gg$ T = ljx(T,4,1);

		x_6	x_5	1
x_3	=	-1	2	12
x_4	=	-1	3	10
x_2	=	-0.33	0.33	4
x_1	=	-0.33	1.33	3
x_7	=	1	-5	15
z	=	1	2	-15

In this tableau, all reduced costs are positive, and so the pivot column selection procedure does not identify an appropriate column. This is as it should be, because this tableau is optimal! For any other feasible point than the one indicated by this tableau, we would have $x_6 \geq 0$ and $x_5 \geq 0$, giving an objective $z = x_6 + 2x_5 - 15 \geq -15$. Hence, we cannot improve z over its current value of -15 by allowing either x_5 or x_6 to enter the basis, and so the tableau is optimal. The values of the basic variables can be read from the last column of the optimal tableau. We are particularly interested in the values of the two variables x_1 and x_2 from the original standard formulation of the problem; they are $x_1 = 3$ and $x_2 = 4$. In general, we have an *optimal tableau* when both the last column and the bottom row are nonnegative. (Note: when talking about the last row or last column, we do not include in our considerations the bottom right element of the tableau, the one indicating the current value of the objective. Its sign is irrelevant to the optimization process.)

Figure 3.1 illustrates Example 3-1-1.

The point labeled "Vertex 1" corresponds to the initial tableau, while "Vertex 2" is represented by the second tableau and "Vertex 3" is represented by the final tableau.

Exercise 3-1-2. Consider the problem

$$\begin{aligned} \min \quad & z = p'x \\ \text{subject to} \quad & Ax \geq b, \quad x \geq 0, \end{aligned}$$

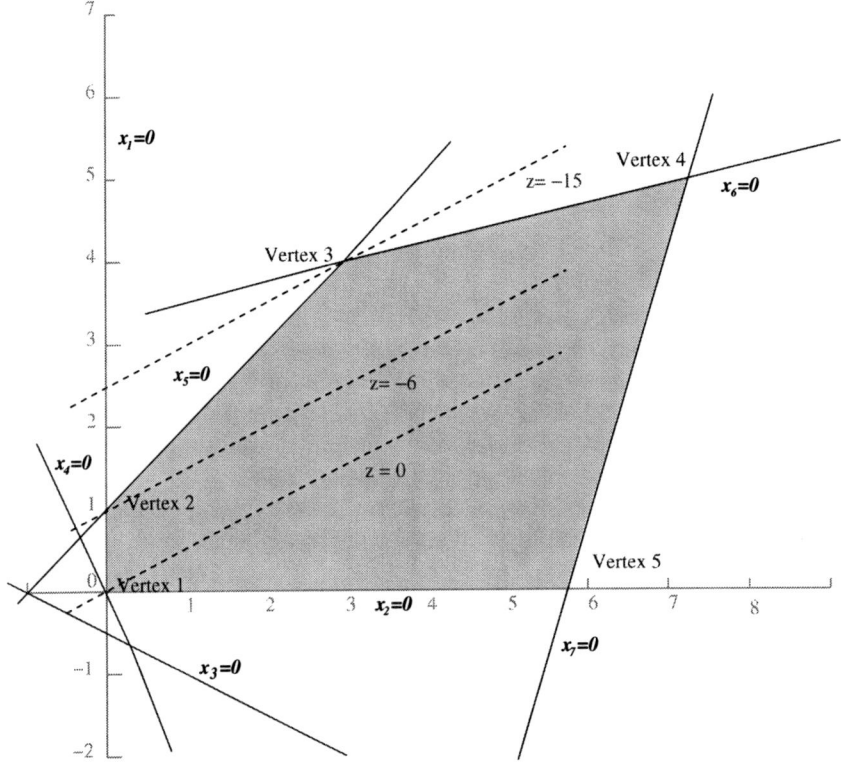

Figure 3.1. *Simplex method applied to Example* $3 - 1 - 1$.

where

$$A = \begin{bmatrix} 0 & -1 \\ -1 & -1 \\ -1 & 2 \\ 1 & -1 \end{bmatrix}, \qquad b = \begin{bmatrix} -5 \\ -9 \\ 0 \\ -3 \end{bmatrix}, \qquad p = \begin{bmatrix} -1 \\ -2 \end{bmatrix}.$$

(i) Draw the feasible region in $\mathbf{R}^2$.

(ii) Draw the contours of $z = -12$, $z = -14$, $z = -16$ and determine the solution graphically.

(iii) Solve the problem in MATLAB using Example 3-1-1 as a template. In addition, trace the path in contrasting color that the simplex method takes on your figure.

3.2 Vertices

The concept of a *vertex* of the feasible region plays an important role in the geometric interpretation of the simplex method. Given the feasible set for (3.1) defined by

$$S := \left\{ x \in \mathbf{R}^n \mid Ax \geq b, x \geq 0 \right\}, \qquad (3.7)$$

a point $x \in S$ is a *vertex* of S if it is not possible to define a line segment lying entirely in S that contains x in its interior. Each vertex can be represented as the intersection of the n hyperplanes defined by $x_i = 0$, for all $i \in N$, where N is the set of nonbasic variables for some feasible tableau. This representation is apparent from Figure 3.1, where for instance Vertex 2 is at the intersection of the lines defined by $x_1 = 0$ and $x_5 = 0$, where x_1 and x_5 are the two nonbasic variables in the tableau corresponding to this vertex.

Definition 3.2.1. For the feasible region S of (3.1) defined by (3.7), let $x_{n+i} := A_i . x - b_i$, $i = 1, 2, \ldots, m$. A *vertex* of S is any point in $(x_1, x_2, \ldots, x_n)' \in S$ that satisfies

$$x_N = 0,$$

where N is any subset of $\{1, 2, \ldots, n + m\}$ containing n elements such that the linear functions defined by x_j, $j \in N$, are linearly independent.

It is important for the n functions in this definition to be linearly independent. If not, then the equation $x_N = 0$ has either zero solutions or infinitely many solutions.

Theorem 3.2.2. *Suppose that $\bar{x}$ is a vertex of S with corresponding index set N. Then if we define*

$$\mathcal{A} := [A \quad - I], \qquad B := \{1, 2, \ldots, n + m\} \backslash N,$$

then $\bar{x}$ satisfies the relationships

$$\mathcal{A}_{.B} x_B + \mathcal{A}_{.N} x_N = b, \qquad x_B \geq 0, \qquad x_N = 0, \qquad (3.8)$$

where $\mathcal{A}_{.B}$ is invertible. Moreover, $\bar{x}$ can be represented by a tableau of the form

$$\begin{array}{rc}
 & \begin{array}{cc} x_N & 1 \end{array} \\
x_B \;=\; & \boxed{\begin{array}{c|c} H & h \end{array}} \\
z \;=\; & \boxed{\begin{array}{c|c} c' & \alpha \end{array}}
\end{array} \qquad (3.9)$$

with $h \geq 0$.

Proof. By the definition of a vertex it follows that

$$\mathcal{A}_{.B} \bar{x}_B + \mathcal{A}_{.N} \bar{x}_N = b, \qquad \bar{x}_B \geq 0, \qquad \bar{x}_N = 0. \qquad (3.10)$$

It remains to prove that $\mathcal{A}_{.B}$ is invertible. Suppose there exists a vector z such that $z' \mathcal{A}_{.B} = 0$. It follows from (3.10) that $z'b = 0$. By definition, the functions x_N satisfy

$$\mathcal{A}_{.B} x_B + \mathcal{A}_{.N} x_N = b, \qquad (3.11)$$

and so $z'A_{.N}x_N = 0$. Since x_N are linearly independent, it follows that $z'A_{.N} = 0$ and hence that $z'A = 0$. Since A has the (negative) identity matrix in its columns, this implies that $z = 0$, and thus $A_{.B}$ is invertible.

Finally, premultiplying (3.11) by $A_{.B}^{-1}$ and rearranging, we see that

$$x_B = -A_{.B}^{-1}A_{.N}x_N + A_{.B}^{-1}b, \tag{3.12}$$

which can be written in tableau form (3.9) with $H = -A_{.B}^{-1}A_{.N}$ and $h = A_{.B}^{-1}b$. The last row of the tableau can be generated by substituting for x_B from (3.12) in the expression $z = p_B'x_B + p_N'x_N$, obtaining $c' = p_N' - p_B'A_{.B}^{-1}A_{.N}$ and $\alpha = p_B'A_{.B}^{-1}b$. Note that $h \geq 0$ since $\bar{x}$ satisfies (3.12) and hence $h = \bar{x}_B$. □

The proof shows that corresponding to each vertex of S there is an invertible square matrix, which we denote by $A_{.B}$ above. This matrix is frequently called the *basis matrix*. It plays an important role in the revised simplex method of Chapter 5.

We illustrate this theorem by considering one of the vertices for the problem in Example 3-1-1. Here we have $n = 2$ and $m = 5$. The point $x = (0, 1)'$ is a vertex of the feasible region, occurring at the intersection of the lines defined by $x_1 = 0$ and $x_5 = 0$. The sets N and B corresponding to this vertex are therefore $N = \{1, 5\}$ and $B = \{2, 3, 4, 6, 7\}$. Thus

$$A_{.B} = \begin{bmatrix} 2 & -1 & 0 & 0 & 0 \\ 1 & 0 & -1 & 0 & 0 \\ -1 & 0 & 0 & 0 & 0 \\ -4 & 0 & 0 & -1 & 0 \\ 1 & 0 & 0 & 0 & -1 \end{bmatrix}, \quad A_{.N} = \begin{bmatrix} 1 & 0 \\ 2 & 0 \\ 1 & -1 \\ 1 & 0 \\ -4 & 0 \end{bmatrix}$$

and simple (MATLAB) calculations show that

$$H = -A_{.B}^{-1}A_{.N} = \begin{bmatrix} 1 & -1 \\ 3 & -2 \\ 3 & -1 \\ -3 & 4 \\ -3 & -1 \end{bmatrix}, \quad h = A_{.B}^{-1}b = \begin{bmatrix} 1 \\ 3 \\ 1 \\ 9 \\ 24 \end{bmatrix}, \tag{3.13}$$

$$c' = p_N' - p_B'A_{.B}^{-1}A_{.N} = \begin{bmatrix} -3 & 6 \end{bmatrix}, \quad \alpha = p_B'A_{.B}^{-1}b = -6,$$

which can be checked against the second tableau given in Example 3-1-1.

To complete our discussion of vertices, we note two more facts.

- The set N in Definition 3.2.1 that corresponds to a particular vertex may not be uniquely defined; that is, the same vertex may be specified by more than one set N. Looking at Figure 3.1 again, we see that Vertex 3 is defined by the unique set $N = \{5, 6\}$ and Vertex 5 is defined uniquely by $N = \{2, 7\}$. However, Vertex 1 can be specified by three possible choices of N: $N = \{1, 2\}$, $N = \{1, 4\}$, or $N = \{2, 4\}$. A vertex that can be specified by more than one set N is sometimes called a *degenerate vertex*.

- Given any set $N \subset \{1, 2, \ldots, n+m\}$ of n indices such that x_N is linearly independent, the point defined by $x_N = 0$ is *not* necessarily a vertex. This is because the point may lie outside the feasible region S. In Figure 3.1, for instance, the point defined by $N = \{3, 5\}$ is not a vertex, since the lines $x_3 = 0$ and $x_5 = 0$ intersect outside the feasible region.

Exercise 3-2-1. Consider the feasible set defined by the following constraints:

$$
\begin{array}{rcrcl}
x_1 & + & 2x_2 & \geq & 2, \\
-2x_1 & - & x_2 & \geq & -4, \\
& & x_1, x_2 & \geq & 0.
\end{array}
$$

1. Plot the feasible region, indicating the vertices on your plot. Indicate all possible choices for the N associated with each vertex.

2. For the vertex $x = (0, 4)'$, write down the sets N and B defined in Theorem 3.2.2, and calculate the quantities H and h in the tableau (3.9) from the formulae (3.13). (Do not use a Jordan exchange.)

3.3 The Phase II Procedure

The simplex method is generally split into two phases. *Phase I* finds a starting point that satisfies the constraints. That is, it finds a tableau of the general form (3.6) such that the last column h is nonnegative. *Phase II* starts with a feasible tableau and applies the pivots needed to move to an optimal tableau, thus solving the linear program.

In the following high-level description of the simplex method, B(r) denotes the label on x corresponding to the rth row of the tableau, and N(s) denotes the label on x corresponding to the sth column of the tableau.

Algorithm 3.1 (Simplex Method).

1. *Construct an initial tableau. If the problem is in standard form (3.1), this process amounts to simply adding slack variables.*

2. *If the tableau is not feasible, apply a Phase I procedure to generate a feasible tableau, if one exists (see Section 3.4). For now we shall assume the origin $x_N = 0$ is feasible.*

3. *Use the pricing rule to determine the pivot column s. If none exists, **stop**; (a): tableau is optimal.*

4. *Use the ratio test to determine the pivot row r. If none exists, **stop**; (b): tableau is unbounded.*

5. *Exchange $x_{B(r)}$ and $x_{N(s)}$ using a Jordan exchange on H_{rs}.*

6. *Go to Step 3.*

Phase II comprises Steps 3 through 6 of the method above—that part of the algorithm that occurs after an initial feasible tableau has been identified.

The method terminates in one of two ways. Stop (a) indicates optimality. This occurs when the last row is nonnegative. In this case, there is no benefit to be obtained by letting any of the nonbasic variables x_N increase away from zero. We can verify this claim mathematically by writing out the last row of the tableau, which indicates that the objective function is

$$
z = c'x_N + \alpha.
$$

When $c \geq 0$ and $x_N \geq 0$, we have $z \geq \alpha$. Therefore, the point corresponding to the tableau (3.6)—$x_B = h$ and $x_N = 0$—is optimal, with objective function value α.

The second way that the method above terminates, Stop (b), occurs when a column with a negative cost c_s has been identified, but the ratio test fails to identify a pivot row. This situation can occur only when all the entries in the pivot column $H_{.s}$ are nonnegative. In this case, by allowing $x_{N(s)}$ to grow larger, without limit, we will be decreasing the objective function to $-\infty$ without violating feasibility. In other words, by setting $x_{N(s)} = \lambda$ for any positive value λ, we have for the basic variables x_B that

$$x_B(\lambda) = H_{.s}\lambda + h \geq h \geq 0,$$

so that the full set of variables $x(\lambda) \in \mathbf{R}^{m+n}$ is defined by the formula

$$x_j(\lambda) = \begin{cases} \lambda & \text{if } j = N(s), \\ H_{is}\lambda + h_i & \text{if } j = B(i), \\ 0 & \text{if } j \in N \setminus \{N(s)\}, \end{cases}$$

which is feasible for all $\lambda \geq 0$. The objective function for $x(\lambda)$ is

$$z = c'x_N(\lambda) + \alpha = c_s\lambda + \alpha,$$

which tends to $-\infty$ as $\lambda \to \infty$. Thus the set of points $x(\lambda)$ for $\lambda \geq 0$ identifies a ray of feasible points along which the objective function approaches $-\infty$. Another way to write this ray is to separate $x(\lambda)$ into a constant vector u and another vector that depends on λ as follows:

$$x(\lambda) = u + \lambda v,$$

where the elements of u and v are defined as follows:

$$u_j = \begin{cases} 0 & \text{if } j \in N, \\ h_i & \text{if } j = B(i), \end{cases} \qquad v_j = \begin{cases} 1 & \text{if } j = N(s), \\ H_{is} & \text{if } j = B(i), \\ 0 & \text{if } j \in N \setminus \{N(s)\}. \end{cases}$$

Example 3-3-1. Show that the following linear program is unbounded below, and find vectors u and v such that $u + \lambda v$ is feasible for all $\lambda \geq 0$. Find a feasible point of this form with objective value -98.

$$\begin{array}{rrcrcrcrl} \min & z = -2x_1 & - & 3x_2 & + & x_3 \\ \text{subject to} & x_1 & + & x_2 & + & x_3 & \geq & -3, \\ & -x_1 & + & x_2 & - & x_3 & \geq & -4, \\ & x_1 & - & x_2 & - & 2x_3 & \geq & -1, \\ & & & & & x_1, x_2, x_3 & \geq & 0. \end{array}$$

```
≫ load ex3-3-1

≫ T = totbl(A,b,p);
```

		x_1	x_2	x_3	1
x_4	=	1	1	1	3
x_5	=	-1	1	-1	4
x_6	=	1	-1	-2	1
z	=	-2	-3	1	0

Since this tableau is feasible, we can start immediately on Phase II (Step 3 of the algorithm above). Selecting column 2 as the pivot column, we find that the ratio test chooses row 3 as the pivot row.

≫ T = ljx(T,3,2);

		x_1	x_6	x_3	1
x_4	=	2	−1	−1	4
x_5	=	0	−1	−3	5
x_2	=	1	−1	−2	1
z	=	−5	3	7	−3

The next pivot column selected must be column 1, but we find that the ratio test fails to select a pivot row, since all the elements in column 1 (except, of course, in the last row) are nonnegative. The tableau is unbounded, and the method therefore terminates at Step 4. By setting $x_1 = \lambda \geq 0$, we have from the tableau that the dependent variables satisfy

$$x_4(\lambda) = 2\lambda + 4, \qquad x_5(\lambda) = 5, \qquad x_2(\lambda) = \lambda + 1,$$

whereas the nonbasic variables x_6 and x_3 are both 0. For the original three variables of the problem x_1, x_2, x_3, we have

$$x(\lambda) = \begin{bmatrix} \lambda \\ \lambda + 1 \\ 0 \end{bmatrix} = \begin{bmatrix} 0 \\ 1 \\ 0 \end{bmatrix} + \lambda \begin{bmatrix} 1 \\ 1 \\ 0 \end{bmatrix},$$

so that by setting $u = (0, 1, 0)'$ and $v = (1, 1, 0)'$ we obtain the direction of unboundedness in the specified form. From the final tableau, we also have $z = -5\lambda - 3$, so that the value $z = -98$ is obtained by setting $\lambda = 19$. The corresponding value of x is then

$$x = u + 19v = \begin{bmatrix} 19 \\ 20 \\ 0 \end{bmatrix}. \quad \blacksquare$$

Exercise 3-3-2. Solve the following linear program. If it is unbounded below, find vectors u and v such that $u + \lambda v$ is feasible for all $\lambda \geq 0$. Find a feasible point of this form with objective value -415.

$$\begin{aligned}
\min \quad & z = x_1 - 2x_2 - 4x_3 + 4x_4 \\
\text{subject to} \quad & x_2 - 2x_3 - x_4 \geq -4, \\
& 2x_1 - x_2 - x_3 + 4x_4 \geq -5, \\
& -x_1 + x_2 - 2x_4 \geq -3, \\
& x_1, x_2, x_3, x_4 \geq 0.
\end{aligned}$$

Linear programs may have more than one solution. In fact, given any collection of solutions $x^1, x^2, \ldots, x^K$, any other point in the *convex hull* of these solutions, defined by

$$\left\{ x \mid x = \sum_{i=1}^{K} \alpha_i x^i, \ \sum_{i=1}^{K} \alpha_i = 1, \ \alpha_i \geq 0, \ i = 1, 2, \ldots, K \right\}, \tag{3.14}$$

is also a solution. To prove this claim, we need to verify that any such x is feasible with respect to the constraints $Ax \geq b, x \geq 0$ in (3.1) and also that x achieves the same objective value as each of the solutions $x^i, i = 1, 2, \ldots, K$. First, note that

$$Ax = \sum_{i=1}^{K} \alpha_i Ax^i \geq \left(\sum_{i=1}^{K} \alpha_i\right) b = b,$$

and so the inequality constraint is satisfied. Since x is a nonnegative combination of the nonnegative vectors x, it is also nonnegative, and so the constraint $x \geq 0$ is also satisfied. Finally, since each x^i is a solution, we have that $p'x^i = z$ for some scalar z_{opt} and all $i = 1, 2, \ldots, K$. Hence,

$$p'x = \sum_{i=1}^{K} \alpha_i p'x^i = \left(\sum_{i=1}^{K} \alpha_i\right) z_{opt} = z_{opt}.$$

Since x is feasible for (3.1) and attains the optimal objective value z_{opt}, we conclude that x is a solution, as claimed.

Phase II can be extended to identify multiple solutions by performing additional pivots on columns with zero reduced costs after an optimal tableau has been identified. We illustrate the technique with the following simple example.

Example 3-3-3.

$$\min_{x_1, x_2, x_3} \quad -x_1 - x_2 - x_3$$

$$\begin{array}{rcrcrcrcr}
\text{subject to} & x_1 & - & x_2 & + & x_3 & \geq & -2, \\
& -x_1 & + & x_2 & + & x_3 & \geq & -3, \\
& x_1 & + & x_2 & - & x_3 & \geq & -1, \\
& -x_1 & - & x_2 & - & x_3 & \geq & -4, \\
& & & x_1, x_2, x_3 & & & \geq & 0.
\end{array}$$

```
≫ load ex3-3-3
≫ T = totbl(A,b,p);
```

		x_1	x_2	x_3	1
x_4	=	1	-1	1	2
x_5	=	-1	1	1	3
x_6	=	1	1	-1	1
x_7	=	-1	-1	-1	4
z	=	-1	-1	-1	0

Since the problem is in standard form, the initial tableau can be created using `totbl`. The right-hand column is nonnegative, and so the resulting tableau is feasible, and no Phase I procedure is needed.

```
≫ T = ljx(T,3,3);
```

		x_1	x_2	x_6	1
x_4	=	2	0	-1	3
x_5	=	0	2	-1	4
x_3	=	1	1	-1	1
x_7	=	-2	-2	1	3
z	=	-2	-2	1	-1

≫ T = ljx(T,4,1);

		x_7	x_2	x_6	1
x_4	=	-1	-2	0	6
x_5	=	0	2	-1	4
x_3	=	-0.5	0	-0.5	2.5
x_1	=	-0.5	-1	0.5	1.5
z	=	1	0	0	-4

At this point we have found a solution to the problem, namely,

$$x = \begin{bmatrix} 1.5 \\ 0 \\ 2.5 \end{bmatrix}.$$

However, it is interesting to note that this is not the only possible solution. If we had chosen a different sequence of pivots, we would have obtained a different solution (with, of course, the same objective value $z = -4$), as we now show.

≫ load ex3-3-3

≫ T = totbl(A,b,p);

≫ T = ljx(T,2,1);

		x_5	x_2	x_3	1
x_4	=	-1	0	2	5
x_1	=	-1	1	1	3
x_6	=	-1	2	0	4
x_7	=	1	-2	-2	1
z	=	1	-2	-2	-3

≫ T = ljx(T,4,3);

		x_5	x_2	x_7	1
x_4	=	0	-2	-1	6
x_1	=	-0.5	0	-0.5	3.5
x_6	=	-1	2	0	4
x_3	=	0.5	-1	-0.5	0.5
z	=	0	0	1	-4

giving the solution

$$x = \begin{bmatrix} 3.5 \\ 0 \\ 0.5 \end{bmatrix}.$$

We can find additional solutions by performing pivots on the columns for which there is a zero in the last row. If we were to take the last tableau above and choose the first column as a pivot column, the ratio test would select row 3 as the pivot row, and the resulting pivot would yield the tableau for the solution that we found first. If, on the other hand, we were to choose the second column as the pivot column, the ratio test would select row 4 as the pivot row as follows:

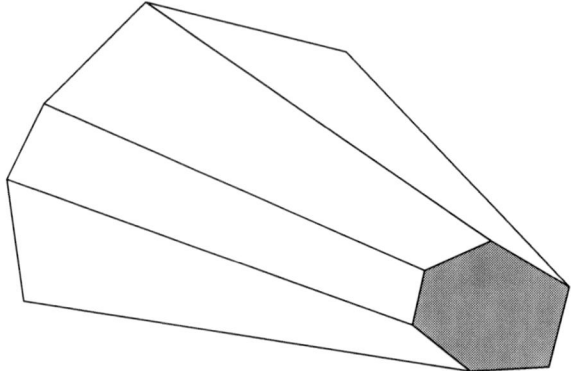

Figure 3.2. *Feasible set in* $\mathbf{R}^3$*, showing an optimal face that contains* 6 *vertices* .

```
>> T = ljx(T,4,2);
```

		x_5	x_3	x_7	1
x_4	=	-1	2	0	5
x_1	=	-0.5	0	-0.5	3.5
x_6	=	0	-2	-1	5
x_2	=	0.5	-1	-0.5	0.5
z	=	0	0	1	-4

yielding the solution $x = (3.5, 0.5, 0)'$. Continuing this process, if we were to choose the second column as the pivot column again, we would return to the previous tableau and hence recover the previous solution. If, on the other hand, we were to select the first column, then the ratio test would select row 1 as the pivot row, and we would obtain

```
>> T = ljx(T,1,1);
```

		x_4	x_3	x_7	1
x_5	=	-1	2	0	5
x_1	=	0.5	-1	-0.5	1
x_6	=	0	-2	-1	5
x_2	=	-0.5	0	-0.5	3
z	=	0	0	1	-4

yielding another solution $x = (1, 3, 0)'$. Proceeding in this fashion, taking care not to double back to a previous solution by choosing the same pivot column twice, we can identify two more distinct solutions: $x = (0, 3, 1)'$ and $x = (3.5, 0, 0.5)'$. These six solutions represent the vertices of an *optimal face*, defined by the convex hull (3.14), like the one illustrated in Figure 3.2. Every point in the optimal face can be associated with a particular choice of coefficients α_i in the formula (3.14).

Note that an algebraic description of the solution set can be derived from any optimal tableau. A solution is a feasible point whose objective value is equal to the optimal value. The equations represented in the tableau, along with $x_i \geq 0$, determine feasibility, while the last row determines the objective value. For example, in the optimal tableau given above, the last row represents the equation

$$z = x_7 - 4.$$

Since $x_7 \geq 0$, it follows that in any solution, we have $x_7 = 0$. The feasibility constraints then amount to $x_3 = \lambda \geq 0$, $x_4 = \mu \geq 0$, and

$$
\begin{array}{rcrcrcr}
0 \leq x_5 & = & -\mu & + & 2\lambda & + & 5, \\
0 \leq x_1 & = & 0.5\mu & - & \lambda & + & 1, \\
0 \leq x_6 & = & & - & 2\lambda & + & 5, \\
0 \leq x_2 & = & -0.5\mu & & & + & 3.
\end{array}
$$

Thus the solution set in this example is

$$\left\{ x = (0.5\mu - \lambda + 1, 3 - 0.5\mu, \lambda)' \geq 0 \mid \lambda, \mu \geq 0, 2\lambda + 5 \geq \mu, 2\lambda \leq 5 \right\}.$$

Note that the solutions generated by each of the six optimal tableaus are all in this set (for particular choice of λ and μ). For concreteness, note that the first optimal tableau found generates the solution given by $\lambda = 2.5$, $\mu = 6$. ■

Exercise 3-3-4. Solve the following problem using MATLAB:

$$
\begin{array}{ll}
\min & z = p'x \\
\text{subject to} & Ax \geq b, \quad x \geq 0,
\end{array}
$$

where

$$
A = - \begin{bmatrix} 1 & 3 & 0 & 1 \\ 2 & 1 & 0 & 0 \\ 0 & 1 & 4 & 1 \end{bmatrix}, \qquad b = - \begin{bmatrix} 4 \\ 3 \\ 3 \end{bmatrix}, \qquad p = - \begin{bmatrix} 2 \\ 4 \\ 1 \\ 1 \end{bmatrix}.
$$

Exercise 3-3-5.   1. Use the simplex procedure to solve

$$
\begin{array}{rcrcrcr}
\min & z = x - y \\
\text{subject to} & -x & + & y & \geq & -2, \\
& -x & - & y & \geq & -6, \\
& & & x, y & \geq & 0.
\end{array}
$$

2. Draw a graphical representation of the problem in x, y space and indicate the path of the simplex steps.

3. Repeat the problem above but using the new objective function $z = -x + y$. This problem has multiple solutions, and so find all the vertex solutions and write down an expression for the full set of solutions.

4. Solve the following problem, and graph the path followed by the simplex method:

$$
\begin{array}{rcrcrcr}
\min & z = -x - y \\
\text{subject to} & 2x & - & y & \geq & -1, \\
& -x & + & y & \geq & -1, \\
& & & x, y & \geq & 0.
\end{array}
$$

Exercise 3-3-6. Solve the following linear program. Is the feasible region unbounded? Is the solution set unbounded? If so, find vectors u and v such that $u + \lambda v$ is optimal for all $\lambda \geq 0$.

$$
\begin{array}{rrrrrrr}
\text{min} & z = 3x_1 & - & 3x_2 & - & 2x_3 & + & 5x_4 \\
\text{subject to} & x_1 & - & x_2 & - & x_3 & - & 0.5x_4 & \geq & -2, \\
& 2.5x_1 & + & 2.25x_2 & - & 0.5x_3 & + & 4.25x_4 & \geq & -4, \\
& -x_1 & + & x_2 & & & - & 2x_4 & \geq & -3, \\
& & & & & & x_1, x_2, x_3, x_4 & \geq & 0.
\end{array}
$$

Exercise 3-3-7. Consider the linear program in three variables given by

$$
\begin{array}{rl}
\text{min} & z = p'x \\
\text{subject to} & x_1 \geq 0, \quad x_2 \geq 0, \quad x_3 \geq 1.
\end{array}
$$

1. Is the feasible region unbounded?

2. When $p = [1 \ 1 \ 1]'$, does the problem have a solution? Is the solution unique?

3. When $p = [1 \ 0 \ 1]'$, does the problem have a solution? Is the solution unique? If not, determine the solution set, and indicate whether the solution set is an unbounded set or not.

4. When $p = [-1 \ 0 \ 1]'$, does the problem have a solution? Is the solution unique?

In all three cases, indicate what properties the final tableau generated by an application of the simplex method to the problem would have. Can you construct an example of a linear program that has multiple solutions but such that the solution set is a bounded set?

3.4 The Phase I Procedure

In all the problems we have examined to date, the linear program has been stated in standard form, and the tableau constructed from the problem data has been feasible. This situation occurs when $x = 0$ is feasible with respect to the constraints, that is, when the right-hand side of all the inequality constraints is nonpositive. In general, however, this need not be the case, and we are often faced with the task of identifying a feasible initial point (that is, a feasible tableau), so that we can go ahead and apply the Phase II procedure described in Section 3.3. The process of identifying an initial feasible tableau is called *Phase I*.

Phase I entails the solution of a linear program that is different from, though closely related to, the problem we actually wish to solve. It is easy to identify an initial feasible tableau for the modified problem, and its eventual solution tells us whether the *original* problem has a feasible tableau or not. If the original problem has a feasible tableau, it can be easily constructed from the tableau resulting from Phase I.

The Phase I problem contains one additional variable x_0, a set of constraints that is the same as the original problem except for the addition of x_0 to some of them, and an objective function of x_0 itself. It can be stated as follows:

$$
\begin{aligned}
\min_{x_0, x} \quad & z_0 = x_0 \\
\text{subject to} \quad & x_{n+i} = A_i.x - b_i + x_0 && \text{if } b_i > 0, \\
& x_{n+i} = A_i.x - b_i && \text{if } b_i \le 0, \\
& x_0, x \ge 0.
\end{aligned}
\tag{3.15}
$$

The variable x_0 is an *artificial variable*. Note that the objective of (3.15) is bounded below by 0, since x_0 is constrained to be nonnegative. We note a number of important facts about this problem:

- We can obtain a feasible point for (3.15) by setting $x_0 = \max\left(\max_{1 \le i \le m} b_i, 0\right)$ and $x_N = 0$ for $N = \{1, \ldots, n\}$. The dependent variables x_B, where $B = \{n+1, n+2, \ldots, n+m\}$, then take the following initial values:

$$
b_i > 0 \implies x_{n+i} = A_i.x - b_i + x_0 = -b_i + \max_{1 \le j \le m, \, b_j > 0} b_j \ge -b_i + b_i = 0,
$$

$$
b_i \le 0 \implies x_{n+i} = A_i.x - b_i = -b_i \ge 0,
$$

  so that $x_B \ge 0$, and these components are also feasible.

- If there exists a point $\bar{x}$ that is feasible for the original problem, then the point $(x_0, x) = (0, \bar{x})$ is feasible for the Phase I problem. (It is easy to check this fact by verifying that $x_{n+i} \ge 0$ for $i = 1, 2, \ldots, m$.)

- Conversely, if $(0, x)$ is a solution of the Phase I problem, then x is feasible for the original problem. We see this by examining the constraint set for the Phase I problem and noting that

$$
b_i > 0 \implies 0 \le x_{n+i} = A_i.x - b_i + x_0 = A_i.x - b_i,
$$

$$
b_i \le 0 \implies 0 \le x_{n+i} = A_i.x - b_i,
$$

  so that $Ax \ge b$.

- If (x_0, x) is a solution of the Phase I problem and x_0 is *strictly positive*, then the original problem must be infeasible. This fact follows immediately from the observations above: If the original problem were feasible, it would be possible to find a feasible point for the Phase I problem with objective zero.

We can set up this starting point by forming the initial tableau for (3.15) in the usual way and performing a "special pivot." We select the x_0 column as the pivot column and choose the pivot row to be a row with the most negative entry in the last column of the tableau.

After the special pivot, the tableau contains only nonnegative entries in its last column, and the simplex method can proceed, using the usual rules for pivot column and row selection. Since the objective of (3.15) is bounded below (by zero), it can terminate only at an optimal tableau. Two possibilities then arise.

- The optimal objective z_0 is *strictly positive*. In this case, we conclude that the original problem (3.1) is *infeasible*, and so we terminate without going to Phase II.

- The optimal objective z_0 is *zero*. In this case, x_0 must also be zero, and the remaining components of x are a feasible initial point for the original problem. We can construct a feasible table for the initial problem from the optimal tableau for the Phase I problem as follows. First, if x_0 is still a dependent variable in the tableau (that is, one of the row labels), perform a Jordan exchange to make it an independent variable. (Since $x_0 = 0$, this pivot will be a degenerate pivot, and the values of the other variables will not change.) Next, delete the column labeled by x_0 and the row labeled by z_0 from the tableau. The tableau that remains is feasible for the original problem (3.1), and we can proceed with Phase II, as described in Section 3.3.

We summarize Phase I and then proceed with an example of the MATLAB implementation.

Algorithm 3.2 (Phase I).

1. *If $b \leq 0$, then $x_\text{B} = -b$, $x_\text{N} = 0$ is a feasible point corresponding to the initial tableau and no Phase I is required. Skip to Phase II.*

2. *If $b \not\leq 0$, introduce the artificial variable x_0 (and objective function $z_0 = x_0$) and set up the Phase I problem (3.15) and the corresponding tableau.*

3. *Perform the "special pivot" of the x_0 column with a row corresponding to the most negative entry of the last column to obtain a feasible tableau for Phase I.*

4. *Apply standard simplex pivot rules until an optimal tableau for the Phase I problem is attained. If the optimal value (for z_0) is positive,* **stop:** *The original problem has no feasible point. Otherwise, perform an extra pivot (if needed) to move x_0 to the top of the tableau.*

5. *Strike out the column corresponding to x_0 and the row corresponding to z_0 and proceed to Phase II.*

The following example shows how to perform the two-phase simplex method.

Example 3-4-1.

$$
\begin{array}{rrcrcr}
\min & 4x_1 & + & 5x_2 & & \\
\text{subject to} & x_1 & + & x_2 & \geq & -1, \\
 & x_1 & + & 2x_2 & \geq & 1, \\
 & 4x_1 & + & 2x_2 & \geq & 8, \\
 & -x_1 & - & x_2 & \geq & -3, \\
 & -x_1 & + & x_2 & \geq & 1, \\
 & & & x_1, x_2 & \geq & 0.
\end{array}
$$

We start by loading the data into a tableau and then adding a column for the artificial variable x_0 and the Phase I objective z_0. We use the MATLAB routines `addrow` and `addcol`. The last argument of each of these routines specifies the position that we wish the new row/column to occupy within the augmented tableau.

```
>> load ex3-4-1
>> T = totbl(A,b,p);
>> neg = [0 1 1 0 1 0]';
>> T = addcol(T,neg,'x0',3);
>> T = addrow(T,[0 0 1 0],'z0',7);
```

		x_1	x_2	x_0	1
x_3	=	1	1	0	1
x_4	=	1	2	1	−1
x_5	=	4	2	1	−8
x_6	=	−1	−1	0	3
x_7	=	−1	1	1	−1
z	=	4	5	0	0
z_0	=	0	0	1	0

We now perform the "special" pivot, using the MATLAB max command to identify the largest element in the vector b (which corresponds to the most negative element in the final column of the tableau). This command also returns the position r occupied by this largest element. The variable s denotes the column position occupied by the x_0 column.

```
>> [maxviol,r] = max(b);
>> s = length(p)+1;
>> T = ljx(T,r,s);
```

		x_1	x_2	x_5	1
x_3	=	1	1	0	1
x_4	=	−3	0	1	7
x_0	=	−4	−2	1	8
x_6	=	−1	−1	0	3
x_7	=	−5	−1	1	7
z	=	4	5	0	0
z_0	=	−4	−2	1	8

We now proceed with the simplex method, using the usual rules for selecting pivot columns and rows. Note that we use the last row in the tableau—the z_0 row—to select the pivot columns, since it is z_0 (not z) that defines the objective for Phase I. However, we modify the entries in row z along with the rest of the tableau.

```
>> T = ljx(T,4,2);
```

		x_1	x_6	x_5	1
x_3	=	0	−1	0	4
x_4	=	−3	0	1	7
x_0	=	−2	2	1	2
x_2	=	−1	−1	0	3
x_7	=	−4	1	1	4
z	=	−1	−5	0	15
z_0	=	−2	2	1	2

```
>> T = ljx(T,3,1);
```

		x_0	x_6	x_5	1
x_3	=	0	−1	0	4
x_4	=	1.5	−3	−0.5	4
x_1	=	−0.5	1	0.5	1
x_2	=	0.5	−2	−0.5	2
x_7	=	2	−3	−1	0
z	=	0.5	−6	−0.5	14
z_0	=	1	0	0	0

At this point, we have solved Phase I and x_0 appears again at the top of the tableau. We now delete the x_0 column and the z_0 row to obtain the following feasible tableau for Phase II:

```
>> T = delrow(T,'z0');
>> T = delcol(T,'x0');
```

		x_6	x_5	1
x_3	=	-1	0	4
x_4	=	-3	-0.5	4
x_1	=	1	0.5	1
x_2	=	-2	-0.5	2
x_7	=	-3	-1	0
z	=	-6	-0.5	14

Although feasible, this tableau is not optimal for Phase II. Simplex rules lead us to perform the following degenerate pivot:

```
>> T = ljx(T,5,1);
```

		x_7	x_5	1
x_3	=	$1/3$	$1/3$	4
x_4	=	1	0.5	4
x_1	=	$-1/3$	$1/6$	1
x_2	=	$2/3$	$1/6$	2
x_6	=	$-1/3$	$-1/3$	0
z	=	2	1.5	14

We have now identified a solution to the original problem: $x_1 = 1$, $x_2 = 2$, $z = 14$. ∎

The Phase I technique is interesting not just as a way to identify a starting point for Phase II but also in its own right as a technique to find a feasible point for a system of inequalities. The approach is the same; the only difference is that the tableau does not have a row for the objective function of the original problem (since there *is* no objective function).

Exercise 3-4-2.   1. Demonstrate that

$$
\begin{aligned}
\min \quad & z = -3x_1 + x_2 \\
\text{subject to} \quad -x_1 \; - \; & x_2 \; \geq \; -2, \\
2x_1 \; + \; & 2x_2 \; \geq \; 10, \\
& x_1, x_2 \; \geq \; 0
\end{aligned}
$$

is infeasible.

2. Demonstrate that

$$
\begin{aligned}
\min \quad & z = -x_1 + x_2 \\
\text{subject to} \quad 2x_1 \; - \; & x_2 \; \geq \; 1, \\
x_1 \; + \; & 2x_2 \; \geq \; 2, \\
& x_1, x_2 \; \geq \; 0
\end{aligned}
$$

is unbounded.

Exercise 3-4-3. Solve the following linear programs using MATLAB. Give an optimal vector and its objective value if solvable; give a direction of unboundedness if unbounded; or verify that the problem is infeasible. Each problem has the form

$$\min \quad z = p'x$$
$$\text{subject to} \quad Ax \geq b, \quad x \geq 0,$$

where the data A, b, and p are given below.

1.

$$A = \begin{bmatrix} -1 & -3 & 0 & -1 \\ -2 & -1 & 0 & 0 \\ 0 & -1 & -4 & -1 \\ 1 & 1 & 2 & 0 \\ -1 & 1 & 4 & 0 \end{bmatrix}, \quad b = \begin{bmatrix} -4 \\ -3 \\ -3 \\ 1 \\ 1 \end{bmatrix}, \quad p = \begin{bmatrix} -2 \\ -4 \\ -1 \\ -1 \end{bmatrix}.$$

2.

$$A = \begin{bmatrix} -1 & 0 & -1 \\ -1 & 1 & 0 \\ 0 & -1 & 1 \end{bmatrix}, \quad b = \begin{bmatrix} 2 \\ -1 \\ 3 \end{bmatrix}, \quad p = \begin{bmatrix} -1 \\ -3 \\ 0 \end{bmatrix}.$$

Exercise 3-4-4. Using the simplex method, find *all* solutions of

$$\min \quad z = 2x_1 + 3x_2 + 6x_3 + 4x_4$$
$$\text{subject to} \quad x_1 + 2x_2 + 3x_3 + x_4 \geq 5,$$
$$x_1 + x_2 + 2x_3 + 3x_4 \geq 3,$$
$$x_1, x_2, x_3, x_4 \geq 0.$$

3.5 Finite Termination

3.5.1 The Nondegenerate Case

We now ask the question, Is the method described in the previous section guaranteed to terminate (either at a solution or by finding a direction of unboundedness), or can we cycle forever between Steps 3 and 6? In this section, we show that termination is guaranteed under certain rather restrictive assumptions. (We relax these assumptions and prove a more general result in Section 3.5.3.)

To set up our result, we define the concept of a *degenerate tableau*.

Definition 3.5.1. A feasible tableau is *degenerate* if the last column contains any zero elements. If the elements in the last column are all strictly positive, the tableau is *nondegenerate*. A linear program is said to be *nondegenerate* if all feasible tableaus for that linear program are nondegenerate.

Geometrically, a tableau is nondegenerate if the vertex it defines is at the intersection of exactly n hyperplanes of the form $x_j = 0$; namely, those hyperplanes defined by $j \in N$. Vertices that lie at the intersection of more than n hyperplanes correspond to degenerate tableaus. (See the examples at the end of Section 3.2, illustrated in Figure 3.1.) Consequently, a linear program is nondegenerate if each of the vertices of the feasible region for that linear program is defined uniquely by a set N.

We encounter degenerate tableaus during the simplex method when there is a tie in the ratio test for selection of the pivot row. After the pivot is performed, zeros appear in the last column of the row(s) that tied but were *not* selected as pivots.

The finite termination of the simplex method under these assumptions is now shown.

Theorem 3.5.2. *If a linear program is* feasible *and* nondegenerate, *then starting at any feasible tableau, the objective function strictly decreases at each pivot step. After a finite number of pivots the method terminates with an* optimal *point or else identifies a direction of* unboundedness.

Proof. At every iteration, we must have a nonoptimal, optimal, or unbounded tableau. In the latter two cases, termination occurs. In the first case, the following transformation occurs when we pivot on an element H_{rs}, for which $h_r > 0$ (nondegeneracy) and $c_s < 0$ (by pivot selection), and $H_{rs} < 0$ (by the ratio test):

$$
\begin{array}{cc}
 & \begin{array}{cc} x_N & 1 \end{array} \\
\begin{array}{c} x_B = \\ z = \end{array} & \begin{array}{|c|c|} \hline H & h \\ \hline c' & \alpha \\ \hline \end{array}
\end{array}
\longrightarrow
\begin{array}{cc}
 & \begin{array}{cc} x_{\tilde{N}} & 1 \end{array} \\
\begin{array}{c} x_{\tilde{B}} = \\ z = \end{array} & \begin{array}{|c|c|} \hline \tilde{H} & \tilde{h} \\ \hline \tilde{c}' & \tilde{\alpha} \\ \hline \end{array}
\end{array}
$$

Here

$$
\tilde{\alpha} = \alpha - \frac{c_s h_r}{H_{rs}} < \alpha,
$$

where the strict inequality follows from the properties of c_s, h_r, and H_{rs}. Hence, we can never return to the tableau with objective α, since this would require us to increase the objective at a later iteration, something the simplex method does not allow. Since we can only visit each possible tableau at most once, and since there are only a finite number of possible tableaus, the method must eventually terminate at either an optimal or an unbounded tableau.   ☐

In fact, a bound on the number of possible tableaus is obtained by determining the number of ways to choose the nonbasic set N (with n indices) from the index set $\{1, 2, \ldots, m+n\}$ which, by elementary combinatorics, is

$$
\binom{m+n}{n} = \frac{(m+n)!}{m!n!}.
$$

3.5.2 Cycling

We start by giving a classic example due to Beale (1955), which shows that for a degenerate linear program, reasonable rules for selecting pivots can fail to produce finite termination. In this example, the simplex method repeats the same sequence of six pivots indefinitely, making no progress toward a solution.

Example 3-5-1 (Beale). Consider the initial tableau

```
>> load beale
>> T = totbl(A,b,p);
```

		x_1	x_2	x_3	x_4	1
x_5	=	-0.5	5.5	2.5	-9	0
x_6	=	-0.5	1.5	0.5	-1	0
x_7	=	-1	0	0	0	1
z	=	-10	57	9	24	0

Let us use the following pivot rules:

1. The pivot column is the one with the most negative entry in the bottom row.

2. If two or more rows tie for being a pivot row in the ratio test, choose the one with the smallest subscript.

Note that the example has zeros in the final column, and so it does not satisfy the nondegeneracy assumption of Section 3.5. Carrying out the simplex method, following the pivot rules above, we obtain the following tableau after six pivots:

```
>> T = ljx(T,1,1);
>> T = ljx(T,2,2);
>> T = ljx(T,1,3);
>> T = ljx(T,2,4);
>> T = ljx(T,1,1);
>> T = ljx(T,2,2);
```

		x_3	x_4	x_1	x_2	1
x_5	=	2.5	-9	-0.5	5.5	0
x_6	=	0.5	-1	-0.5	1.5	0
x_7	=	0	0	-1	0	1
z	=	9	24	-10	57	0

Notice that apart from a column reordering, this tableau is identical to the initial tableau. If we continue to apply the pivot rules above, we will continue to cycle through the same six pivots. The positions of the columns may change at the end of every cycle, but the objective function will never change and we will make no progress toward a solution. This phenomenon is known as *cycling*. Cycling can occur only for degenerate problems; a nondegenerate pivot cannot be part of a cycle since it yields a decrease in the objective function. ∎

3.5.3 The General Case

We now revisit the issue of finite termination of the simplex method. The main result shows that a particular variant of the simplex method, when started from a feasible tableau, is guaranteed to produce either a solution or a direction of unboundedness within a finite number of iterations—without any nondegeneracy assumption such as the one used in Section 3.5. Finite termination of the two-phase simplex method—that is, determination of infeasibility, optimality, or unboundedness within a finite number of simplex pivots—follows as a simple consequence. Finite termination depends crucially on the rule used to select pivot columns (in the event of more than one negative entry in the last row) and on the rule for selecting the pivot row (in the event of a tie in the ratio test). As shown above, even apparently reasonable rules can fail to produce finite termination.

We now modify the pivot selection rule of the simplex method to overcome this problem. This rule was introduced by Bland (1977) and is commonly called Bland's rule or the *smallest-subscript rule*.

1. *Pricing* (pivot column selection): The pivot column is the smallest $N(s)$ of nonbasic variable indices such that column s has a negative element in the bottom row (reduced cost).

2. *Ratio Test* (pivot row selection): The pivot row is the smallest $B(r)$ of basic variable indices such that row r satisfies

$$-h_r/H_{rs} = \min_i \{-h_i/H_{is} \mid H_{is} < 0\}.$$

In other words, among all possible pivot columns (those with negative reduced costs), we choose the one whose label has the smallest subscript. Among all possible pivot rows (those that tie for the minimum in the ratio test), we again choose the one whose label has the smallest subscript.

Example 3-5-2. We apply the smallest-subscript rule to Beale's problem.

```
>> load beale
>> T = totbl(A,b,p);
```

		x_1	x_2	x_3	x_4	1
x_5	=	-0.5	5.5	2.5	-9	0
x_6	=	-0.5	1.5	0.5	-1	0
x_7	=	-1	0	0	0	1
z	=	-10	57	9	24	0

The pivot column must be the column labeled x_1, since it is the only one with a negative reduced cost. In the ratio tests, the rows labeled x_5 and x_6 tie for minimum ratio; the smallest-subscript rule chooses x_5. By continuing in this manner, we generate the following sequence of five pivots, ending with the tableau shown below.

```
>> T = ljx(T,1,1);
>> T = ljx(T,2,2);
>> T = ljx(T,1,3);
>> T = ljx(T,2,4);
>> T = ljx(T,1,1);
```

		x_3	x_6	x_1	x_2	1
x_5	=	-2	9	4	-8	0
x_4	=	0.5	-1	-0.5	1.5	0
x_7	=	0	0	-1	0	1
z	=	21	-24	-22	93	0

At this point, the smallest-subscript rule chooses the column labeled x_1 as the pivot column, whereas the "most negative" rule described earlier would have chosen x_6. By continuing with the next pivots, we obtain the following tableau:

```
>> T = ljx(T,2,3);
>> T = ljx(T,3,1);
```

		x_7	x_6	x_4	x_2	1
x_5	=	-2	5	-4	-2	2
x_1	=	-1	0	0	0	1
x_3	=	-1	2	2	-3	1
z	=	1	18	42	30	-1

Finally, we have arrived at an optimal tableau. The optimal value is -1, achieved when $x = (1, 0, 1, 0, 2, 0, 0)$. ■

The following theorem establishes finiteness of the simplex method using the smallest-subscript rule, without any nondegeneracy assumption. The proof closely follows the one given by Chvátal (1983).

Theorem 3.5.3. *If a linear program is* feasible, *then starting at any feasible tableau, and using the smallest-subscript anticycling rule, the simplex method terminates after a finite number of pivots at an* optimal *or* unbounded *tableau.*

Proof. Since there are at most $\binom{m+n}{m}$ ways of choosing m basic variables from $n + m$ variables, some choice of basic variables (and the tableau that corresponds to these variables) must repeat if the simplex method cycles. We will show, by contradiction, that the method cannot cycle when the smallest-subscript anticycling rule is used.

Suppose for contradiction that some tableau is repeated. Then there is a sequence of degenerate pivots that leads from some tableau T_0 back to itself. We denote this sequence of tableaus as follows:

$$T_0, T_1, \ldots, T_k = T_0.$$

If any one of these pivots were nondegenerate, the objective function would decrease and the cycle would be broken. Hence, all pivots in the above sequence must be degenerate.

A variable will be called *fickle* if it is nonbasic in some of the tableaus $T_0, T_1, \ldots, T_k = T_0$ and basic in others. *Every* fickle variable must become basic at least once in the cycle *and* become nonbasic at least once in the cycle since the set of basic and nonbasic variables is the same at the beginning and end of the cycle. Furthermore, the value of a fickle variable is always zero, since every pivot is degenerate.

Among all fickle variables, let x_l have the largest subscript. Then among $T_1, \ldots, T_k$ there is some T where x_l is the basic variable chosen to become nonbasic. Let x_s be the (fickle) nonbasic variable that is chosen to become basic at this pivot. Our logic will show that there is another fickle variable $x_r, r < l$, that is eligible to become nonbasic at this pivot step as well, contradicting our choice of x_l as the variable to become nonbasic. The tableau T looks like this so far when we order according to subscripts:

$$
T: \qquad\qquad\qquad\qquad
\begin{array}{c|cc|c}
 & & x_s & 1 \\
\hline
x_r = & & H_{rs} & \\
x_l = & & H_{ls} < 0 & 0 \\
\hline
z = & \geq 0 & c_s < 0 & \alpha
\end{array}
\qquad\qquad (3.16)
$$

For the purposes of this proof, we abuse notation and assume that the entries of the tableau h, c, and H are indexed not by $1, 2, \ldots, m$ and $1, 2, \ldots, n$ but by the corresponding basic and nonbasic labels B and N. Note therefore that $c_s < 0$ since s is chosen to enter at this pivot step. Since x_l is fickle, it must also be chosen to become basic at some other tableau $\tilde{T}$ in the sequence $T_1, T_2, \ldots, T_k$. Since the objective value is the same for all tableaus, the bottom row of $\tilde{T}$ represents the equation

$$z = \tilde{c}'_{\tilde{N}} x_{\tilde{N}} + \alpha, \qquad\qquad (3.17)$$

where $\tilde{N}$ (and $\tilde{B}$) correspond to the nonbasic (and basic) variables in $\tilde{T}$. By defining $\tilde{c}_{\tilde{B}} = 0$, we obtain

$$z = \tilde{c}'_{\tilde{B}} x_{\tilde{B}} + \tilde{c}'_{\tilde{N}} x_{\tilde{N}} + \alpha = \tilde{c}'x + \alpha.$$

For any $\lambda \in \mathbf{R}$, the tableau T (3.16) above defines the following relationships between the variables:

$$x_i = \begin{cases} \lambda & \text{if } i = s, \\ H_{js}\lambda + h_j & \text{if } i = B(j), \\ 0 & \text{if } i \in N \setminus \{s\}, \end{cases} \tag{3.18}$$

$$z = c_s\lambda + \alpha.$$

By Theorem 2.1.1, the value $z = c_s\lambda + \alpha$ must be the same as the value of z obtained by substituting (3.18) into (3.17). Hence,

$$c_s\lambda + \alpha = \tilde{c}_{\tilde{N}} x_{\tilde{N}} + \alpha = \tilde{c}_s\lambda + \sum_{i \in B} \tilde{c}_i (H_{is}\lambda + h_i) + \alpha,$$

which implies that

$$\left(c_s - \tilde{c}_s - \sum_{i \in B} \tilde{c}_i H_{is}\right)\lambda = \sum_{i \in B} \tilde{c}_i h_i.$$

Since this relationship holds for *any* $\lambda \in \mathbf{R}$, we must have

$$c_s - \tilde{c}_s - \sum_{i \in B} \tilde{c}_i H_{is} = 0. \tag{3.19}$$

Since x_s becomes basic in T, we have from (3.16) that $c_s < 0$. Also, x_s does not become basic in $\tilde{T}$ (x_l does), and so because $s < l$, we must have $\tilde{c}_s \geq 0$. Thus from (3.19), we have

$$\sum_{i \in B} \tilde{c}_i H_{is} < 0,$$

and so there must be some $r \in B$ such that

$$\tilde{c}_r H_{rs} < 0. \tag{3.20}$$

This is the r we require for our contradiction. Note the following properties of x_r:

1. x_r is fickle: From (3.20), $\tilde{c}_r \neq 0$, and so $\tilde{c}_{\tilde{B}} = 0$ implies that $r \in \tilde{N}$. On the other hand, $r \in B$, and so x_r is fickle. Since l is the largest fickle index, we must have $r \leq l$.

2. $r < l$: Since x_l becomes nonbasic in T, $H_{ls} < 0$; see (3.16). Also $\tilde{c}_l < 0$ because x_l becomes basic in $\tilde{T}$. It follows from (3.20) that $r \neq l$.

3. $H_{rs} < 0$: Since (as established above) $r \in \tilde{N}$, and since $r < l$, we must have that $\tilde{c}_r \geq 0$ (otherwise r, not l, would be chosen to become basic at $\tilde{T}$). It now follows from (3.20) that $H_{rs} < 0$.

By combining all this information, we conclude that T has the following structure:

		x_s	1
x_r	$=$	$H_{rs} < 0$	0
x_l	$=$	$H_{ls} < 0$	0
z	$=$	$\geq 0 \quad c_s < 0$	α

According to this tableau, x_r is chosen to become nonbasic, and $r < l$ contradicts our hypothesis that x_l is chosen to become nonbasic in T.   □

We can now use Theorem 3.5.3 to show that the two-phase simplex method identifies a solution, or else indicates infeasibility or unboundedness, after a finite number of pivots.

Theorem 3.5.4. *For a linear program (3.1), the two-phase simplex method with the smallest-subscript anticycling rule terminates after a finite number of pivots with a conclusion that the problem is* infeasible, *or at an* optimal *or* unbounded *tableau.*

Proof. Consider first Phase I. After the initial pivot, we have a feasible tableau for Phase I, and since the Phase I problem is bounded below, it must have an optimal solution. Theorem 3.5.3 indicates that this solution is found after a finite number of pivots. If $x_0 > 0$ at optimality, then the original problem is infeasible. If $x_0 = 0$, then we can construct a feasible tableau for Phase II. By applying Theorem 3.5.3 again, we find that an optimal or unbounded tableau is reached for Phase II after a finite number of pivots.   □

Exercise 3-5-3. Consider the following tableau:

		x_2	x_3	1
x_1	$=$	1	-2	2
x_5	$=$	2	-1	1
x_4	$=$	2	-1	2
z	$=$	1	$\mu - 3$	-1

Note that the variables $x_1, x_2, \ldots, x_5$ are all nonnegative. Indicate for each of the three cases $\mu < 3$, $\mu > 3$, $\mu = 3$ whether the tableau is

- infeasible;

- feasible but not optimal (do not do any pivoting);

- optimal with a unique solution (write it down);

- optimal with a nonunique solution (write down at least two solutions).

Exercise 3-5-4. Read the following statements carefully to see whether it is *true* or *false*. Explain your answers briefly in each case.

1. If a linear program has more than one optimal solution, it must have an infinite number of optimal solutions.

2. The Phase I problem in the two-phase simplex method can be unbounded.

3. In solving a linear program by the simplex method, starting with a feasible tableau, a different feasible point is generated after every pivot step.

Although anticycling rules are needed to guarantee finite termination of the simplex method, it has been observed that cycling is a rare phenomenon in practice (Kotiah & Steinberg (1977), Kotiah & Steinberg (1978)). In practice, anticycling rules are not applied at every pivot, since they tend to increase the number of pivots required (Gass (1985), p. 183). Practical codes often "turn on" the anticycling rules if the simplex method stalls at the same vertex for a large number of pivots.

3.6 Linear Programs in Nonstandard Form

We know from Theorem 3.5.4 that the two-phase simplex method can solve any linear program that is posed in standard form. In this section, we show that *any* linear program, including problems with equality constraints and free variables, can be rewritten in standard form. Hence the simplex method can be applied to any linear program whatsoever, and the results of Theorem 3.5.4 apply.

We discuss two approaches. In the first, known as *Scheme I*, we use explicit transformations and substitutions to express a nonstandard linear program in standard form. We can then apply the two-phase simplex method. Once the solution is obtained, we can recover the solution of the original problem by substituting into the transformations we made. The second approach, *Scheme II*, takes a more direct route. Here we handle some of the nonstandard aspects of the formulation by performing some preliminary pivots on the tableau, *prior* to starting the simplex method.

We describe these two techniques in turn, after outlining some of the transformations needed to change the form of the constraints.

3.6.1 Transforming Constraints and Variables

We first note that any maximization problem can be converted into an equivalent minimization problem by taking the negative of the objective. Specifically, given any function f and constraint set S, any vector $\bar{x}$ that solves the problem $\max_{x \in S} f(x)$ is also a solution of the problem $\min_{x \in S} (-f(x))$, and vice versa. To solve the first (maximization) problem, we find the solution of the second (minimization) problem and then negate the optimal objective value.

Second, we can convert any less-than inequalities into greater-than inequalities by simply multiplying both sides of the inequality by -1. For instance, the constraint $x_1 - 2x_3 \leq 4$ is equivalent to $-x_1 + 2x_3 \geq -4$.

As an example, we revisit the Professor's diary problem, introduced in Section 1.1. The following problem was solved using a graphical method:

$$
\begin{array}{rrrrr}
\max_{x,y} & z = 5x + 4y & & & \\
\text{subject to} & x & & \leq & 6, \\
& .25x & + \quad y & \leq & 6, \\
& 3x & + \quad 2y & \leq & 22, \\
& & x, y & \geq & 0.
\end{array}
$$

To process this problem using Phase II of the simplex method, we need to convert the maximization into a minimization and reverse all the inequalities. The first step enters the modified data and then forms the initial tableau.

```
>> p = [5; 4]; b = [6; 6; 22];
>> A = [1 0; 0.25 1; 3 2];
>> T = totbl(-A,-b,-p);
>> T = relabel(T,'x1','x','x2','y');
```

		x	y	1
x_3	$=$	-1	0	6
x_4	$=$	-0.25	-1	6
x_5	$=$	-3	-2	22
z	$=$	-5	-4	0

The relabel command simply replaces labels in the obvious manner. Since this tableau is feasible, we simply apply Phase II of the simplex method, resulting in the following two pivots:

```
>> T = ljx(T,2,2);
>> T = ljx(T,3,1);
```

		x_5	x_4	1
x_3	$=$	0.4	-0.8	2
y	$=$	0.1	-1.2	5
x	$=$	-0.4	0.8	4
z	$=$	1.6	0.8	-40

Note that the final solution has $x = 4$ and $y = 5$ and that the final objective value is 40 (where we convert the value -40 since the original problem is a maximization).

These techniques are further demonstrated in the following more general example.

Example 3-6-1.

$$
\begin{array}{rrrrrrr}
\max & z = x_1 + x_2 - x_3 & & & & & \\
\text{subject to} & x_1 & + & x_2 & + & x_3 & \leq & 4, \\
& 2x_1 & - & x_2 & & & \geq & 2, \\
& -x_1 & - & x_2 & + & x_3 & \leq & 10, \\
& x_1 & + & 2x_2 & - & x_3 & \geq & 1, \\
& & & & x_1, x_2, x_3 & \geq & 0.
\end{array}
$$

```
>> load ex3-6-1
>> A(1,:) = -A(1,:); b(1) = -b(1);
>> A(3,:) = -A(3,:); b(3) = -b(3);
>> T = totbl(A,b,-p);
```

		x_1	x_2	x_3	1
x_4	$=$	-1	-1	-1	4
x_5	$=$	2	-1	0	-2
x_6	$=$	1	1	-1	10
x_7	$=$	1	2	-1	-1
z	$=$	-1	-1	1	0

Since this tableau is in standard form, we can apply the two-phase simplex method to solve it. The details are left to the reader. ∎

Exercise 3-6-2. Use the two-phase simplex procedure to solve

$$
\begin{array}{rlrcl}
\max & z = 2x_1 + 3x_2 & & & \\
\text{subject to} & -4x_1 & + \ 3x_2 & \leq & 12, \\
& 2x_1 & + \ \ x_2 & \leq & 6, \\
& x_1 & + \ \ x_2 & \geq & 3, \\
& 5x_1 & + \ \ x_2 & \geq & 4, \\
& & x_1, x_2 & \geq & 0.
\end{array}
$$

Using the technique above to transform less-than constraints, we can convert any set of general constraints (equalities, less-than inequalities, greater-than inequalities) into a set of equalities and greater-than inequalities. We now consider how to eliminate all bounds on the variables except nonnegativity bounds.

Suppose first that a variable x_i has both a lower and an upper bound; that is,

$$
\bar{l}_i \leq x_i \leq \bar{u}_i. \tag{3.21}
$$

(In Section 5.2.1 of Chapter 5, we describe a variant of the simplex method that handles constraints of this type directly. However, it is instructive to see how such constraints can also be handled using our current version of simplex.)

Substitution Method.   There are two methods to remove these general bound constraints. The first method uses substitution.

When x_i has only a lower bound $x_i \geq \bar{l}_i$ (when $\bar{l}_i \neq 0$), we can apply the following shifting procedure: Define a new variable w_i to be $x_i - \bar{l}_i$, replace x_i by $w_i + \bar{l}_i$ in the formulation, and declare w_i to be a nonnegative variable (this latter restriction being equivalent to $x_i \geq \bar{l}_i$). After solving the problem, we recover the optimal x_i by setting $x_i = w_i + \bar{l}_i$.

Similarly, when x_i has only an upper bound $x_i \leq \bar{u}_i$, we can make the substitution $w_i = \bar{u}_i - x_i$ (that is, make the substitution $x_i = \bar{u}_i - w_i$ throughout) and declare w_i to be a nonnegative variable. After solving the problem, we recover the optimal x_i by setting $x_i = \bar{u}_i - w_i$.

For the more general case with two bounds, if we happen to have $\bar{l}_i = 0$, we can simply add $-x_i \geq -\bar{u}_i$ to the set of general constraints and declare x_i to be a nonnegative variable in the formulation. If $l_i \neq 0$, we make the substitution $w_i = x_i - \bar{l}_i$ as above, and the constraints on w_i are then

$$
0 \leq w_i \leq \bar{u}_i - \bar{l}_i.
$$

We can add the upper bound $w_i \leq \bar{u}_i - \bar{l}_i$ to the list of general constraints, make the substitution $x_i = w_i + \bar{l}_i$ to replace x_i in the other constraints and in the objective, and declare w_i to be a nonnegative variable. After solving the problem, we recover the optimal x_i by setting $x_i = w_i + \bar{l}_i$.

Example 3-6-3. Given the linear program

$$\begin{aligned}
\min \quad & 3x_1 + 2x_2 - x_3 \\
\text{subject to} \quad & x_1 + x_2 = 5, \\
& -1 \le x_1 \le 6, \\
& x_2 \ge 3, \\
& 0 \le x_3 \le 5,
\end{aligned}$$

we use the substitutions $w_1 = x_1 + 1$, $w_2 = x_2 - 3$ to generate the following problem:

$$\begin{aligned}
\min \quad & 3w_1 + 2w_2 - x_3 + 3 \\
\text{subject to} \quad & -w_1 && && \ge \;\; -7, \\
& && - x_3 && \ge \;\; -5, \\
& w_1 \; + \; w_2 && && = \;\;\;\;\; 3, \\
& && w_1, w_2, x_3 && \ge \;\;\;\;\; 0. \quad \blacksquare
\end{aligned}$$

Free Variable Method. An alternative technique for dealing with (3.21) that avoids any substitution is to add the following two inequality constraints to the general constraint set:

$$x_i \ge \bar{l}_i, \qquad -x_i \ge -\bar{u}_i,$$

and declare x_i to be a free variable (that is, do not apply a nonnegativity constraint explicitly).

When we have a simple lower bound constraint, we can add $x_i \ge \bar{l}_i$ to the list of general constraints and declare x_i to be a free variable in the formulation. Similarly, for simple upper bound constraints, we can add $-x_i \ge -\bar{u}_i$ to the list of general constraints and declare x_i to be free.

Using either of the simple reductions described above, we can obtain a formulation in which all the general constraints are either equalities or greater-than inequalities and in which all the variables are either nonnegative variables or are free. We can therefore transform any linear programming problem to the following "general" form:

$$\begin{aligned}
\min_{x,y} \quad & p'x + q'y \\
\text{subject to} \quad & Bx + Cy \;\ge\; d, \\
& Ex + Fy \;=\; g, \\
& x \ge 0, \; y \text{ free.}
\end{aligned} \tag{3.22}$$

For expository purposes, we have gathered all the remaining nonnegative variables into a vector x, and all the free variables into a vector y, and rearranged the variables so that the nonnegative variables are listed before the free variables. Similarly, the equality constraints are listed after all the inequalities.

Example 3-6-4. For the linear program data of Example 3-6-3, we reformulate using the free variable method as a problem with only equalities, greater-than inequalities, nonnegative variables, and free variables as shown:

$$
\begin{array}{rrcrcr}
\min & -x_3 + 3x_1 + 2x_2 & & & & \\
\text{subject to} & x_1 & & & \geq & -1, \\
& -x_1 & & & \geq & -6, \\
& & & x_2 & \geq & 3, \\
& -x_3 & & & \geq & -5, \\
& x_1 & + & x_2 & = & 5, \\
\end{array}
$$
$$x_3 \geq 0,\ x_1, x_2 \text{ free.}$$

Note that the transformed problem has five "general" constraints. The formulation using substitution given previously, while slightly more complicated to carry out, results in a problem with the same number of variables but only three general constraints. ∎

3.6.2   Scheme I

We can reformulate (3.22) in standard form by (i) replacing the free variables with a difference between their positive and negative parts, that is,

$$y \text{ free} \iff y = y^+ - y^-, \qquad y^+, y^- \geq 0;$$

and (ii) replacing the equations by two inequalities as shown:

$$
Ex + Fy = g \iff \begin{array}{l} Ex + Fy \geq g, \\ Ex + Fy \leq g. \end{array}
$$

By making both these substitutions in (3.22), we obtain the following:

$$
\begin{array}{rrcrcrcrcr}
\min_{x,y^+,y^-} & p'x + q'y^+ - q'y^- & & & & & & & \\
\text{subject to} & Bx & + & Cy^+ & - & Cy^- & \geq & d, \\
& Ex & + & Fy^+ & - & Fy^- & \geq & g, \\
& -Ex & - & Fy^+ & + & Fy^- & \geq & -g, \\
& & & x, y^+, y^- & & & \geq & 0.
\end{array}
$$

By collecting the variables x, y^+, and y^- into a vector t and defining H, h, and c by

$$
H = \begin{bmatrix} B & C & -C \\ E & F & -F \\ -E & -F & F \end{bmatrix}, \qquad
h = \begin{bmatrix} d \\ g \\ -g \end{bmatrix}, \qquad
c = \begin{bmatrix} p \\ q \\ -q \end{bmatrix}, \qquad (3.23)
$$

we see that the above problem can then be rewritten in standard form as

$$
\begin{array}{rrcl}
\min_{t} & c't & & \\
\text{subject to} & Ht & \geq & h, \\
& t & \geq & 0.
\end{array} \qquad (3.24)
$$

In *Scheme I*, we perform this reduction to standard form explicitly and solve the problem (3.24) explicitly using the two-phase simplex method. In a "postprocessing" phase, we recover the optimal values for the original variables by "undoing" the transformations that we applied to obtain the variables t.

Example 3-6-5.

$$
\begin{aligned}
\min \quad & 2x_1 - x_2 + x_3 \\
\text{subject to} \quad & x_1 \;-\; x_2 \;+\; 4x_3 \;\geq\; -1, \\
& x_1 \;-\; x_2 \;-\; x_3 \;\geq\; 2, \\
& x_1 \;+\; 3x_2 \;+\; 2x_3 \;=\; 3, \\
& x_1, x_2 \geq 0.
\end{aligned}
$$

Note that x_3 is a free variable, and the third constraint is an equality. We identify this problem with the formulation (3.22) by defining the data as follows:

$$
B = \begin{bmatrix} 1 & -1 \\ 1 & -1 \end{bmatrix}, \quad
C = \begin{bmatrix} 4 \\ -1 \end{bmatrix}, \quad
E = \begin{bmatrix} 1 & 3 \end{bmatrix}, \quad
F = \begin{bmatrix} 2 \end{bmatrix}, \quad
d = \begin{bmatrix} -1 \\ 2 \end{bmatrix},
$$

$$
g = \begin{bmatrix} 3 \end{bmatrix}, \quad
p = \begin{bmatrix} 2 \\ -1 \end{bmatrix}, \quad
q = \begin{bmatrix} 1 \end{bmatrix}, \quad
x = \begin{bmatrix} x_1 \\ x_2 \end{bmatrix}, \quad
y = \begin{bmatrix} x_3 \end{bmatrix}.
$$

The following MATLAB code defines this data, assembles the Scheme I problem as defined in (3.24), (3.23), and defines the initial tableau:

```
>> B = [1 -1; 1 -1]; C = [4; -1];
>> E = [1 3]; F = 2;
>> d = [-1; 2]; g = 3;
>> p = [2; -1]; q = 1;
>> H = [B C -C; E F -F; -E -F F];
>> h = [d; g; -g];
>> c = [p; q; -q];
>> T = totbl(H,h,c,0,'t');
```

	t_1	t_2	t_3	t_4	1
$t_5 =$	1	-1	4	-4	1
$t_6 =$	1	-1	-1	1	-2
$t_7 =$	1	3	2	-2	-3
$t_8 =$	-1	-3	-2	2	3
$z =$	2	-1	1	-1	0

Note that we have used a five-argument form of `totbl`. The fourth argument represents the initial value to be placed in the lower right cell of the tableau (representing the constant term in the objective function, if one exists), while the fifth argument defines the text part of the labels to be used for rows and columns (if we required something other than the default x).

We now proceed with Phase I of the two-phase simplex method.

```
>> T = addcol(T,[0 1 1 0 0]',
   'x0',5);
>> w = [0 0 0 0 1 0];
>> T = addrow(T,w,'z0',6);
```

	t_1	t_2	t_3	t_4	x_0	1
$t_5 =$	1	-1	4	-4	0	1
$t_6 =$	1	-1	-1	1	1	-2
$t_7 =$	1	3	2	-2	1	-3
$t_8 =$	-1	-3	-2	2	0	3
$z =$	2	-1	1	-1	0	0
$z_0 =$	0	0	0	0	1	0

The next step is the special pivot that generates a feasible tableau for Phase I.

≫ T = ljx(T,3,5);

		t_1	t_2	t_3	t_4	t_7	1
t_5	=	1	−1	4	−4	0	1
t_6	=	0	−4	−3	3	1	1
x_0	=	−1	−3	−2	2	1	3
t_8	=	−1	−3	−2	2	0	3
z	=	2	−1	1	−1	0	0
z_0	=	−1	−3	−2	2	1	3

A further pivot leads to an optimal Phase I tableau.

≫ T = ljx(T,3,1);

		x_0	t_2	t_3	t_4	t_7	1
t_5	=	−1	−4	2	−2	1	4
t_6	=	0	−4	−3	3	1	1
t_1	=	−1	−3	−2	2	1	3
t_8	=	1	0	0	0	−1	0
z	=	−2	−7	−3	3	2	6
z_0	=	1	0	0	0	0	0

Striking out the x_0 column and the z_0 row, we note that the resulting tableau is feasible but not optimal. Two more pivots lead to an optimal tableau.

≫ T = delcol(T,'x0');
≫ T = delrow(T,'z0');
≫ T = ljx(T,2,1);
≫ T = ljx(T,1,3);

		t_6	t_3	t_5	t_7	1
t_4	=	0.2	1	−0.2	0	0.6
t_2	=	−0.1	0	−0.15	0.25	0.7
t_1	=	0.7	0	0.05	0.25	2.1
t_8	=	0	0	0	−1	0
z	=	1.3	0	0.45	0.25	2.9

We now need to convert the optimal values for t back into optimal values for x. Note that $x_1 = t_1 = 2.1$, $x_2 = t_2 = 0.7$, $x_3 = t_3 - t_4 = -0.6$ (since we replaced x_3 by the difference of its positive part t_3 and its negative part t_4 in the formulation). The optimal objective value is $z = 2.9$. ∎

We have shown in Section 3.6.1 that *any* linear program can be reformulated in the general form (3.22), while in this section we have shown that (3.22) can be reformulated in standard form and then solved with the simplex method of this chapter. We combine these two observations with Theorem 3.5.4 to obtain the following result.

Theorem 3.6.1. *Given any linear program, suppose we apply Scheme I together with the two-phase simplex method using the smallest-subscript anticycling rule. Then, after a finite number of pivots, the algorithm either terminates with a conclusion that the problem is infeasible or else arrives at an* optimal *or* unbounded *tableau.*

Exercise 3-6-6. Convert the following problem into standard form using Scheme I and solve using the simplex method:

$$
\begin{array}{lrcrcrcrcl}
\min & x + 2y + 3z \\
\text{subject to} & x & - & y & + & 3z & \leq & 3, \\
& 4x & + & y & & & \geq & 1, \\
& & & & & z & \geq & 0.
\end{array}
$$

Exercise 3-6-7. Convert the following problem to standard form and solve using the two-phase simplex method:

$$
\begin{array}{lrcrcrcrcl}
\max & -2x_1 - x_2 - x_3 - 2x_4 \\
\text{subject to} & x_1 & - & x_2 & + & x_3 & - & x_4 & = & -1, \\
& -x_1 & - & x_2 & - & x_3 & - & x_4 & = & -3, \\
& -x_1 & + & x_2 & - & x_3 & + & x_4 & \leq & 1, \\
& x_1 & + & x_2 & - & x_3 & - & x_4 & \leq & -1, \\
& -x_1 & - & x_2 & + & x_3 & + & x_4 & \leq & 1, \\
& -x_1 & + & x_2 & + & x_3 & - & x_4 & \leq & -2, \\
& & & & & & x_1, x_4 & \geq & 0.
\end{array}
$$

(Note that the variables x_2, x_3 are free.)

In Example 3-6-5, we added only one equation and one variable to the problem during conversion to standard form. When there are multiple free variables and multiple equations, we end up increasing the size of the problem significantly, as seen in Exercise 3-6-7. By using a variant on the Scheme I technique, we can generate a standard-form problem by just adding one extra variable and one extra constraint. The role of the extra variable is to absorb the maximum negativity of the free variables y. We replace y by a set of nonnegative variables $\hat{y}$ and the extra variable η as follows:

$$
y \text{ free} \quad \Longleftrightarrow \quad y = \hat{y} - e\eta, \qquad \hat{y} \geq 0, \qquad \eta \geq 0.
$$

For the equality constraints, we make the following substitution:

$$
Ex + Fy = g \quad \Longleftrightarrow \quad
\begin{array}{rcl}
Ex + Fy & \geq & g, \\
e'(Ex + Fy - g) & \leq & 0.
\end{array}
$$

Here $e = (1, 1, \ldots, 1)'$ is a vector of ones of appropriate dimension. By making these two substitutions into the general form (3.22), we obtain the following standard-form linear program:

$$
\begin{array}{ll}
\min_{x, \hat{y}, \eta} & p'x + q'(\hat{y} - e\eta) \\
\text{subject to} &
\begin{array}{rcrcl}
Bx & + & C(\hat{y} - e\eta) & \geq & d, \\
Ex & + & F(\hat{y} - e\eta) & \geq & g, \\
-e'Ex & - & e'F(\hat{y} - e\eta) & \geq & -e'g, \\
& & x, \hat{y}, \eta & \geq & 0.
\end{array}
\end{array}
$$

Exercise 3-6-8. Use the above approach to solve the problem given in Exercise 3-6-7.

Exercise 3-6-9. Solve the following problem:

$$
\begin{array}{rrrrrrrrl}
\max & 2y_1 & - & y_3 & & & & & \\
\text{subject to} & 5y_1 & - & 2y_2 & + & y_3 & - & y_4 & = & 36, \\
& y_1 & & & + & y_3 & & & \geq & 4, \\
& y_1 & + & 3y_2 & + & y_3 & & & \geq & 1, \\
& & & y_1 \leq 8, & y_2 \geq 0, & y_3 \leq 0, & y_4 \geq 0. & & &
\end{array}
$$

Does the problem have a unique optimal solution? Justify.

Exercise 3-6-10. Use the simplex method to solve the following problem:

$$
\begin{array}{rrrrrrrl}
\max & 2x_1 + 5x_2 - x_3 & & & & & & \\
\text{subject to} & 5x_1 & & & + & x_3 & \geq & 1, \\
& -2x_1 & & & + & x_3 & \leq & 22, \\
& -4x_1 & + & x_2 & - & x_3 & \leq & -6, \\
& & & x_1 \leq 0, & x_2 \geq 3, & x_3 \geq 0. & &
\end{array}
$$

3.6.3 Scheme II

We now describe an alternative approach to solving the general formulation (3.22) that avoids transformation to standard form and avoids adding extra variables and equations to the formulation. This approach, known as *Scheme II*, handles the rows and columns of the tableau that correspond to free variables and to equality constraints in a special way, as we now describe.

Equality constraints could be handled by introducing artificial variables that represent the violation of these constraints and requiring these variables to be zero at the solution. Accordingly, we aim to manipulate the tableau *prior* to applying the simplex method to pivot all the artificial variables associated with equality constraints to the top of the tableau. Once they are at the top of the tableau, we are able to assign them the value 0. Since we want to ensure that these variables are fixed at zero and remain nonbasic for the rest of the computation, we can simply delete their columns from the tableau.

We deal with all the free variables, which are on top of the initial tableau, by performing some initial pivots to move them to the side of the tableau. Once they are row labels, we do not wish to consider them as possible row pivots during the subsequent application of the simplex method. In fact, we do not care if these variables take on negative values at subsequent iterations. The best practical strategy for dealing with these rows in the MATLAB implementation is to move them to the bottom of the tableau and ignore them during subsequent selection of row and column pivots. Once the simplex method has terminated, we can read off the values of these variables by referring to the entries in the final column of the tableau.

Example 3-6-11. We now apply Scheme II to Example 3-6-5.

```
≫ load ex3-6-5
≫ T = totbl(A,b,p);
```

	x_1	x_2	x_3	1
$x_4 =$	1	-1	4	1
$x_5 =$	1	-1	-1	-2
$x_6 =$	1	3	2	-3
$z =$	2	-1	1	0

We now remove the artificial variable x_6 that corresponds to the equality constraint by using a Jordan exchange to move it to the top of the tableau and then delete its column. The equality constraint represented by x_6 will be satisfied for all subsequent tableaus.

```
>> T = ljx(T,3,1);
>> T = delcol(T,'x6');
```

		x_2	x_3	1
x_4	=	-4	2	4
x_5	=	-4	-3	1
x_1	=	-3	-2	3
z	=	-7	-3	6

We now move the free variable x_3 to the side of the tableau by means of the following pivot:

```
>> T = ljx(T,2,2);
```

		x_2	x_5	1
x_4	=	-6.67	-0.67	4.67
x_3	=	-1.33	-0.33	0.33
x_1	=	-0.33	0.67	2.33
z	=	-3	1	5

Note that since x_3 is free, the second row of the tableau can be viewed as a defining relationship for x_3. The Jordan exchange has used this relationship to substitute x_3 out of all the remaining equations in the tableau. Since we need not consider the row labeled by x_3 further as a possible pivot row, we use the command permrows to move it out of the way, to the bottom of the tableau. We can then ignore it for the remainder of the calculations.

```
>> T = permrows(T,[1 3 4 2]);
```

		x_2	x_5	1
x_4	=	-6.67	-0.67	4.67
x_1	=	-0.33	0.67	2.33
z	=	-3	1	5
x_3	=	-1.33	-0.33	0.33

(Note that permrows reorders the rows as described in its second argument.) At this stage, all the variables left in the tableau x_2, x_5, x_4, and x_1 are nonnegative in the formulation, and so we can proceed with the two-phase simplex method. In this particular example, since the tableau is already feasible, we can dispense with Phase I and move directly to Phase II.

```
>> T = ljx(T,1,1);
```

		x_4	x_5	1
x_2	=	-0.15	-0.1	0.7
x_1	=	0.05	0.7	2.1
z	=	0.45	1.3	2.9
x_3	=	0.2	-0.2	-0.6

The final tableau is optimal, with $x_1 = 2.1$, $x_2 = 0.7$, and the free variable x_3 having the value -0.6. The optimal objective value is $z = 2.9$, as we calculated earlier in Example 3-6-5.

The pivots that are performed to bring the artificial variables associated with equality constraints to the top of the tableau and the free variables to the side are best thought of as a "preprocessing" step; they are not part of the simplex method proper. We have a great deal of latitude in selecting these pivots; any legal pivots can be used. In fact, we can solve the above problem using the following alternative approach:

```
>> load ex3-6-5
>> T = totbl(A,b,p);
>> T = ljx(T,3,3);
>> T = delcol(T,'x6');
>> T = permrows(T,[1 2 4 3]);
```

		x_1	x_2	1
x_4	=	-1	-7	7
x_5	=	1.5	0.5	-3.5
z	=	1.5	-2.5	1.5
x_3	=	-0.5	-1.5	1.5

We have used a single pivot to *both* move the artificial variable to the top of the tableau and move the free variable to the side. Unlike the previous choice of pivots, however, we need to start the simplex method with a Phase I. By adding the Phase I variable x_0 and objective z_0 in the usual way, we obtain

```
>> T = addcol(T,[0 1 0 0]','x0',3);
>> T = addrow(T,[0 0 1 0],'z0',5);
```

		x_1	x_2	x_0	1
x_4	=	-1	-7	0	7
x_5	=	1.5	0.5	1	-3.5
z	=	1.5	-2.5	0	1.5
x_3	=	-0.5	-1.5	0	1.5
z_0	=	0	0	1	0

The Phase I procedure then proceeds in the usual way.

```
>> T = ljx(T,2,3);
>> T = ljx(T,2,1);
>> T = delrow(T,'z0');
>> T = delcol(T,'x0');
```

		x_2	x_5	1
x_4	=	-6.67	-0.67	4.67
x_1	=	-0.33	0.67	2.33
z	=	-3	1	5
x_3	=	-1.33	-0.33	0.33

Note that the last MATLAB operations above just remove the row and the column corresponding to the Phase I variables. A single Phase II pivot `ljx(T,1,1)` leads to the same optimal tableau as given earlier. ∎

Exercise 3-6-12. Carry out Scheme II on the problem given in Exercise 3-6-7. Check that you obtain the same solutions as before.

Scheme II can run into difficulties when zero pivots make it impossible to pivot the free variables to the side of the tableau and the equality constraints to the top. We now discuss how to handle these cases and what they indicate about the problem.

The equality constraints present the easier case. If we are unable to pivot the artificial variable x_{n+i} for an equality constraint to the top of the tableau, then the whole row of the tableau must contain zeros, except possibly for the entry in the last column, which gives the value of x_{n+i} for this tableau. (This situation can happen when the equality constraint is linearly dependent on other constraints in the system.) We have two cases:

(a) If the right-hand side is zero, the equality $x_{n+i} = 0$ is consistent, and we can delete this row from the tableau and continue.

(b) If the right-hand side is nonzero, the equality $x_{n+i} = 0$ is inconsistent with the other constraints, and so the problem is infeasible.

The case in which a free variable cannot be pivoted to the side of the tableau is a little more complicated. Scheme II runs into trouble when all possible pivots in column j (corresponding to the free variable $x_{N(j)}$) are zero. (We exclude from consideration here the reduced cost c_j for $x_{N(j)}$ and also the rows corresponding to other free variables which have already been pivoted to the side of the tableau.) By examining the reduced cost c_j, we can recognize two cases immediately.

(a) If $c_j = 0$, we can leave the column in place and proceed with further Scheme II steps and with the simplex method. By examining the Jordan exchange formulae (2.5) and (2.6) closely, we see that the eligible pivots in this column (and the reduced cost) remain at zero at all subsequent steps. If an optimal tableau is eventually found, we can obtain a solution by setting $x_{N(j)}$ to an arbitrary value and calculating the values of the free variables which *were* successfully pivoted to the side by reading off the tableau in the usual manner. (If we are interested in finding just one solution of the linear program, we can simply set $x_{N(j)} = 0$ and read the values of the remaining free variables from the last column of the tableau.)

(b) If the reduced cost c_j is nonzero and the tableau is feasible, then we can drive the objective to $-\infty$ by either increasing $x_{N(j)}$ to ∞ (if $c_j < 0$) or decreasing it to $-\infty$ (if $c_j > 0$), while maintaining feasibility of the other variables. In this case the problem is unbounded.

One important case remains unresolved: the case in which $c_j \neq 0$ but the tableau is not feasible. In this case, we need to perform more work before determining whether the problem is unbounded or infeasible. By examining the Jordan exchange formula (2.6) closely, we can exclude the possibility that the problem has an optimal solution in this case. Any subsequent pivots applied to the tableau will not change the values in column j; the possible pivots will remain at zero and c_j will not change. If the further pivots lead to a feasible tableau (for instance via subsequent Scheme II pivots, or by the use of Phase I), then we are in the situation of case (b) above, and we can declare the problem to be unbounded. If, however, we determine that no feasible tableau can be obtained (for instance if Phase I terminates with a positive objective), then we declare the problem to be infeasible.

We now discuss some simple examples that illustrate these cases.

Example 3-6-13. Consider first the following problem:

$$\begin{array}{lll} \min_{x_1, x_2} & 2x_1 - x_2 & \\ \text{subject to} & x_1 - 2x_2 & \geq -2, \\ & x_1 & = 4, \\ & 2x_1 & = 6, \\ & x_1 & \geq 0, \end{array}$$

for which the tableau is as follows:

```
>> load ex3-6-13-1
>> T = totbl(A,b,p);
```

		x_1	x_2	1
x_3	$=$	1	-2	2
x_4	$=$	1	0	-4
x_5	$=$	2	0	-6
z	$=$	2	-1	0

By pivoting x_4 to the top and eliminating the corresponding column, we obtain the following tableau:

```
>> T = ljx(T,2,1);
>> T = delcol(T,'x4');
```

		x_2	1
x_3	$=$	-2	6
x_1	$=$	0	4
x_5	$=$	0	2
z	$=$	-1	8

We cannot pivot x_5, the artificial variable corresponding to the other equality constraint, to the top of the tableau, because there is a zero in the only eligible pivot position. Since the last column contains a nonzero in this row, we conclude that the problem is infeasible. If the third constraint in the original problem had been $2x_1 = 8$, then the last column would have contained a zero in this position, and we would have been able to delete the row corresponding to x_5 and then proceed with the algorithm.

In the next examples we have a nonnegative variable $x_1 \geq 0$, a free variable x_2, a slack variable x_3 associated with an inequality, and an artificial variable x_4 associated with an equality:

$$\min_{x_1,x_2} \quad 2x_1 - x_2$$
$$\text{subject to} \quad \begin{aligned} x_1 &\geq -6, \\ -x_1 &= -4, \\ x_1 &\geq 0. \end{aligned}$$

This problem corresponds to the following tableau:

```
>> load ex3-6-13-2
>> T = totbl(A,b,p);
```

		x_1	x_2	1
x_3	$=$	1	0	6
x_4	$=$	-1	0	4
z	$=$	2	-1	0

We would like to perform pivots to move x_4 to the top of the tableau and x_2 to the side of the tableau. Note that we cannot pivot x_2 to the side, as all elements in its column (except the reduced cost) are zero. Since the reduced cost is nonzero, we cannot simply delete the column and proceed. Nor can we yet determine that the problem is unbounded, since the tableau is not yet feasible. (It yields the value $x_4 = 4$, which is not feasible since we require this artificial variable to be zero.) Hence we proceed with pivoting, choosing the $(2, 1)$ element as the pivot element to move the x_4 element to the top, and obtain the following:

```
>> T = ljx(T,2,1);
```

		x_4	x_2	1
x_3	$=$	-1	0	10
x_1	$=$	-1	0	4
z	$=$	-2	-1	8

This tableau is now feasible: x_4 is zero, and x_1 and x_3 are both positive, as required. At this point, we can conclude that the problem is unbounded. (It is easy to see that if we let x_2 go to $+\infty$, x_1 and x_3 remain feasible while z goes to $-\infty$.)

Suppose that we change the right-hand side in the first constraint to 6, that is,

$$\min_{x_1, x_2} \quad 2x_1 - x_2$$
$$\text{subject to} \quad \begin{aligned} x_1 &\geq 6, \\ -x_1 &= -4, \\ x_1 &\geq 0. \end{aligned}$$

```
>> b(1)=6;
>> T = totbl(A,b,p);
```

		x_1	x_2	1
x_3	$=$	1	0	-6
x_4	$=$	-1	0	4
z	$=$	2	-1	0

Again, we note that we cannot pivot x_2 to the side, and so we perform the pivot on the $(2, 1)$ element as before and delete the column corresponding to x_4 to obtain the following:

```
>> T=ljx(T,2,1);
>> T = delcol(T,'x4');
```

		x_2	1
x_3	$=$	0	-2
x_1	$=$	0	4
z	$=$	-1	8

The resulting tableau is still not feasible, and so we still cannot determine whether the problem is infeasible or unbounded. We now proceed with the usual Phase I as follows:

```
>> T = addcol(T,[1 0 0]','x0',2);
>> T = addrow(T,[0 1 0],'z0',4);
>> T = ljx(T,1,2);
```

		x_2	x_3	1
x_0	$=$	0	1	2
x_1	$=$	0	0	4
z	$=$	-1	0	8
z_0	$=$	0	1	2

We see now that the problem is infeasible, since Phase I has terminated with a nonzero objective value. ∎

Exercise 3-6-14. Consider the linear program

$$
\begin{array}{rrrrrrrrrr}
\min & -x_1 + x_2 - x_3 - 4x_4 \\
\text{subject to} & x_1 & - & x_2 & + & 2x_3 & - & x_4 & = & -4, \\
& -x_1 & + & x_2 & - & x_3 & - & x_4 & = & 4, \\
& x_1 & + & x_2 & - & x_3 & - & x_4 & \geq & 2, \\
& -x_1 & - & x_2 & + & x_3 & - & 2x_4 & \geq & -4, \\
& x_1 & + & x_2 & + & 2x_3 & + & x_4 & \geq & -3, \\
& & & x_1, x_2 \geq 0, & & x_3, x_4 \text{ free.}
\end{array}
$$

1. Use Scheme I to reformulate and solve the above problem as a standard form linear program with 6 variables and 7 constraints.

2. Use Scheme II to solve the original problem in MATLAB. (Follow the procedure of Example 3-6-11.) Be sure to recover the solution of the original problem.

Exercise 3-6-15. Convert the following problem into standard form (using Scheme II) and solve:

$$
\begin{array}{rrrrrrr}
\max & x_1 + 4x_2 + x_3 \\
\text{subject to} & 2x_1 & + & 2x_2 & + & x_3 & = & 4, \\
& x_1 & & & - & x_3 & = & 1, \\
& & & x_2 \geq 0, & & x_3 \geq 0.
\end{array}
$$

3.6.4 Summary

The complete process for solving a general linear program involves the following steps:

1. Convert maximization problems into minimization problems.

2. Transform less-than inequalities into greater-than inequalities.

3. Use substitution to convert generally bounded variables into nonnegative or free variables.

4. Apply Scheme I or Scheme II to remove free variables and equations from the formulation.

5. If the tableau is infeasible, apply the Phase I method to generate a feasible tableau. If Phase I terminates with a positive objective function value, stop and declare the problem infeasible.

6. Apply Phase II pivots to determine unboundedness or an optimal tableau.

7. Recover the values of the original variables if substitution was applied.

8. If the problem was originally a maximization, negate the objective value in the final tableau to give the optimal value of the original problem.

Exercise 3-6-16. Solve the following problem:

$$
\begin{array}{rl}
\min & z = -3x_1 + x_2 + 3x_3 - x_4 \\
\text{subject to} & x_1 + 2x_2 - x_3 + x_4 = 0, \\
& 2x_1 - 2x_2 + 3x_3 + 3x_4 = 9, \\
& x_1 - x_2 + 2x_3 - x_4 = 6, \\
& x_1, x_2, x_3, x_4 \geq 0.
\end{array}
$$

Exercise 3-6-17. Solve the following problem in MATLAB:

$$
\begin{array}{rl}
\min & c'x \\
\text{subject to} & Ax = b, \quad x \geq 0,
\end{array}
$$

where

$$
A = \begin{bmatrix} 1 & 2 & 0 & 1 \\ 1 & 2 & 1 & 1 \\ 1 & 3 & -1 & 2 \\ 1 & 1 & 1 & 0 \end{bmatrix}, \qquad
b = \begin{bmatrix} 6 \\ 7 \\ 7 \\ 5 \end{bmatrix}, \qquad
c = \begin{bmatrix} 2 \\ 6 \\ 1 \\ 1 \end{bmatrix}.
$$

Chapter 4

Duality

Duality is a fundamental concept in linear algebra and mathematical programming that arises from examining a problem from two different viewpoints. To be specific, a system of linear relations defined in terms of a matrix $A \in \mathbf{R}^{m \times n}$ can be interpreted either in terms of the column space of A or in terms of its row space. These alternative, *dual* interpretations lead us to some fundamental results, such as equality of row and column ranks of a matrix and equality of the optimal values of a primal-dual pair of linear programs. In the first part of this chapter, we derive these two results via a constructive approach utilizing the Jordan exchange. After discussing the Karush–Kuhn–Tucker (KKT) optimality conditions for linear programming, we derive the *dual simplex method* by applying the simplex method of the previous chapter to the dual formulation of the standard-form linear program. Finally, we show how to construct the dual of a linear program stated in *general form* and derive some fundamental results about linear programming by making use of duality relationships.

4.1   Duality and Rank in Linear Systems

Given any linear system $y = Ax$, we define the *dual system* as follows: $v = -A'u$. For the purposes of this chapter, we refer to the tableau for $y = Ax$ as the *primal tableau* and to the tableau for $v = -A'u$ as the *dual tableau*. Note that by repeating the transformation that we used to get from the primal system $y = Ax$ to its dual $v = -A'u$ (that is, replacing the matrix by its transpose and negating it), we recover the original system. Hence, each of these systems is the dual of the other. We refer to the two corresponding tableaus as a pair of *dual tableaus* and represent them as follows:

$$
\begin{array}{cc}
 & x \\
y = & \boxed{A}
\end{array}
\qquad
\begin{array}{cc}
 & v = \\
-u & \boxed{A}
\end{array}
\tag{4.1}
$$

Note that since $A \in \mathbf{R}^{m \times n}$, the primal tableau independent variables x are in $\mathbf{R}^n$, while the independent variables u of the dual tableau are in $\mathbf{R}^m$.

The simple relationship described above lies at the heart of the concept of duality, which is powerful both in constructing practical solution techniques for linear programming and in proving fundamental results about linear algebra and linear programming problems.

89

The immediate significance of the relationship between the two tableaus above is that a Jordan exchange on the primal system $y = Ax$ is equivalent to a corresponding exchange on the dual system $v = -A'u$, as we now show.

Theorem 4.1.1 (Dual Transformation). *A Jordan exchange with pivot element A_{rs} has two equivalent interpretations:*

1. *(Primal): Solve $y_r = \sum_{j=1}^{n} A_{rj}x_j$ for x_s and substitute for x_s in the remaining $y_i = \sum_{j=1}^{n} A_{ij}x_j$, $i \neq r$.*

2. *(Dual): Solve $v_s = -\sum_{i=1}^{m} A_{is}u_i$ for u_r and substitute for u_r in the remaining $v_j = -\sum_{i=1}^{m} A_{ij}u_i$, $j \neq s$.*

Proof. The first statement follows immediately from our derivation of the Jordan exchange in Chapter 2. For the second part, we write out the equation $v = -A'u$ componentwise as follows:

$$v_j = -\sum_{i=1}^{m} A_{ij}u_i, \qquad j \neq s,$$

$$v_s = -\sum_{i \neq r} A_{is}u_i - A_{rs}u_r.$$

By rearranging the last equation above and substituting into the other equations, we have

$$u_r = -\left(\frac{1}{A_{rs}}v_s + \sum_{i \neq r} \frac{A_{is}}{A_{rs}}u_i \right)$$

$$v_j = -A_{rj}u_r - \sum_{i \neq r} A_{ij}u_i, \qquad j \neq s$$

$$= -\left(\frac{-A_{rj}}{A_{rs}}v_s + \sum_{i \neq r} \left(A_{ij} - \frac{A_{rj}A_{is}}{A_{rs}} \right) u_i \right), \qquad j \neq s.$$

These are exactly the equations we would obtain by writing out the tableau for $v = -A'u$, with labels for the rows v_j and columns u_i, and performing the Jordan exchange with pivot element A_{rs} as described in Chapter 2. $\square$

Example 4-1-1. Here is a simple example of the dual interpretation of Jordan exchanges. The primal system

$$\begin{aligned}
y_1 &= x_1 - x_2 + 2x_3 - x_4, \\
y_2 &= -x_1 - x_2 - x_3 + x_4, \\
y_3 &= x_1 + x_2 - x_3 - x_4
\end{aligned}$$

corresponds to the following dual system:

$$\begin{aligned}
v_1 &= -u_1 + u_2 - u_3, \\
v_2 &= u_1 + u_2 - u_3, \\
v_3 &= -2u_1 + u_2 + u_3, \\
v_4 &= u_1 - u_2 + u_3.
\end{aligned}$$

The tableau representation (complete with dual labels) can be easily constructed, using the following MATLAB code:

```
>> load ex4-1-1
>> T = totbl(A);
>> T = dualbl(T);
```

$$
\begin{array}{cccc}
v_1 = & v_2 = & v_3 = & v_4 = \\
x_1 & x_2 & x_3 & x_4
\end{array}
$$

$$
\begin{array}{rr}
-u_1 & y_1 = \\
-u_2 & y_2 = \\
-u_3 & y_3 =
\end{array}
\left[
\begin{array}{rrrr}
1 & -1 & 2 & -1 \\
-1 & -1 & -1 & 1 \\
1 & 1 & -1 & -1
\end{array}
\right]
$$

Note that $u_1 = u_2 = u_3 = 1$ implies that $v_1 = -1$, $v_2 = 1$, $v_3 = 0$, and $v_4 = 1$. A Jordan exchange on the $(2, 3)$ element gives

```
>> T = ljx(T,2,3);
```

$$
\begin{array}{cccc}
v_1 = & v_2 = & u_2 = & v_4 = \\
x_1 & x_2 & y_2 & x_4
\end{array}
$$

$$
\begin{array}{rr}
-u_1 & y_1 = \\
-v_3 & x_3 = \\
-u_3 & y_3 =
\end{array}
\left[
\begin{array}{rrrr}
-1 & -3 & -2 & 1 \\
-1 & -1 & -1 & 1 \\
2 & 2 & 1 & -2
\end{array}
\right]
$$

Setting the dual independent variables $u_1 = 1$, $v_3 = 0$, and $u_3 = 1$ gives $v_1 = -1$, $v_2 = 1$, $u_2 = 1$, and $v_4 = 1$. The values of these variables remain the same after the Jordan exchange, suggesting that the relationship between the variables has not changed as a result of the exchange. To verify this observation, we can read off the following linear relations from the dual labels on the tableau above: $v_1 = u_1 + v_3 - 2u_3$, $v_2 = 3u_1 + v_3 - 2u_3$, $u_2 = 2u_1 + v_3 - u_3$, and $v_4 = -u_1 - v_3 + 2u_3$. By verifying that these relations are consistent with the original dual relation $v = -A'u$, we can see that Theorem 4.1.1 holds for this particular Jordan exchange. ∎

In fact, all the results established for the tableau for the primal system $y = Ax$ have precise analogues in the dual setting. We will use these dual results in the remainder of the chapter without further proof.

We turn now to a discussion of the *rank* of a matrix A.

Definition 4.1.2. The rank of a matrix $A \in \mathbf{R}^{m \times n}$ is the maximum number of linearly independent rows that are present in A.

Rank is a fundamental concept in linear algebra that plays an important role in linear programming. A simple constructive method that uses Jordan exchanges to determine the rank of a matrix is given in the following theorem. We show in particular that the rank determined from this method does not depend on the choice of pivot sequence in the method.

Theorem 4.1.3. *Given $A \in \mathbf{R}^{m \times n}$, form the tableau $y := Ax$. Using Jordan exchanges, pivot as many of the y's to the top of the tableau as possible. The rank of A is equal to the number of y's pivoted to the top.*

Proof. We show directly that the maximum number of y_i's that can be pivoted to the top is independent of our choice of y_i's and of the order in which they are selected for pivoting. Suppose for contradiction that this claim is false. That is, there is one subset y_I of the y_i's that can be pivoted to the top of the tableau before no further such pivots are possible, and another subset y_K with the same property, except that $|K| > |I|$. For any $j \notin I$, we can examine the tableau that remains after the components of y_I have been pivoted and write $y_j = z^j y_I$ for some row vector z^j. (Note in particular that y_j cannot depend on any of the x components that remain at the top of the tableau, because if it did, we would be able to pivot y_j to the top of the tableau as well.) Therefore, for every x, we can use relations from the original tableau to write

$$A_{j.}x = y_j = z^j y_I = z^j A_{I.}x.$$

Since this relation holds for *every* x, we can write $A_{j.} = z^j A_{I.}$ for $j \notin I$. For the remaining indices $i \in I$, we can also write $A_{i.} = z^i A_{I.}$ by setting z^i to be the unit vector with a 1 in the position corresponding to row i and zeros elsewhere.

We now consider the indices $l \in K$. Since $A_{l.} = z^l A_{I.}$ for some vector z^l and all $l \in K$, we have by assembling the z^l into a matrix Z that

$$A_{K.} = Z A_{I.},$$

where Z has dimension $|K| \times |I|$. Since $|K| > |I|$, it follows from Proposition 2.2.2 that the rows of Z are linearly dependent. Therefore,

$$v'Z = 0 \qquad \text{for some } v \neq 0,$$

and so

$$v'A_{K.} = v'Z A_{I.} = 0 \cdot A_{I.} = 0 \quad \text{for some } v \neq 0.$$

It follows that the rows of $A_{K.}$ are linearly dependent. However, by applying the Steinitz theorem (Theorem 2.2.3) to $A_{K.}$, we know that since all the components of y_K can be pivoted to the top of this tableau, the rows of $A_{K.}$ are linearly independent. Hence, we have a contradiction. We conclude that the set y_K with the given properties cannot exist and that our specific choice of pivots does not affect the maximum number of y_i's that can be pivoted to the top of the tableau $y := Ax$.

The conclusion of the theorem now follows immediately from the Steinitz theorem (Theorem 2.2.3). $\square$

The dual interpretation given in Theorem 4.1.1 can be combined with Theorem 4.1.3 to prove the classical result that the maximum number of linearly independent rows of A precisely equals the maximum number of linearly independent columns of A, as we now show. Note that this result is not particularly intuitive. For example, given a set of 91 numbers, arranged as a 7×13 matrix, it does not seem obvious that the "row rank" and "column rank" should be the same.

Theorem 4.1.4. *Let $A \in \mathbf{R}^{m \times n}$; then* $\mathrm{rank}(A)$—*the number of linearly independent rows of A—is equal to the number of linearly independent columns of A.*

Proof. We prove the result by construction. First, set up the primal tableau $y := Ax$ and then use Theorem 4.1.1 to add the dual variable labels. Now exchange as many y_i's to the top as possible. After possibly reordering rows and columns, the tableau will now have the following form:

$$
\begin{array}{cccc}
 & & u_{I_1} = & v_{J_2} = \\
 & & y_{I_1} & x_{J_2} \\
-v_{J_1} & x_{J_1} = & \boxed{\begin{array}{cc} B_{I_1 J_1} & B_{I_1 J_2} \end{array}} \\
-u_{I_2} & y_{I_2} = & B_{I_2 J_1} \quad 0
\end{array}
$$

By applying Theorem 4.1.3 to this tableau, we find that the maximum number of linearly independent rows is equal to $|I_1|$. However, because of the correspondence between Jordan exchanges on the primal and dual tableaus, we also can apply Theorem 4.1.3 to the dual tableau $v := -A'u$ to find that the maximum number of linearly independent rows of $-A'$ (and therefore the maximum number of linearly independent columns of A) is equal to $|J_1|$. Since $B_{I_1 J_1}$ is a square matrix, we must have that $|I_1| = |J_1|$. □

Example 4-1-2. Determine the rank of

$$
A = \begin{bmatrix} 1 & 0 & 2 \\ 0 & -1 & 1 \\ 1 & 1 & 1 \end{bmatrix}.
$$

We construct a tableau from this matrix as follows:

```
>> load ex4-1-2
>> T = totbl(A);
>> T = dualbl(T);
```

$$
\begin{array}{ccccc}
 & & v_1 = & v_2 = & v_3 = \\
 & & x_1 & x_2 & x_3 \\
-u_1 & y_1 = & \boxed{1} & 0 & 2 \\
-u_2 & y_2 = & 0 & -1 & 1 \\
-u_3 & y_3 = & 1 & 1 & 1
\end{array}
$$

After two pivots, we are blocked; no more y_i's can be pivoted to the top.

```
>> T = ljx(T,1,1);
```

$$
\begin{array}{ccccc}
 & & u_1 = & v_2 = & v_3 = \\
 & & y_1 & x_2 & x_3 \\
-v_1 & x_1 = & \boxed{1} & 0 & -2 \\
-u_2 & y_2 = & 0 & -1 & 1 \\
-u_3 & y_3 = & 1 & 1 & -1
\end{array}
$$

```
>> T = ljx(T,2,2);
```

$$
\begin{array}{ccccc}
 & & u_1 = & u_2 = & v_3 = \\
 & & y_1 & y_2 & x_3 \\
-v_1 & x_1 = & \boxed{1} & 0 & -2 \\
-v_2 & x_2 = & 0 & -1 & 1 \\
-u_3 & y_3 = & 1 & -1 & 0
\end{array}
$$

We have that $v_3 = 2v_1 - v_2$, $y_3 = y_1 - y_2$ (that is, $A_{\cdot 3} = 2A_{\cdot 1} - A_{\cdot 2}$, $A_{3 \cdot} = A_{1 \cdot} - A_{2 \cdot}$), and rank$(A) = 2$.

Of course, in MATLAB, we could use the built-in function `rank`. Indeed, this function may give a more accurate result than the Jordan-exchange technique used above, since numerical rounding is treated better within the built-in MATLAB function.

Exercise 4-1-3. Suppose the matrix $A \in \mathbf{R}^{m \times n}$ has linearly independent rows and that for some $p < m$, the first p columns $A_{.1}, A_{.2}, \ldots, A_{.p}$ are linearly independent. Show that $m - p$ of the remaining $n - p$ columns of A can be added to these first p columns to form a set of m linearly independent vectors. (Hint: use the dual interpretation as in the proof of Theorem 4.1.4.)

4.2 Duality in Linear Programming

Associated with the standard linear program

$$\begin{array}{ll} \min_{x} & z = p'x \\ \text{subject to} & Ax \geq b, \quad x \geq 0, \end{array} \tag{4.2}$$

is the *dual linear program*

$$\begin{array}{ll} \max_{u} & w = b'u \\ \text{subject to} & A'u \leq p, \quad u \geq 0. \end{array} \tag{4.3}$$

We refer to these two problems as a *primal-dual pair* of linear programs. The two problems are intimately related in a number of ways but chiefly by the following key duality result: The primal problem has a solution if and only if the dual problem has a solution, in which case the optimal objective values of the two problems are equal. In addition, *both* problems can be represented *simultaneously* by a single tableau as follows:

			$u_{m+1} =$ x_1	$\cdots$	$u_{m+n} =$ x_n	$w =$ 1
$-u_1$	x_{n+1}	=	A_{11}	$\cdots$	A_{1n}	$-b_1$
	$\vdots$		$\vdots$	$\ddots$	$\vdots$	$\vdots$
$-u_m$	x_{n+m}	=	A_{m1}	$\cdots$	A_{mn}	$-b_m$
1	z	=	p_1	$\cdots$	p_n	0

As shown in Theorem 4.1.1, each Jordan exchange performed on this tableau can be viewed from both the primal and the dual perspective. In consequence, the simplex method, which consists of a sequence of Jordan exchanges, can be construed as solving not only the primal problem (4.2) but also its dual counterpart (4.3). The following example shows that if we apply the simplex method to solve the primal problem, the solution for the dual problem can be read from the optimal tableau.

Example 4-2-1. Consider the primal problem

$$\begin{array}{rlrcrcr} \min_{x_1, x_2} & z = -2x_1 & & & & & \\ \text{subject to} & -x_1 & - & x_2 & \geq & -1, \\ & -2x_1 & - & x_2 & \geq & -2, \\ & & & x_1, x_2 & \geq & 0, \end{array}$$

for which the corresponding dual problem is

$$\max_{u_1, u_2} \quad w = -u_1 - 2u_2$$

$$\text{subject to} \quad \begin{array}{rcrcr} -u_1 & - & 2u_2 & \leq & -2, \\ -u_1 & - & u_2 & \leq & 0, \\ & & u_1, u_2 & \geq & 0. \end{array}$$

We construct the tableau for this primal-dual pair as follows:

```
>> load ex4-2-1
>> T = totbl(A,b,p);
>> T = dualbl(T);
```

		$u_3 =$ x_1	$u_4 =$ x_2	$w =$ 1
$-u_1$	$x_3 =$	-1	-1	1
$-u_2$	$x_4 =$	-2	-1	2
1	$z =$	-2	0	0

Note that we have used labels u_3 and u_4 for the dual tableau in place of v in (4.1). This labeling is analogous to the use of x_3 and x_4 in place of y in the primal tableau in (4.1), the purpose of which is to make the tableau amenable to the simplex method.

If we perform a Jordan exchange using the (1,1) element as pivot, we obtain the following:

```
>> T = ljx(T,1,1);
```

		$u_1 =$ x_3	$u_4 =$ x_2	$w =$ 1
$-u_3$	$x_1 =$	-1	-1	1
$-u_2$	$x_4 =$	2	1	0
1	$z =$	2	2	-2

This tableau is now primal optimal, with solution $x_1 = 1$, $x_2 = 0$ and objective value $z = -2$. By setting $u_2 = u_3 = 0$, we have from the tableau that $u_1 = 2$ and $u_4 = 2$. It is clear that this is a feasible point for the dual problem and also that it is a solution point. (It achieves the objective value -2, and for any other dual feasible point, we have from the objective and the first constraint that $w = -u_1 - 2u_2 \leq -2$.) Hence, we conclude that the tableau is dual optimal, with solution $u_1 = 2$, $u_2 = 0$, and $w = -2$. In solving the primal problem, we have also found a solution to the dual problem. ■

Exercise 4-2-2. Solve the following linear program using the simplex method:

$$\min \quad z = 2x_1 + 9x_2 + 3x_3$$

$$\text{subject to} \quad \begin{array}{rcrcrcr} -x_1 & - & 6x_2 & & & \geq & -3, \\ x_1 & + & 4x_2 & + & x_3 & \geq & 1, \\ -2x_1 & - & 14x_2 & & & \geq & -5, \\ & & & x_1, x_2, x_3 & & \geq & 0. \end{array}$$

Formulate the dual of this problem and read off an optimal solution of the dual problem from the final tableau.

We now show how a primal linear program and its dual are intimately related by a number of theoretical and computational results.

Table 4.1. *Data for the "snack" problem.*

	Chocolate	Sugar	Cream cheese	Cost
Brownie	3	2	2	50
Cheesecake	0	4	5	80
Requirements	6	10	8	

4.3   Interpretation of Linear Programming Duality

We present a small example illustrating the relationship between primal and dual linear programs that shows that these two programs arise from two different perspectives of the same problem. The example is a simple instance of the diet problem described in Section 1.3.1 and is a modification of the example in Winston & Venkataramanan (2003, Section 6.6).

A student is deciding what to purchase from a bakery for a tasty afternoon snack. There are two choices of food: brownies, which cost 50 cents each, and mini-cheesecakes, which cost 80 cents. The bakery is service-oriented and is happy to let the student purchase a fraction of an item if she wishes. The bakery requires three ounces of chocolate to make each brownie (no chocolate is needed in the cheesecakes). Two ounces of sugar are needed for each brownie and four ounces of sugar for each cheesecake. Finally, two ounces of cream cheese are needed for each brownie and five ounces for each cheesecake. Being health-conscious, the student has decided that she needs at least six total ounces of chocolate in her snack, along with ten ounces of sugar and eight ounces of cream cheese. She wishes to optimize her purchase by finding the least expensive combination of brownies and cheesecakes that meet these requirements. The data is summarized in Table 4.1.

The student's problem can be formulated as the classic diet problem as follows:

$$
\begin{aligned}
\min_{x_1, x_2} \quad & 50x_1 + 80x_2 \\
\text{subject to} \quad & 3x_1 && \geq && 6, \\
& 2x_1 + 4x_2 && \geq && 10, \\
& 2x_1 + 5x_2 && \geq && 8, \\
& x_1, x_2 && \geq && 0,
\end{aligned}
\tag{4.4}
$$

where x_1 and x_2 represent the number of brownies and cheesecakes purchased, respectively. By applying the simplex method of the previous chapter, we find that the unique solution is $x = (2, 3/2)'$, with optimal cost of \$2.20.

We now adopt the perspective of the wholesaler who supplies the baker with the chocolate, sugar, and cream cheese needed to make the goodies. The baker informs the supplier that he intends to purchase at least six ounces of chocolate, ten ounces of sugar, and eight ounces of cream cheese to meet the student's minimum nutritional requirements. He also shows the supplier the other data in Table 4.1. The supplier now solves the following optimization problem: How can I set the prices per ounce of chocolate, sugar, and cream cheese (u_1, u_2, u_3) so that the baker will buy from me, and so that I will maximize my revenue? The baker will buy only if the total cost of raw materials for brownies is below 50 cents; otherwise he runs the risk of making a loss if the student opts to buy brownies. This restriction imposes the following constraint on the prices:

$$3u_1 + 2u_2 + 2u_3 \leq 50.$$

Similarly, he requires the cost of the raw materials for each cheesecake to be below 80 cents, leading to a second constraint:

$$4u_2 + 5u_3 \leq 80.$$

Clearly, all the prices must be nonnegative. Moreover, the revenue from the guaranteed sales is $6u_1 + 10u_2 + 8u_3$. In summary, the problem that the supplier solves to maximize his guaranteed revenue from the student's snack is as follows:

$$
\begin{array}{rrcl}
\max_{u_1, u_2, u_3} & 6u_1 + 10u_2 + 8u_3 & & \\
\text{subject to} & 3u_1 + 2u_2 + 2u_3 & \leq & 50, \\
& 4u_2 + 5u_3 & \leq & 80, \\
& u_1, u_2, u_3 & \geq & 0.
\end{array} \tag{4.5}
$$

The solution of this problem is $u = (10/3, 20, 0)'$, with an optimal revenue of \$2.20.

It may seem strange that the supplier charges nothing for the cream cheese ($u_3 = 0$), especially since, once he has announced his prices, the baker actually takes delivery of 11.5 ounces of it, rather than the required minimum of 8 ounces. A close examination of the problem (4.5) shows, however, that his decision is a reasonable one. If he had decided to charge a positive amount for the cream cheese (that is, $u_3 > 0$), he would have had to charge less for the sugar (u_2) and possibly also for the chocolate (u_1) in order to meet the pricing constraints, and his total revenue would have been lower. It is better to supply the cream cheese for free and charge as much as possible for the sugar!

Note that the student's problem is precisely in the form of the primal, and the supplier's problem has the dual form. Furthermore, the dual variables have the interpretation of prices on the ingredients. For this reason, in the economics literature, the dual variables are typically called "shadow prices."

We will return to this example in the next two sections, as we develop the key results concerning the relationship between primal and dual linear programs.

4.4 Duality Theory

We begin with an elementary theorem that bounds the objectives of the primal-dual pair of linear programs.

Theorem 4.4.1 (Weak Duality Theorem). *If x is primal feasible and u is dual feasible, then the dual objective function evaluated at u is less than or equal to the primal objective function evaluated at x, that is,*

$$
\left.
\begin{array}{ll}
Ax \geq b, & x \geq 0 \\
A'u \leq p, & u \geq 0
\end{array}
\right\} \implies b'u \leq p'x.
$$

Proof. Note first that for any two vectors s and t of the same size, for which $s \geq 0$ and $t \geq 0$, we have $s't \geq 0$. By applying this observation to the feasibility relationships above, we have

$$p'x = x'p \geq x'A'u = u'Ax \geq u'b = b'u,$$

where the first inequality follows from $p - A'u \geq 0$ and $x \geq 0$, and the second inequality follows from $Ax - b \geq 0$, $u \geq 0$.   □

Relating this result to the example of Section 4.3, it is easy to see that $(5, 0)'$ is feasible for the student's problem (the primal), while $(0, 0, 10)'$ is feasible for the supplier's problem (the dual). For these values, the objective value of the primal at the given feasible point is 250, which is greater than the objective value of the dual at the given point 80. Note that 80 provides a lower bound on the optimal value of the primal problem, and 250 provides an upper bound on the optimal value of the dual problem.

Exercise 4-4-1. If the following linear program is solvable, find a lower bound on the optimal value of its objective function, *without* using the simplex method. If it is unsolvable, explain why.

$$
\begin{array}{rrrrrrl}
\min & -47x_1 & + & 13x_2 & + & 22x_3 & \\
\text{subject to} & -4x_1 & + & x_2 & - & 17x_3 & \geq \ 2, \\
& -x_1 & + & x_2 & + & 39x_3 & \geq \ 1, \\
& & & & x_1, x_2, x_3 & \geq \ 0.
\end{array}
$$

Exercise 4-4-2. Consider the (primal) problem

$$
\begin{array}{rrrrrl}
\min & x_1 & - & 2x_2 & \\
\text{subject to} & x_1 & - & x_2 & \geq & 1, \\
& x_1 & + & 2x_2 & \geq & -3, \\
& & & x_1, x_2 & \geq & 0.
\end{array}
$$

1. Solve the primal problem or produce an unbounded ray.

2. Write down the dual problem.

3. What can you say about the dual problem? Be sure to quote any results you use to justify your comments.

We now discuss one of the most fundamental theorems of linear programming.

Theorem 4.4.2 (Strong Duality or Trichotomy Theorem). *Exactly one of the following three alternatives holds:*

(i) Both *primal and dual problems are* feasible *and consequently both have optimal solutions with equal extrema.*

(ii) *Exactly* one *of the problems is* infeasible *and consequently the other problem has an unbounded objective function in the direction of optimization on its feasible region.*

(iii) Both *primal and dual problems are* infeasible.

Proof. (i) If both problems are feasible, then both objectives are bounded by weak duality, Theorem 4.4.1. Specifically, if $\bar{u}$ is any dual feasible point, we have from Theorem 4.4.1 that $p'x \geq b'\bar{u}$ for all primal feasible x; a bound on the dual objective

$b'u$ can be obtained similarly. Hence the simplex method with the smallest-subscript rule applied to the tableau

			$u_{m+1} =$	$\cdots$	$u_{m+n} =$	$w =$
			x_1	$\cdots$	x_n	1
$-u_1$	x_{n+1}	$=$	A_{11}	$\cdots$	A_{1n}	$-b_1$
$\vdots$			$\vdots$	$\ddots$	$\vdots$	$\vdots$
$-u_m$	x_{n+m}	$=$	A_{m1}	$\cdots$	A_{mn}	$-b_m$
1	z	$=$	p_1	$\cdots$	p_n	0

cannot cycle and must terminate at a primal optimal tableau as follows:

$$(4.6)$$

			$u_{\hat{B}} =$	$w =$
			x_N	1
$-u_{\hat{N}}$	x_B	$=$	H	h
1	z	$=$	c'	α

It follows from this tableau that $x_B = h \geq 0$, $x_N = 0$ is an optimal solution and $c' \geq 0$, $z = \alpha$. Moreover, we have from Theorem 4.1.1 that $u_{\hat{B}} = c$, $u_{\hat{N}} = 0$ is dual feasible and also that the dual objective w equals α at this point. Since the primal and dual objectives have the same value, we conclude from weak duality (Theorem 4.4.1) that both objective achieve their bounds and hence are optimal.

(ii) Consider the case in which exactly one of the problems is infeasible. Suppose that the other problem has a bounded objective. Then the simplex method with the smallest-subscript rule will terminate at a primal-dual optimal tableau (as in (i)), contradicting infeasibility. Hence, the feasible problem cannot have a bounded objective.

(iii) We can illustrate this case by setting $A = 0$, $b = 1$, and $p = -1$. $\square$

For the snack problem example of Section 4.3, we have already seen that both problems are feasible, and therefore the strong duality theorem dictates that their optimal values should be equal—so it was no coincidence that both the student's problem and the supplier's problem had an optimal value of $2.20. It is interesting to note that if both the student and the supplier make their optimal choices, the baker is "squeezed" and makes no profit at all.

Exercise 4-4-3. Let z^* be the optimal value of

$$\begin{array}{ll} \max & c'x \\ \text{subject to} & Ax \leq b, \quad x \geq 0, \end{array}$$

and let y^* be any optimal solution of the corresponding dual problem. Prove that

$$c'x \leq z^* + t'y^*$$

for every feasible point x of

$$\begin{array}{ll} \max & c'x \\ \text{subject to} & Ax \leq b + t, \quad x \geq 0. \end{array}$$

4.5 KKT Optimality Conditions

In this section we derive necessary and sufficient optimality conditions for linear programming—a set of algebraic relationships that are satisfied by primal and dual variables if and only if these variables are solutions of a primal-dual pair of linear programs. These conditions are commonly called the KKT conditions, after their originators (Karush (1939), Kuhn & Tucker (1951)). We state them as follows:

$\bar{x}$ is a solution to the primal problem and $\bar{u}$ is a solution to the dual if and only if $\bar{x}$ and $\bar{u}$ satisfy the following relationships:

$$A\bar{x} \geq b, \qquad \bar{x} \geq 0, \qquad A'\bar{u} \leq p, \qquad \bar{u} \geq 0, \qquad (4.7)$$

and

$$\bar{u}_i(A_{i.}\bar{x} - b_i) = 0, \qquad \bar{x}_j(-A'_{.j}\bar{u} + p_j) = 0 \qquad \forall i, j. \qquad (4.8)$$

This claim follows from the equivalence of (b) and (d) in the following theorem.

Theorem 4.5.1 (KKT Conditions). *The following conditions are equivalent:*

(a) *$\bar{x}$ solves the primal problem*

$$\min_{x} \quad z = p'x$$
$$\text{subject to} \quad Ax - b \geq 0, \quad x \geq 0.$$

(b) *$\bar{x}$ is primal feasible, some $\bar{u}$ is dual feasible, and $p'\bar{x} = b'\bar{u}$; that is, the primal objective value at some primal feasible point is equal to the dual objective value at some dual feasible point. Furthermore, $\bar{x}$ is primal optimal and $\bar{u}$ is dual optimal.*

(c) *$\bar{x}$ is primal feasible, some $\bar{u}$ is dual feasible, and $p'\bar{x} \leq b'\bar{u}$; that is, the primal objective value is less than or equal to the dual objective value.*

(d) *There exist $\bar{x}$ and $\bar{u}$ satisfying $A\bar{x} \geq b$, $\bar{x} \geq 0$, $A'\bar{u} \leq p$, $\bar{u} \geq 0$, and*

$$\bar{u}_i(A_{i.}\bar{x} - b_i) = 0, \qquad \bar{x}_j(-A'_{.j}\bar{u} + p_j) = 0 \qquad \forall i, j.$$

(e) *There exist $\bar{x}$ and $\bar{u}$ satisfying $A\bar{x} \geq b$, $\bar{x} \geq 0$, $A'\bar{u} \leq p$, $\bar{u} \geq 0$, and*

$$\bar{u}'(A\bar{x} - b) + \bar{x}'(-A'\bar{u} + p) = 0.$$

Proof.

(a) $\implies$ (b) This implication follows from strong duality, Theorem 4.4.2.

(b) $\implies$ (c) This implication is trivial.

(c) $\implies$ (e) We have the following relationships:

$$0 \leq \bar{u}'(A\bar{x} - b) + \bar{x}'(-A'\bar{u} + p) = -\bar{u}'b + \bar{x}'p \leq 0.$$

The first inequality follows from the feasibility conditions and the second inequality from (c). Since the left and right sides of the above inequality are equal, we must have equality throughout, proving (e).

(e) $\implies$ (d) This implication is trivial since each component of each sum in (e) is nonnegative.

(d) $\implies$ (a) Let x be any primal feasible point, and let $\bar{x}$ and $\bar{u}$ be as in (d). Since $\bar{x}$ is primal feasible, all we need to show is that $p'x \geq p'\bar{x}$. Since $\bar{u}$ is dual feasible, it follows from the weak duality theorem (Theorem 4.4.1) that $p'x \geq b'\bar{u}$. From (d), it follows that $\bar{x}'p = \bar{x}'A'\bar{u}$ and $\bar{x}'A'\bar{u} = b'\bar{u}$, so that $p'\bar{x} = \bar{x}'p = b'\bar{u}$. Hence $p'x \geq p'\bar{x}$, as required. $\quad\Box$

We can write the KKT conditions (4.7) and (4.8) equivalently and more succinctly as

$$0 \leq A\bar{x} - b \perp \bar{u} \geq 0, \tag{4.9a}$$
$$0 \leq p - A'\bar{u} \perp \bar{x} \geq 0, \tag{4.9b}$$

where the orthogonality notation $\perp$, defined in (A.1), stands here for $\bar{u}'(A\bar{x} - b) = 0$ and $\bar{x}'(p - A'\bar{u}) = 0$, respectively.

The conditions (4.8) are frequently referred to as the *complementary slackness* or *complementarity* conditions. They imply the following complementarity relationships:

$$x_j = 0 \quad \text{or} \quad p_j = (A'_{.j}u), \qquad j = 1, 2, \ldots, n, \tag{4.10}$$

and

$$u_i = 0 \quad \text{or} \quad (A_{i.}x) = b_i, \qquad i = 1, 2, \ldots, m. \tag{4.11}$$

In other words, each primal variable or its corresponding dual slack variable is zero, and each dual variable or its corresponding primal slack variable is zero. When *exactly one* of the equalities in each pair of conditions in (4.10) and (4.11) holds, we say that the solutions x and u satisfy *strict complementary slackness* or *strict complementarity*.

Referring back to the example of Section 4.3, we see that at the solution of the primal, the third constraint in (4.4) has a positive slack variable (that is, $2x_1 + 5x_2 - 8 > 0$), and so it follows that the optimal dual variable on that constraint (namely u_3) must be 0. Furthermore, since $x_1 > 0$ at the optimal solution of the primal, the complementarity conditions guarantee that the first constraint of the dual problem (4.5) must be satisfied as an equality at the dual solution. We leave the reader to confirm this fact.

These complementarity relationships are satisfied for every primal solution and *any* dual solution, not just for a particular pair of primal-dual optimal solutions. We shall discuss complementarity in more detail in Chapter 7.

Example 4-5-1. Use the complementary slackness theorem to check whether $u = (7, 0, 2.5, 0, 3, 0, 0.5)'$ is an optimal solution of

$$
\begin{array}{lllllllll}
\max & w = u_1 + 2u_2 + u_3 - 3u_4 + u_5 + u_6 - u_7 \\
\text{subject to} & u_1 & + & u_2 & & - & u_4 & & + & 2u_6 & - & 2u_7 & \leq & 6, \\
& & & u_2 & & - & u_4 & + & u_5 & - & 2u_6 & + & 2u_7 & \leq & 4, \\
& & & u_2 & + & u_3 & & & & + & u_6 & - & u_7 & \leq & 2, \\
& & & u_2 & & & - & u_4 & & & - & u_6 & + & u_7 & \leq & 1, \\
& & & & & & & & u_1, u_2, u_3, u_4, u_5, u_6, u_7 & \geq & 0.
\end{array}
$$

Note that this problem is in the standard form of a dual linear program (4.3), with

$$
A = \begin{bmatrix}
1 & 0 & 0 & 0 \\
1 & 1 & 1 & 1 \\
0 & 0 & 1 & 0 \\
-1 & -1 & 0 & -1 \\
0 & 1 & 0 & 0 \\
2 & -2 & 1 & -1 \\
-2 & 2 & -1 & 1
\end{bmatrix}, \qquad
b = \begin{bmatrix}
1 \\
2 \\
1 \\
-3 \\
1 \\
1 \\
-1
\end{bmatrix}, \qquad
p = \begin{bmatrix}
6 \\
4 \\
2 \\
1
\end{bmatrix}.
$$

We attempt to determine an x that together with this u satisfies the KKT conditions (4.7) and (4.8). First, it is easy to verify that $A'u \le p$ and $u \ge 0$. Next, we use the complementary slackness conditions to determine the components i for which $A_{i.}\bar{x} = b_i$ must be satisfied by x. Since $u_1 > 0$, $u_3 > 0$, $u_5 > 0$, and $u_7 > 0$, we have

$$
A_{1.}x = b_1, \qquad A_{3.}x = b_3, \qquad A_{5.}x = b_5, \qquad \text{and} \quad A_{7.}x = b_7.
$$

Taken together, these constraints imply that $x = (1, 1, 1, 0)'$.

The final step is to check that $Ax \ge b$ and $x \ge 0$ are satisfied and that $x_j(-A'u+p)_j = 0$ for all $j = 1, 2, \ldots, n$. Since these conditions hold, we have that (4.7) and (4.8) are both satisfied with x replacing $\bar{x}$ and u replacing $\bar{u}$. Hence, by the KKT conditions, x solves the primal problem and u solves the dual. ∎

Exercise 4-5-2. Without using any simplex pivots, show that $x = (0, 1.8, 0.1)'$ solves the following linear program:

$$
\begin{array}{rrcrcrcr}
\min & -12x_1 & + & 10x_2 & + & 2x_3 & & \\
\text{subject to} & -4x_1 & + & x_2 & - & 8x_3 & \ge & 1, \\
& -x_1 & + & x_2 & + & 12x_3 & \ge & 3, \\
& & & & & x_1, x_2, x_3 & \ge & 0.
\end{array}
$$

4.6   Dual Simplex Method

By making use of the duality results just established we can solve any linear program by applying the simplex method to its dual. Rather than formulating the dual explicitly, however, we can use an implicit approach that involves applying Jordan exchanges to the original primal tableau—an approach known as the *dual simplex method*. This approach has a significant advantage when the origin $u = 0 \in \mathbf{R}^m$ is feasible in the dual problem (that is, $p \ge 0$), while the origin $x = 0 \in \mathbf{R}^n$ is not primal feasible (that is, $b \not\le 0$). In this situation, we can apply the dual simplex method, rather than applying Phase I to compute a feasible initial point for the primal problem. This situation arises frequently in practice, for instance, in the approximation problems of Chapter 9.

Consider the following primal-dual pair:

$$
\begin{array}{cl}
\min\limits_{x} & z = p'x \\
\text{subject to} & Ax - b \ge 0, \quad x \ge 0,
\end{array}
$$

and

$$\max_{u} \quad w = b'u$$
$$\text{subject to} \quad -A'u + p \geq 0, \quad u \geq 0.$$

We add primal slack variables $x_{n+1}, \ldots, x_{n+m}$ and dual slack variables $u_{m+1}, \ldots, u_{m+n}$ to obtain the following tableau:

			$u_{m+1} =$	$\cdots$	$u_{m+s} =$	$\cdots$	$u_{m+n} =$	$w =$
			x_1	$\cdots$	x_s	$\cdots$	x_n	1
$-u_1$	x_{n+1}	=	A_{11}	$\cdots$	A_{1s}	$\cdots$	A_{1n}	$-b_1$
			$\vdots$	$\vdots$	$\vdots$	$\vdots$	$\vdots$	$\vdots$
$-u_r$	x_{n+r}	=	A_{r1}	$\cdots$	A_{rs}	$\cdots$	A_{rn}	$-b_r$
			$\vdots$	$\vdots$	$\vdots$	$\vdots$	$\vdots$	$\vdots$
$-u_m$	x_{n+m}	=	A_{m1}	$\cdots$	A_{ms}	$\cdots$	A_{mn}	$-b_m$
1	z	=	p_1	$\cdots$	p_s	$\cdots$	p_n	0

This tableau is a special case of the general tableau

			$u_{\hat{B}} =$	$w =$
			x_N	1
$-u_{\hat{N}}$	x_B	=	H	h
1	z	=	c'	α

where the sets B and N form a partition of $\{1, 2, \ldots, n + m\}$ (containing m and n indices, respectively), while the sets $\hat{B}$ and $\hat{N}$ form a partition of $\{1, 2, \ldots, m + n\}$ (containing n and m indices, respectively). The initial tableau above has $B = \{n + 1, \ldots, n + m\}$ and $\hat{B} = \{m + 1, \ldots, m + n\}$.

Let us describe the dual simplex method by assuming that $c \geq 0$ in the general tableau above. We can then define a dual feasible point by setting $u_{\hat{B}} = c$, $u_{\hat{N}} = 0$. We attempt to find a dual *optimal* point by performing Jordan exchanges to make the last column nonnegative, while maintaining dual feasibility (nonnegativity of the last row). (This approach is analogous to Phase II of the primal simplex method, in which we start with a nonnegative last *column* and perform Jordan exchanges to obtain a nonnegative last *row*.) For this new approach, the pivot selection rules are as follows:

1. (Pivot Row Selection): The pivot row is any row r with $h_r < 0$. If none exist, the current tableau is dual optimal.

2. (Pivot Column Selection): The pivot column is any column s such that

$$c_s / H_{rs} = \min_{j} \left\{ c_j / H_{rj} \mid H_{rj} > 0 \right\}.$$

If $H_{rj} \leq 0$ for all j, the dual objective is unbounded above.

The justification for the column selection rule is that if we let $u_{\hat{N}(r)} = \lambda \geq 0$, $u_{\hat{N}(i)} = 0$, $i \neq r$, then we have that

$$u_{\hat{B}(j)} = -H_{rj}\lambda + c_j$$

for $j = 1, 2, \ldots, n$. For $u_{\hat{N}(j)} \geq 0$ we need $\lambda \leq c_j / H_{rj}$ for each $H_{rj} > 0$, $j = 1, 2, \ldots, n$, and thus the minimum ratio determines the dual basic variable $u_{\hat{N}(s)}$ which blocks increase of $u_{\hat{N}(r)}$. For the case in which $H_{rj} \leq 0$ for all j, we have that $u_{\hat{N}(j)} = -H_{rj}\lambda + c_j \geq 0$, for every $\lambda \geq 0$ and $w = -h_r \lambda + \alpha \to \infty$ as $\lambda \to \infty$, verifying unboundedness.

The smallest-subscript rule can be used to avoid cycling in the dual simplex method, just as in the original (primal) simplex method (see Section 3.5). If there are multiple rows r with $h_r < 0$ in Step 1, we choose the one with the smallest subscript on the dual label for that row. If there are multiple columns that achieve the minimum ratio in Step 2, we again choose the one with the smallest subscript on the dual label.

Example 4-6-1. Consider the problem

$$
\begin{aligned}
\min \quad & x_1 + x_2 \\
\text{subject to} \quad & 3x_1 + x_2 \geq 2, \\
& 3x_1 + 4x_2 \geq 5, \\
& 4x_1 + 2x_2 \geq 8, \\
& x_1, x_2 \geq 0,
\end{aligned}
$$

with dual feasible initial tableau constructed using the following code:

```
>> load ex4-6-1
>> T = totbl(A,b,p);
>> T = dualbl(T);
```

		$u_4 =$	$u_5 =$	$w =$
		x_1	x_2	1
$-u_1$	$x_3 =$	3	1	-2
$-u_2$	$x_4 =$	3	4	-5
$-u_3$	$x_5 =$	4	2	-8
1	$z =$	1	1	0

The pivot row selection rule would allow any of the three rows to be chosen, since all have negative elements in the last column. If we were applying the smallest-subscript rule, we would choose row 1, but let us choose row 3 here. By applying the pivot column selection rule, we obtain column 1 as the pivot column. The resulting Jordan exchange leads to the following tableau:

```
>> T = ljx(T,3,1);
```

		$u_3 =$	$u_5 =$	$w =$
		x_5	x_2	1
$-u_1$	$x_3 =$	0.75	-0.5	4
$-u_2$	$x_4 =$	0.75	2.5	1
$-u_4$	$x_1 =$	0.25	-0.5	2
1	$z =$	0.25	0.5	2

Since the last column of this tableau is nonnegative (and since nonnegativity of the last row has been maintained), the tableau is optimal. Thus in one step of the dual simplex method, we have obtained a primal optimal solution $x_1 = 2$, $x_2 = 0$, $z = 2$ and a dual optimal solution $u_1 = 0$, $u_2 = 0$, $u_3 = 0.25$, $w = 2$. Note that if the primal simplex were employed, we would have had to apply a Phase I procedure first, and the computational effort would have been greater. ∎

Exercise 4-6-2. Solve the following problem in MATLAB by both the dual simplex method and the two-phase (primal) simplex method. Use the `addrow` and `addcol` routines to set up the appropriate Phase I problem.

$$
\begin{array}{lrcrcrcrcr}
\min & & & x_2 & + & x_3 & & & & \\
\text{subject to} & x_1 & + & x_2 & - & 2x_3 & \geq & 1, & & \\
& -x_1 & + & x_2 & & & \geq & 2, & & \\
& & - & x_2 & & & \geq & -6, & & \\
& & & x_1, x_2, x_3 & & & \geq & 0. & &
\end{array}
$$

The steps of the dual simplex method are identical to those that would be taken if we were to apply the primal simplex method to the dual problem, using consistent pivoting rules. This fact is illustrated by the following simple example.

Example 4-6-3. Consider the problem

$$
\begin{array}{lrcrcr}
\min & & x_1 & + & 2x_2 & \\
\text{subject to} & -x_1 & + & x_2 & \geq & 1, \\
& x_1 & + & x_2 & \geq & 3, \\
& & x_1, x_2 & \geq & 0. &
\end{array}
$$

The dual simplex method obtains an optimal tableau in two steps:

```
>> load ex4-6-3
>> T = totbl(A,b,p);
>> T = dualbl(T);
```

		$u_3 =$	$u_4 =$	$w =$
		x_1	x_2	1
$-u_1$	$x_3 =$	-1	1	-1
$-u_2$	$x_4 =$	1	1	-3
1	$z =$	1	2	0

```
>> T = ljx(T,2,1);
```

		$u_2 =$	$u_4 =$	$w =$
		x_4	x_2	1
$-u_1$	$x_3 =$	-1	2	-4
$-u_3$	$x_1 =$	1	-1	3
1	$z =$	1	1	3

```
>> T = ljx(T,1,2);
```

		$u_2 =$	$u_1 =$	$w =$
		x_4	x_3	1
$-u_4$	$x_2 =$	0.5	0.5	2
$-u_3$	$x_1 =$	0.5	-0.5	1
1	$z =$	1.5	0.5	5

The corresponding dual problem is

$$
\begin{array}{lrcrcr}
\max & & u_1 & + & 3u_2 & \\
\text{subject to} & -u_1 & + & u_2 & \leq & 1, \\
& u_1 & + & u_2 & \leq & 2, \\
& & u_1, u_2 & \geq & 0. &
\end{array}
$$

When reformulated into standard primal form, the steps of the primal simplex method give the following tableaus applied to the dual problem. Note that the optional arguments to `totbl` are used to set the objective value α to 0 and change the primal labels from "x" and "z" to "u" and "w," respectively.

```
>> load ex4-6-3
>> T = totbl(-A',-p,-b,0,'u','-w');
>> T = dualbl(T);
```

			$x_3 =$	$x_4 =$	$z =$
			u_1	u_2	1
$-x_1$	u_3	$=$	1	-1	1
$-x_2$	u_4	$=$	-1	-1	2
1	$-w$	$=$	-1	-3	0

```
>> T = ljx(T,1,2);
```

			$x_3 =$	$x_1 =$	$z =$
			u_1	u_3	1
$-x_4$	u_2	$=$	1	-1	1
$-x_2$	u_4	$=$	-2	1	1
1	$-w$	$=$	-4	3	-3

```
>> T = ljx(T,2,1);
```

			$x_2 =$	$x_1 =$	$z =$
			u_4	u_3	1
$-x_4$	u_2	$=$	-0.5	-0.5	1.5
$-x_3$	u_1	$=$	-0.5	0.5	0.5
1	$-w$	$=$	2	1	-5

After appropriate changes of signs and a transposition operation, these three tableaus are identical to the three tableaus obtained in the dual simplex method. This example illustrates the claim that the primal simplex method applied to the dual problem is identical to the dual simplex method applied to the primal problem. ∎

Exercise 4-6-4.   1. Form the dual of the following problem and solve the dual using the (primal) simplex method, using the most negative reduced cost to choose the variable to leave the basis.

$$\begin{array}{llll}
\min & 3x_1 + 4x_2 + 5x_3 \\
\text{subject to} & x_1 + 2x_2 + 3x_3 & \geq & 5, \\
& 2x_1 + 2x_2 + x_3 & \geq & 6, \\
& x_1, x_2, x_3 & \geq & 0.
\end{array}$$

2. Apply the dual simplex method to the problem given above. Use the most negative element of the last column to indicate the pivot row. Verify that the tableaus obtained in this approach are identical to those in part 1, after transposition and sign changes.

4.7 General Linear Programs

The duality results derived above were relevant to primal problems in standard form. The primal-dual simplex tableau was used to obtain various duality results. We now consider linear programs in "general" form, in which both inequality and equality constraints and both nonnegative and free variables are present. We show how the duals of such problems can be generated via conversion to standard form.

The "general" form of the linear programming problem is as follows:

$$
\begin{aligned}
\min_{x,y} \quad & p'x + q'y \\
\text{subject to} \quad & Bx + Cy \;\geq\; d, \\
& Ex + Fy \;=\; g, \\
& Hx + Jy \;\leq\; k, \\
& x \qquad\quad\; \geq\; 0.
\end{aligned}
\tag{4.12}
$$

Note the presence of equality constraints and free variables. The dual of this problem can be constructed by means of a three-step process:

- put into standard form;

- construct the dual for the resulting standard form problem;

- simplify the resulting dual problem.

By applying the first step of this procedure to the problem above, replacing the free variable y by the difference $y^+ - y^-$ with y^+, y^- both nonnegative, we obtain

$$
\min_{x,y^+,y^-} \quad
\begin{bmatrix} p' & q' & -q' \end{bmatrix}
\begin{bmatrix} x \\ y^+ \\ y^- \end{bmatrix}
$$

$$
\text{subject to} \quad
\begin{bmatrix} B & C & -C \\ E & F & -F \\ -E & -F & F \\ -H & -J & J \end{bmatrix}
\begin{bmatrix} x \\ y^+ \\ y^- \end{bmatrix}
\geq
\begin{bmatrix} d \\ g \\ -g \\ -k \end{bmatrix}, \qquad
\begin{bmatrix} x \\ y^+ \\ y^- \end{bmatrix}
\geq 0.
$$

Proceeding with the second step, the dual of this problem is as follows:

$$
\max_{u,s,t} \quad
\begin{bmatrix} d' & g' & -g' & -k' \end{bmatrix}
\begin{bmatrix} u \\ r \\ s \\ t \end{bmatrix}
$$

$$
\text{subject to} \quad
\begin{bmatrix} B' & E' & -E' & -H' \\ C' & F' & -F' & -J' \\ -C' & -F' & F' & J' \end{bmatrix}
\begin{bmatrix} u \\ r \\ s \\ t \end{bmatrix}
\leq
\begin{bmatrix} p \\ q \\ -q \end{bmatrix}, \qquad
\begin{bmatrix} u \\ r \\ s \\ t \end{bmatrix}
\geq 0.
$$

For the third step, we define the free variable $v = r - s$ and the nonpositive variable $w = -t$ and simplify as follows:

$$
\begin{aligned}
\max \quad & d'u + g'v + k'w \\
\text{subject to} \quad & B'u + E'v + H'w \leq p, \\
& C'u + F'v + J'w = q, \\
& u \geq 0, \quad w \leq 0.
\end{aligned}
\tag{4.13}
$$

We could have added nonpositive variables to (4.12) as well, but to simplify the exposition we leave this as an exercise to the reader. We summarize this construction with the following correspondence between elements of the primal and dual linear programs in a primal-dual pair:

Min problem	Max problem
Nonnegative variable $\geq$	Inequality constraint $\leq$
Nonpositive variable $\leq$	Inequality constraint $\geq$
Free variable	Equality constraint $=$
Inequality constraint $\geq$	Nonnegative variable $\geq$
Inequality constraint $\leq$	Nonpositive variable $\leq$
Equality constraint $=$	Free variable

We can apply Theorems 4.4.1, 4.4.2, and 4.5.1 to a general primal-dual pair of linear programs. In particular, for Theorem 4.5.1, we have that the KKT conditions for the general primal-dual pair are as follows: $(\bar{x}, \bar{y})$ is a solution to the primal problem (4.12) and $(\bar{u}, \bar{v}, \bar{w})$ is a solution to the dual (4.13) if these vectors satisfy the following relationships:

$$
\begin{aligned}
B\bar{x} + C\bar{y} \geq d, \quad E\bar{x} + F\bar{y} = g, \quad H\bar{x} + J\bar{y} \leq k, \quad \bar{x} \geq 0, \\
B'\bar{u} + E'\bar{v} + H'\bar{w} \leq p, \quad C'\bar{u} + F'\bar{v} + J'\bar{w} = q, \quad \bar{u} \geq 0, \quad \bar{w} \leq 0,
\end{aligned}
\tag{4.14}
$$

and

$$
\begin{aligned}
\bar{x}_i (B'\bar{u} + E'\bar{v} + H'\bar{w} - p)_i &= 0 \quad \forall i, \\
\bar{u}_j (B\bar{x} + C\bar{y} - d)_j &= 0 \quad \forall j, \\
\bar{u}_l (H\bar{x} + J\bar{y} - k)_l &= 0 \quad \forall l.
\end{aligned}
\tag{4.15}
$$

Alternatively, using more succinct notation like that of (4.9), we can restate these conditions equivalently as follows:

$$0 \leq p - B'\bar{u} - E'\bar{v} - H'\bar{w} \perp \bar{x} \geq 0, \tag{4.16a}$$

$$q - C'\bar{u} - F'\bar{v} - J'\bar{w} = 0, \tag{4.16b}$$

$$0 \leq B\bar{x} + C\bar{y} - d \perp \bar{u} \geq 0, \tag{4.16c}$$

$$E\bar{x} + F\bar{y} = g, \tag{4.16d}$$

$$0 \geq H\bar{x} + J\bar{y} - k \perp \bar{w} \leq 0. \tag{4.16e}$$

Exercise 4-7-1. Consider the standard-form (primal) problem

$$
\begin{aligned}
\min \quad & p'x \\
\text{subject to} \quad & Ax \geq b, \quad x \geq 0,
\end{aligned}
$$

whose dual problem is

$$\begin{array}{ll} \max & b'u \\ \text{subject to} & A'u \leq p, \quad u \geq 0. \end{array}$$

1. Reformulate the dual problem to put it in standard form.

2. Construct the dual problem of the problem from part 1.

3. How is this "dual of the dual" related to the primal problem?

Note that the conclusion of the above exercise allows us to start from either side of the table above when constructing the dual of a given problem. Thus, if we have a minimization problem, we read from left to right, whereas for a maximization problem, we read from right to left.

Example 4-7-2. The canonical-form linear program has the form

$$\begin{array}{ll} \min_{x} & p'x \\ \text{subject to} & \mathcal{A}x = b, \quad x \geq 0. \end{array}$$

Using the construction above, the dual problem has a free variable u associated with the equality constraint $\mathcal{A}x = b$ and a dual inequality constraint $\leq$ associated with the nonnegative variable x. Hence the dual problem is

$$\begin{array}{ll} \max_{x} & b'u \\ \text{subject to} & \mathcal{A}'u \leq p. \quad \blacksquare \end{array}$$

Example 4-7-3. We wish to find the dual of the following linear program:

$$\begin{array}{ll} \min & p'x \\ \text{subject to} & Ax \leq b, \quad Cx = d, \quad 0 \leq x \leq f. \end{array}$$

First, we put this problem into general form by some simple transformations:

$$\begin{array}{ll} \min & p'x \\ \text{subject to} & \begin{bmatrix} -A \\ -I \end{bmatrix} x \geq \begin{bmatrix} b \\ -f \end{bmatrix}, \quad Cx = d, \quad x \geq 0. \end{array}$$

We can now substitute into the formula for the dual of general form to obtain

$$\begin{array}{ll} \max & -b'u - f'w + d'v \\ \text{subject to} & -A'u - w + C'v \leq p, \quad u \geq 0, \quad w \geq 0. \quad \blacksquare \end{array}$$

Exercise 4-7-4. Write down the dual of

$$\begin{array}{ll} \min_{x} & c'x \\ \text{subject to} & Ax = b, \quad l \leq x \leq f, \end{array}$$

where l and f are fixed lower and upper bounds on the variables x. Simplify the dual so that it has the fewest possible number of variables and constraints.

Exercise 4-7-5. Consider the following problem:

$$\text{min} \qquad b'r - b's$$
$$\text{subject to} \qquad A'r - A's - t \;=\; c,$$
$$\qquad\qquad\qquad r, s, t \;\geq\; 0.$$

1. Construct the dual of this problem and simplify it.

2. Simplify the problem first and *then* construct the dual.

Exercise 4-7-6. Consider the linear program

$$\text{max} \qquad x_1 - 2x_2 - 4x_3 - 2x_4$$
$$\text{subject to} \quad \begin{array}{rcrcrcrcl} x_1 & + & 2x_2 & - & x_3 & + & x_4 & \geq & 0, \\ 4x_1 & + & 3x_2 & + & 4x_3 & - & 2x_4 & \leq & 3, \\ -x_1 & - & x_2 & + & 2x_3 & + & x_4 & = & 1, \\ & & & & x_2, x_3, x_4 & & & \geq & 0. \end{array}$$

1. Solve the problem, explicitly justifying why you carry out the particular pivots.

2. Write down the dual of this problem.

3. Without using any further pivots, write down a solution of the dual problem. Be sure to quote any results you use and prove that this point is indeed a solution.

Exercise 4-7-7. Consider the primal linear programming problem

$$\text{max} \qquad c'x$$
$$\text{subject to} \quad Ax = b, \quad x \geq 0.$$

Suppose this problem and its dual are feasible, and let λ denote a known optimal solution to the dual.

1. If the kth row of the matrix A and the kth element of the right-hand side b are multiplied by $\mu \neq 0$, determine an optimal solution w to the dual of this modified problem.

2. Suppose that, in the original primal, we add μ times the kth equation to the rth equation. What is a solution w to the corresponding dual problem?

3. Suppose, in the original primal, we add μ times the kth row of A to the cost vector c. What is a solution to the corresponding dual problem?

4.8   Big M Method

In Chapters 3 and 5, we discuss the use of Phase I to obtain a feasible point for the original linear program, from which we can apply the Phase II simplex method. We demonstrate here an alternative technique that avoids Phase I, by constructing a new linear program from which the solution of the original linear program can easily be extracted, and for which it is easy to find an initial point. This approach is known as the "big M" method. *Note:* In this section, we use the notation M to denote a large scalar. This should not be confused with the notation of Chapter 7, in which M denotes the coefficient matrix in a linear complementarity problem.

Consider the linear program in standard form and its dual:

$$\min_{x} \quad z = p'x$$
$$\text{subject to} \quad Ax \geq b, \quad x \geq 0, \tag{4.17}$$

$$\max_{u} \quad w = b'u$$
$$\text{subject to} \quad A'u \leq p, \quad u \geq 0. \tag{4.18}$$

Given positive constant M, we now add an extra variable ξ to the problem and modify the dual accordingly as follows:

$$\min_{x} \quad p'x + M\xi$$
$$\text{subject to} \quad Ax + e_m\xi \geq b, \quad x, \xi \geq 0, \tag{4.19}$$

$$\max_{u} \quad w = b'u$$
$$\text{subject to} \quad A'u \leq p, \quad e'_m u \leq M, \quad u \geq 0, \tag{4.20}$$

where e_m is the vector of length m whose entries are all 1. In the following result, we show that for large enough M, the solution of (4.19) yields a solution of the original problem in (4.17).

Theorem 4.8.1. *If $\bar{x}$ solves (4.17), then there exists $\bar{M} \geq 0$ such that $(\bar{x}, 0)$ solves (4.19) for some or all $M \geq \bar{M}$. Conversely, if $(\bar{x}, 0)$ solves (4.19) for some M, then $\bar{x}$ solves (4.17).*

Proof. We start with the first statement of the theorem. If $\bar{x}$ solves the primal problem in (4.17), then the strong duality theorem posits the existence of a solution $\bar{u}$ to the dual problem (4.18). Let $\bar{M} = \|\bar{u}\|_1$. For all $M \geq \bar{M}$, $\bar{u}$ will be feasible for (4.20). Since $(\bar{x}, \bar{u})$ satisfies the KKT conditions for (4.17), (4.18), it is easy to see that $(x, \xi, u) = (\bar{x}, 0, \bar{u})$ satisfies the KKT conditions for (4.19), (4.20). Hence, by sufficiency of the KKT conditions, $(\bar{x}, 0)$ is optimal for (4.19), for $M \geq \bar{M}$, as claimed.

For the second statement of the theorem, let $(\bar{x}, 0)$ solve (4.19) some M. By strong duality, there is a solution $\bar{u}(M)$ to the corresponding dual problem (4.20) such that $p'\bar{x} + M0 = b'\bar{u}(M)$. Note that $\bar{x}$ and $\bar{u}$ are feasible for the primal and dual problems in (4.17) and (4.18), respectively, and so by weak duality, we have $b'\bar{u}(M) \leq p'\bar{x}$. By putting these two relations together, we have

$$b'\bar{u}(M) \leq p'\bar{x} = p'\bar{x} + M0 = b'\bar{u}(M),$$

which implies that $b'\bar{u}(M) \leq p'\bar{x}$ and hence that $\bar{x}$ and $\bar{u}$ are in fact optimal for (4.17) and (4.18). □

The proof shows just how large M needs to be for the big M method to work: M must exceed $\|u\|_1$, where u is a solution of the dual problem (4.18). Unfortunately, this value is not known a priori. Hence, the big M method is usually implemented by choosing an increasing sequence of values for M, stopping when we find an M for which the solution of (4.19) has $\xi = 0$. The general scheme is as follows.

Algorithm 4.1 (Big M Method).

1. *Make an initial guess of M.*

2. *Solve* (4.19).

3. *If $\xi = 0$, STOP: x is a solution of* (4.17).

4. *Otherwise, increase M and go to* 2.

Note that this scheme will fail when the original primal problem in (4.17) is infeasible. In this case, for all M large enough, the solution of (4.19) will have $\xi = \min_x \|b - Ax\|_\infty > 0$. The solution of (4.19) will be such that x and ξ violate the constraint $Ax \geq b$ as little as possible in the ∞-norm sense. The following theorem formalizes this claim.

Theorem 4.8.2. *If there exists an $\bar{x}$ such that $(\bar{x}, \|(-A\bar{x} + b)_+\|_\infty)$ solves* (4.19) *for all $M \geq \bar{M}$, then $\bar{x}$ solves*

$$\min_x \left\{ p'x \mid Ax + e_m\bar{\xi} \geq b, x \geq 0 \right\},$$

where $\bar{\xi} = \min_x \|(-Ax + b)_+\|_\infty$. In particular, if $\{x \mid Ax \geq b, x \geq 0\} \neq \emptyset$, then $\bar{\xi} = 0$ and $\bar{x}$ solves (4.17).

Proof. See Mangasarian (1999*b*, Theorem 2.1). ☐

Exercise 4-8-1. Show that if the objective function of the primal big M method corresponding to a given feasible LP is unbounded below, then the objective function of the original primal LP itself is unbounded below on its feasible region.

4.9   Applications of Duality

We now present some fundamental results that are simple consequences of strong duality, Theorem 4.4.2. We begin with the Farkas lemma, which is key in deriving the optimality conditions of linear and nonlinear programming. The Farkas lemma can also be used to derive the strong duality result, Theorem 4.4.2, itself.

Theorem 4.9.1 (Farkas Lemma). *Let $A \in \mathbf{R}^{m \times n}$ and $b \in \mathbf{R}^m$. Then exactly one of the following two systems has a solution:*

(I) $Ax = b, x \geq 0$.

(II) $A'u \geq 0, b'u < 0$.

Proof. Consider the linear program

$$\begin{array}{ll} \max & 0'x \\ \text{subject to} & Ax = b, \quad x \geq 0, \end{array}$$

and its dual

$$\begin{array}{ll} \min & b'u \\ \text{subject to} & A'u \geq 0. \end{array}$$

If (I) holds, then the primal is feasible, and its optimal objective is obviously zero. By applying the weak duality result (Theorem 4.4.1), we have that any dual feasible vector u (that is, one which satisfies $A'u \geq 0$) must have $b'u \geq 0$. Hence, the inequalities $A'u \geq 0$, $b'u < 0$ cannot simultaneously hold, and so (II) is not satisfied. If, on the other hand, (I) does not hold, then the primal is infeasible. The dual is, however, always feasible, since $u = 0$ satisfies $A'u \geq 0$ trivially. By applying strong duality, Theorem 4.4.2, we deduce that the dual objective is unbounded below, and hence $A'u \geq 0$, $b'u < 0$ has a solution.   □

This result is an example of a *theorem of the alternative*. Many other examples can be proved by a similar technique, for example the Gordan theorem, which states that exactly one of the following two systems has a solution:

(I) $Ax > 0$.

(II) $A'y = 0, 0 \neq y \geq 0$.

Exercise 4-9-1. Prove the Gordan theorem of the alternative.

Exercise 4-9-2. Use the Farkas lemma to prove that the set $\{Ax \mid x \geq 0\}$ is closed.

The following result is an example of a separation lemma. In layman's terms it states that if the two given sets do not intersect, then a hyperplane $w^T x = \gamma$ has all the points from one set on one side and all the points from the other set on the other side. (One possible value for γ is $\frac{1}{2}(\min_i(A'w)_i + \max_j(B'w)_j)$.)

Exercise 4-9-3. Let $A_{.j}, j = 1, \ldots, n$, be n points in $\mathbf{R}^m$, and $B_{.j}, j = 1, \ldots, k$, be k points in $\mathbf{R}^m$. Show that the set $\text{conv}\{A_{.j} \mid j = 1, \ldots, n\}$ does not intersect the set $\text{conv}\{B_{.j} \mid j = 1, \ldots, k\}$ if and only if there exists a $w \in \mathbf{R}^m$ such that

$$\min_i(A'w)_i > \max_j(B'w)_j.$$

Note that the convex hull of a finite set of points is defined as the set of all convex combinations of those points, that is,

$$\text{conv}\{A_{.j} \mid j = 1, \ldots, n\} := \left\{\sum_{j=1}^n A_{.j}\lambda_j \mid \sum_{j=1}^n \lambda_j = 1, \lambda \geq 0\right\}.$$

A second example of the usefulness of duality theory for linear programming is the following classical result from linear algebra.

Theorem 4.9.2. *Let $A \in \mathbf{R}^{m \times n}$. Any $c \in \mathbf{R}^n$ can be uniquely decomposed as*

$$c = u + w, \qquad Au = 0, \qquad w = A'v. \tag{4.21}$$

Thus,

$$\mathbf{R}^n = \ker A \oplus \operatorname{im} A',$$

where

$$\ker A := \{u \mid Au = 0\}, \qquad \operatorname{im} A' := \{w \mid w = A'v\}.$$

Proof. Consider the dual linear programs

$$
\begin{array}{ll}
\max\limits_{u,v} & 0'u + 0'v \\
\text{subject to} & u + A'v = c, \\
& Au = 0,
\end{array}
\qquad\qquad
\begin{array}{ll}
\min\limits_{x,y} & c'x \\
\text{subject to} & x + A'y = 0, \\
& Ax = 0.
\end{array}
$$

Note that

$$
\begin{aligned}
0 = x + A'y, \quad Ax = 0 \\
\implies \quad AA'y = 0 \\
\implies \quad y'AA'y = 0 \\
\implies \quad A'y = 0 \\
\implies \quad x = -A'y = 0.
\end{aligned}
\tag{4.22}
$$

It follows from (4.22) that $x = 0$ for all dual feasible points and the dual minimum is zero. Hence, by strong duality (Theorem 4.4.2), the primal problem is also solvable for any c, and (4.21) holds. If c has two decompositions (u^1, w^1) and (u^2, w^2), then

$$
0 = (u^1 - u^2) + A'(v^1 - v^2), \qquad A(u^1 - u^2) = 0.
$$

By using (4.22), we conclude that $u^1 = u^2$ and $w^1 = A'v^1 = A'v^2 = w^2$, thereby verifying uniqueness of the decomposition (4.21). $\square$

Our final result concerns the existence of solutions to dual linear programs that satisfy strict complementarity.

Theorem 4.9.3 (Existence of Strictly Complementary Solutions). *Suppose that the primal problem (4.2) has an optimal solution. Then there exist vectors $\hat{x}$ and $\hat{u}$ such that*

(i) *$\hat{x}$ is an optimal solution of (4.2),*

(ii) *$\hat{u}$ is an optimal solution of (4.3), and*

(iii) *$\hat{u} + (A\hat{x} - b) > 0$, $\hat{x} + (-A'\hat{u} + p) > 0$.*

Proof. Since (4.2) has an optimal solution (say $\bar{x}$), we have from Theorem 4.4.2 that the dual problem (4.3) also has an optimal solution (say $\bar{u}$). Consider now the linear program

$$
\begin{array}{ll}
\max\limits_{x,u,\epsilon} & \epsilon \\
\text{subject to} & -Ax \le -b, \quad x \ge 0, \\
& A'u \le p, \quad u \ge 0, \\
& p'x = b'u, \\
& \epsilon e \le u + (Ax - b), \\
& \epsilon e \le x + (-A'u + p), \\
& \epsilon \le 1.
\end{array}
$$

By the KKT conditions (4.7) and (4.8), the point $(x, u, \epsilon) = (\bar{x}, \bar{u}, 0)$ is feasible for this problem. Since the problem is bounded (its optimal objective value can be no larger than 1),

Theorem 4.4.2 implies there exists an optimal solution to both this problem and its dual, which is

$$\min_{r,s,t,v,\xi,\psi} \quad -b'(r+v) + p'(s+t) + \psi$$

$$\begin{aligned}
\text{subject to} \quad -A'(r+v) + p\xi - t &\geq 0, \\
A(s+t) - b\xi - v &\geq 0, \\
e'(t+v) + \psi &= 1, \\
r, s, t, v, \psi &\geq 0.
\end{aligned}$$

Suppose the optimal objective value of these dual problems is zero, and let (r, s, t, v, ξ, ψ) be an optimal solution of the dual. Then, by premultiplying the first dual constraint by $(t+s)'$ and the second dual constraint by $(r+v)'$ and adding, we see that

$$(b\xi + v)'(r+v) + (t - p\xi)'(t+s) \leq 0. \tag{4.23}$$

Since the objective of the primal is zero at the optimum, we have $\epsilon = 0$, and so the last constraint in the primal is inactive. Hence, by complementarity, the corresponding dual variable ψ is zero, and so we have by the fact that the objective of the dual is also zero that

$$0 = -b'(r+v) + p'(s+t) + \psi = -b'(r+v) + p'(s+t).$$

Hence, from (4.23), we have

$$v'(r+v) + t'(t+s) \leq 0.$$

Since all the variables appearing in the last inequality are nonnegative, we can rearrange this expression to obtain

$$v'v + t't \leq -v'r - t's \leq 0,$$

and so we must have $v = 0$ and $t = 0$. These values, along with $\psi = 0$ as already noted, contradict the dual constraint $e'(t+v) + \psi = 1$. We conclude that the primal linear program above has a solution with $\epsilon > 0$, proving the result. $\quad\square$

Note that by Theorem 4.5.1 and (4.8), we have that $u_i(Ax-b)_i = 0, x_j(-A'u+p)_j = 0$, for all $i = 1, 2, \ldots, m$ and $j = 1, 2, \ldots, n$, for every solution x of the primal problem and u of the dual problem. Theorem 4.9.3 shows further that for some solution $\hat{x}$ of the primal problem and $\hat{u}$ of the dual problem, we have

$$\hat{u}_i = 0 \text{ and } (A\hat{x} - b)_i > 0, \quad \text{or} \quad \hat{u}_i > 0 \text{ and } (A\hat{x} - b)_i = 0, \qquad \forall i = 1, 2, \ldots, m,$$

and

$$\hat{x}_j = 0 \text{ and } (-A'\hat{u}+p)_j > 0, \quad \text{or} \quad \hat{x}_j > 0 \text{ and } (-A'\hat{u}+p)_j = 0, \qquad \forall j = 1, 2, \ldots, n.$$

The notion of strict complementarity will be useful for the interior-point methods of Chapter 8.

Exercise 4-9-4. Prove that $Ax \leq b, 0 \leq x \leq e$, has no solution if and only if there exists $u \geq 0$ such that for all x satisfying $0 \leq x \leq e$, the single inequality $u'Ax \leq u'b$ has no solution.

Chapter 5

Solving Large Linear Programs

To make the simplex approach of Chapter 3 suitable for solving large linear programming problems, we need to pay careful attention to several important issues. Chief among these are pricing (the choice of variable to enter the basis) and efficient implementation of the linear algebra at each simplex step. In this chapter, we describe the *revised simplex method*, a powerful and economical approach that forms the basis of practical implementations of the simplex method that are suitable for solving large linear programs. We also discuss network flow problems, a special class of linear programs that comprise a large fraction of practical linear programming problems. We mention in particular how the simplex approach can be adapted to take advantage of the special structure of these problems.

In Chapter 3, we considered linear programming problems in the standard form

$$\begin{aligned} \min \quad & z = \bar{p}'x \\ \text{subject to} \quad & Ax \geq b, \quad x \geq 0, \end{aligned} \tag{5.1}$$

where $x, \bar{p} \in \mathbf{R}^n$, $b \in \mathbf{R}^m$, and $A \in \mathbf{R}^{m \times n}$. We showed that if we define slack variables $x_{n+1}, x_{n+2}, \ldots, x_{n+m}$ by

$$x_{n+i} := A_i.x - b_i,$$

then it follows that

$$A \begin{bmatrix} x_1 \\ \vdots \\ x_n \end{bmatrix} - I \begin{bmatrix} x_{n+1} \\ \vdots \\ x_{n+m} \end{bmatrix} = b.$$

Letting $x \in \mathbf{R}^l$ include the original variables and the slack variables, we define synonymous extensions of the cost vector $\bar{p}$ and coefficient matrix A as follows:

$$p := \begin{bmatrix} \bar{p} \\ 0 \end{bmatrix}, \qquad \mathcal{A} := [A \quad -I]. \tag{5.2}$$

It follows that the standard form above can be rewritten as the following *canonical form*:

$$\begin{aligned} \min \quad & z = p'x \\ \text{subject to} \quad & \mathcal{A}x = b, \quad x \geq 0, \end{aligned} \tag{5.3}$$

where $x, p \in \mathbf{R}^l$ for $l = m + n$, $b \in \mathbf{R}^m$, and $\mathcal{A} \in \mathbf{R}^{m \times l}$. In this chapter, we work with the canonical form. For generality, we do not assume that $\mathcal{A}$ and p were necessarily derived from the standard form in the manner described above but simply that the rows of $\mathcal{A}$ are linearly independent.

5.1 Foundations

In this section, we introduce the notions of a basis matrix and a basic feasible solution for the constraint set of the linear program (5.3). We relate these ideas to the geometric interpretation of linear programming discussed in Chapter 3. These ideas will be combined in the next section with our knowledge of the simplex method to derive the revised simplex method.

5.1.1 Basic Feasible Solutions and Basis Matrices

We make extensive use of the concept of linear independence of sets of (row and column) vectors in this discussion; see Chapter 2 for definitions of linear dependence and independence. By convention, we say that an empty set of vectors is linearly independent.

Definition 5.1.1. Let $\mathcal{A} \in \mathbf{R}^{m \times l}$, $b \in \mathbf{R}^m$, and consider the constraints $\mathcal{A}x = b$, $x \geq 0$. A *basic solution* is a vector $\bar{x} \in \mathbf{R}^l$ that satisfies $\mathcal{A}\bar{x} = b$, where for some $J \subset \{1, 2, \ldots, l\}$, $\mathcal{A}_J$ has linearly independent columns and $\bar{x}_j = 0$ for $j \notin J$.

We say that $\bar{x}$ is a *basic feasible solution* if $\bar{x}$ is a basic solution and in addition it satisfies the nonnegativity condition $\bar{x} \geq 0$.

Example 5-1-1. Given the matrix and vector

$$\mathcal{A} = \begin{bmatrix} -1 & 2 & 2 \\ 0 & 1 & 0 \end{bmatrix}, \qquad b = \begin{bmatrix} 3 \\ 0 \end{bmatrix},$$

we seek basic solutions of the system $\mathcal{A}x = b$. Since $l = 3$, there are $2^3 = 8$ possible choices for the index subset J. We consider each of these subsets in turn and see whether they lead to a vector $\bar{x}$ that satisfies Definition 5.1.1.

For the choice $J = \emptyset$, we must have $\bar{x}_j = 0$ for all $j = 1, 2, 3$, so that $\mathcal{A}x = b$ cannot be satisfied.

For $J = \{1\}$, it is easy to see that $\bar{x} = (-3, 0, 0)'$ satisfies $\mathcal{A}x = b$ and also $\bar{x}_j = 0$ for $j \notin J$. Moreover, the column submatrix $\mathcal{A}_J$ is the single vector $(-1, 0)'$, which has linearly independent columns. Hence $\bar{x} = (-3, 0, 0)'$ is a basic solution with $J = \{1\}$. Similarly, the choice $J = \{3\}$ leads to the basic solution $\bar{x} = (0, 0, 1.5)'$. For $J = \{2\}$, it is not possible to find an $\bar{x}$ with $\bar{x}_1 = \bar{x}_3 = 0$ that satisfies $\mathcal{A}\bar{x} = b$.

For $J = \{1, 2\}$, we find that $\bar{x} = (-3, 0, 0)$ has $\mathcal{A}x = b$ and also $\bar{x}_j = 0$ for $j \notin J$. Moreover, the column submatrix corresponding to this choice of J is

$$\begin{bmatrix} -1 & 2 \\ 0 & 1 \end{bmatrix},$$

whose columns are linearly independent. For similar reasons, the set $J = \{2, 3\}$ leads to the basic solution $\bar{x} = (0, 0, 1.5)'$. However, the set $J = \{1, 3\}$ does not lead to a basic solution, as the column submatrix for this set has linearly dependent columns. For the same reason, the set $J = \{1, 2, 3\}$ also does not lead to a basic solution.

Finally, we note that $\bar{x} = (0, 0, 1.5)'$ is the only basic *feasible* solution of this system, since the other basic solution $(-3, 0, 0)$ has a negative component. ∎

The following fundamental result shows that a basic feasible solution exists whenever a feasible point exists.

Theorem 5.1.2. *Let $b \in \mathbf{R}^m$ and $\mathcal{A} \in \mathbf{R}^{m \times l}$ be given. If $\mathcal{A}x = b$, $x \geq 0$, has a solution, then it has a basic feasible solution $\bar{x} \in \mathbf{R}^l$. That is, there is a set $J \subseteq \{1, \dots, l\}$ such that*

$$\sum_{j \in J} \mathcal{A}_{.j} \bar{x}_j = b, \quad \bar{x}_j \geq 0,$$

and $\mathcal{A}_{.j}$ has linearly independent columns.

Proof. We prove this result using a constructive approach via the simplex method. Since the system $\mathcal{A}x = b$, $x \geq 0$, has a solution, the linear program

$$\begin{aligned} \min \quad & e'(\mathcal{A}x - b) \\ \text{subject to} \quad & \mathcal{A}x \geq b, \ x \geq 0 \end{aligned}$$

must have optimal objective value 0. We can solve this linear program by setting up an initial tableau as follows:

		x_{J_1}	x_{J_2}	1
y_{I_1}	=	$\mathcal{A}_{I_1 J_1}$	$\mathcal{A}_{I_1 J_2}$	$-b_{I_1}$
y_{I_2}	=	$\mathcal{A}_{I_2 J_1}$	$\mathcal{A}_{I_2 J_2}$	$-b_{I_2}$
z	=	$e'\mathcal{A}$		$-e'b$

where we use y's to indicate the slack variables we have added. Note that $y = \mathcal{A}x - b$. In this tableau, we have partitioned the x and y components according to where they appear in the final tableau, which we obtain by solving with the two-phase simplex method and which has the following form:

		y_{I_1}	x_{J_2}	1
x_{J_1}	=	$B_{I_1 J_1}$	$B_{I_1 J_2}$	h_{I_1}
y_{I_2}	=	$B_{I_2 J_1}$	$B_{I_2 J_2}$	0
z	=	c'		0

The zeros in the last column follow from $e'(\mathcal{A}x - b) = e'y = 0$ and $y \geq 0$, which together imply that $y = 0$ at the optimum. Note that the final tableau can be obtained by simply performing a block pivot on (I_1, J_1) in the initial tableau, so that $B_{I_1 J_1} = \mathcal{A}_{I_1 J_1}^{-1}$ (see (2.9) and (2.10)). It follows that $\mathcal{A}_{I_1 J_1}^{-1}$ is nonsingular, which implies that $\mathcal{A}_{.J_1}$ has linearly independent columns. By substituting the optimal values of x and y into the original tableau, we obtain

$$0 = \begin{bmatrix} y_{I_1} \\ y_{I_2} \end{bmatrix} = \begin{bmatrix} \mathcal{A}_{I_1 J_1} \\ \mathcal{A}_{I_2 J_1} \end{bmatrix} h_{I_1} + \begin{bmatrix} \mathcal{A}_{I_1 J_2} \\ \mathcal{A}_{I_2 J_2} \end{bmatrix} 0 - b.$$

We complete the proof by setting $\bar{x} = \begin{bmatrix} h_{I_1} \\ 0 \end{bmatrix}$ and $J = J_1$. ☐

Note that if $b = 0$, the theorem holds with $J = \emptyset$.

Exercise 5-1-2. Consider the following linear programming problem with a single equality constraint and nonnegative variables:

$$
\begin{aligned}
\max \quad & c'x \\
\text{subject to} \quad & a'x = 1, \quad x \geq 0,
\end{aligned}
$$

where $c_i > 0$ and $a_i > 0$ for all $i = 1, 2, \ldots, l$. Characterize all the basic feasible solutions for this problem and develop a method for obtaining a solution of this problem directly.

In Theorem 3.2.2, we showed that vertices of the feasible region could be represented by tableaus of the following form:

$$
\begin{array}{c}
 \quad\quad x_N \quad\ 1 \\
\begin{array}{rcl}
x_B & = & \boxed{\begin{array}{c|c} H & h \end{array}} \\
z & = & \boxed{\begin{array}{c|c} c' & \alpha \end{array}}
\end{array}
\end{array}
\tag{5.4}
$$

Furthermore, for a given tableau/vertex that was generated from the canonical form problem, we showed that the system

$$
\mathcal{A}_{.B} x_B + \mathcal{A}_{.N} x_N = b, \qquad x_B \geq 0, \qquad x_N = 0,
\tag{5.5}
$$

had a solution, with $B \subseteq \{1, 2, \ldots, l\}$ having exactly m elements and $\mathcal{A}_{.B}$ being invertible. In fact, we can see the correspondence between (5.4) and (5.5) by simply setting $H = -\mathcal{A}_{.B}^{-1}\mathcal{A}_{.N}$ and $h = \mathcal{A}_{.B}^{-1}b$. Motivated by these observations, we define a basis matrix formally as follows.

Definition 5.1.3. Given $\mathcal{A} \in \mathbf{R}^{m \times l}$, consider the column submatrix $\mathcal{A}_{.B}$ for some subset $B \subseteq \{1, 2, \ldots, l\}$ containing m elements. If $\mathcal{A}_{.B}$ is invertible, it is called a *basis matrix*.

Since $\mathcal{A}_{.B}$ contains a subset of the columns of $\mathcal{A}$, it follows from Theorem 4.1.4 that if a basis matrix exists, then the rows of $\mathcal{A}$ are linearly independent.

We now show that every basic feasible solution can be associated with a basis matrix whenever $\mathcal{A}$ has linearly independent rows.

Proposition 5.1.4. *Let $\mathcal{A} \in \mathbf{R}^{m \times l}$ and $b \in \mathbf{R}^m$, where $\mathcal{A}$ has linearly independent rows. Suppose $\bar{x}$ is a basic feasible solution to $\mathcal{A}x = b$, $x \geq 0$, with $\bar{x}_j = 0$ for $j \notin J$, where $\mathcal{A}_{.J}$ has linearly independent columns. Then the set J can be extended to a set B with m elements such that $\mathcal{A}_{.B}$ is a basis matrix.*

Proof. This result can be proved by a simple application of Exercise 4-1-3.   □

The following definition applies to the case of a basic feasible solution in which fewer than m components are nonzero.

Definition 5.1.5. A basic feasible solution $\bar{x}$ is a *degenerate basic feasible solution* if there is some $j \in J$ for which $\bar{x}_j = 0$, where J is some index set for which $\bar{x}$ satisfies Definition 5.1.1.

Degenerate basic feasible solutions correspond to the degenerate tableaus we defined in Definition 3.5.1 and to degenerate vertices like those illustrated in Figure 3.1.

Note that B may not be uniquely determined by Proposition 5.1.4. That is, there may be more than one way to extend the set J to a set B such that $\mathcal{A}_{.B}$ is nonsingular.

Example 5-1-3.

$$\mathcal{A} = \begin{bmatrix} 1 & 3 & 3 & 4 \\ 0 & 4 & 0 & 5 \end{bmatrix}, \qquad b = \begin{bmatrix} 1 \\ 0 \end{bmatrix}.$$

The system $\mathcal{A}x = b$, $x \geq 0$ has a basic feasible solution corresponding to J = {1}, namely

$$\mathcal{A}_J = \begin{bmatrix} 1 \\ 0 \end{bmatrix}, \qquad \bar{x} = (1, 0, 0, 0)'.$$

We can extend J to B in two ways to produce a basis matrix, namely B = {1, 2} and B = {1, 4}. In both cases, $\mathcal{A}_{.B}$ is invertible and $\bar{x}$ is the same as given above. (Note that B = {1, 3} does not correspond to a basis matrix since $\mathcal{A}_{.B}$ has linearly dependent columns.) In fact, $\bar{x}$ is a degenerate basic feasible solution, since for the choices J = {1, 2} and J = {1, 4}, both of which satisfy Definition 5.1.1, we have $\bar{x}_j = 0$ for some $j \in J$. ∎

Exercise 5-1-4. Show, by means of an example, that a degenerate basic feasible solution may be optimal without the last row of the tableau having all its entries nonnegative.

Exercise 5-1-5. Consider the following problem:

$$
\begin{array}{rrrcl}
\min & z = 3x_1 + 5x_2 \\
\text{subject to} & 3x_1 + & x_2 & \leq & 6, \\
& x_1 + & x_2 & \leq & 4, \\
& x_1 + & 2x_2 & \leq & 6, \\
& & x_1, x_2 & \geq & 0.
\end{array}
$$

(i) Sketch the set S of feasible points.

(ii) Find the vertices of S and the corresponding tableaus.

(iii) The problem can be converted to canonical form by introducing a slack variable into each inequality. For each tableau in (ii), write down the corresponding basis matrix from the canonical form.

(iv) For the given objective function find the optimal solution(s).

5.1.2 Geometric Viewpoint

We turn now to a geometric interpretation of the basis matrix $\mathcal{A}_{.B}$ and its usefulness in finding directions of movement from the corresponding basic feasible solution to other adjacent basic feasible solutions. As in our description of the simplex method, such directions are obtained by allowing one of the components in the nonbasic set N to increase away from zero (while holding the other nonbasic variables fixed at zero) and determining the effect of this change on the basic components. Geometrically, these directions point along an edge of the feasible region that links one vertex to an adjacent vertex.

We illustrate by considering the feasible region defined by $\mathcal{A}x = b$, $x \geq 0$, with

$$\mathcal{A} = \begin{bmatrix} 1 & 3 & 0 & 4 \\ 0 & 4 & 2 & 5 \end{bmatrix}, \qquad b = \begin{bmatrix} 12 \\ 20 \end{bmatrix}. \tag{5.6}$$

Consider in particular the basic feasible solution $x = (12, 0, 10, 0)'$. For this x we have $B = \{1, 3\}$ and $N = \{2, 4\}$, with

$$A_{\cdot B} = \begin{bmatrix} 1 & 0 \\ 0 & 2 \end{bmatrix}, \qquad A_{\cdot N} = \begin{bmatrix} 3 & 4 \\ 4 & 5 \end{bmatrix}.$$

By partitioning x into its B and N components, we describe the dependence of x_B on x_N by rearranging (5.5), and using nonsingularity of $A_{\cdot B}$, as follows:

$$x_B = -A_{\cdot B}^{-1} A_{\cdot N} x_N + A_{\cdot B}^{-1} b. \tag{5.7}$$

In the present case, this expression becomes

$$\begin{bmatrix} x_1 \\ x_3 \end{bmatrix} = \begin{bmatrix} -3 & -4 \\ -2 & -2.5 \end{bmatrix} \begin{bmatrix} x_2 \\ x_4 \end{bmatrix} + \begin{bmatrix} 12 \\ 10 \end{bmatrix}.$$

Consider the effect of allowing x_2 to increase away from zero, while maintaining $x_4 = 0$. From (5.7), we obtain

$$\begin{bmatrix} x_1 \\ x_3 \end{bmatrix} = \begin{bmatrix} -3 \\ -2 \end{bmatrix} x_2 + \begin{bmatrix} 12 \\ 10 \end{bmatrix}.$$

In particular, setting $x_2 = \lambda$, where λ is some nonnegative scalar, results in a move to the following point

$$(12 - 3\lambda, \lambda, 10 - 2\lambda, 0)',$$

which we can write as $x + \lambda d^1$, where

$$x = (12, 0, 10, 0)', \qquad d^1 = (-3, 1 - 2, 0)'.$$

By taking $\lambda = 4$ (the maximum value of λ for which the nonnegativity condition $x + \lambda d^1 \geq 0$ is satisfied), we arrive at the adjacent basic feasible solution $(0, 4, 2, 0)'$.

Similarly, if we allow x_4 to increase away from 0, while maintaining $x_2 = 0$, we step to the point $(12 - 4\lambda, 0, 10 - 2.5\lambda, 1)'$ for some $\lambda \geq 0$. We express this point as $x + \lambda d^2$, where x is as before and $d^2 = (-4, 0, -2.5, 1)'$.

We can assemble the possible directions of movement toward adjacent basic feasible solutions into a matrix D. For our example, we have

$$D = \begin{bmatrix} d^1 & d^2 \end{bmatrix} = \begin{bmatrix} -3 & -4 \\ 1 & 0 \\ -2 & -2.5 \\ 0 & 1 \end{bmatrix}.$$

Recalling the notation $N(k)$ for the kth element of N and $B(i)$ for the ith element of B, we find that D is defined in the general case as follows:

$$D_{j\cdot} := \begin{cases} I_{k\cdot} & \text{if } j = N(k), \\ -(A_{\cdot B}^{-1} A_{\cdot N})_{i\cdot} & \text{if } j = B(i). \end{cases} \tag{5.8}$$

Exercise 5-1-6.

(i) For the second column d^2 of the matrix D above, find the positive value of λ such that $x + \lambda d^2$ is a basic feasible solution.

(ii) From the basic feasible solution $(0, 4, 2, 0)$ construct a new matrix D (in the same manner as indicated above) that corresponds to the two edge moves that are possible. Is there a direction that leads to a basic feasible solution that is different from all those previously found?

(iii) Why do the index sets $B = \{1, 2\}$, $B = \{1, 4\}$, and $B = \{2, 4\}$ not correspond to basic feasible solutions?

Each of the directions defined by columns of D in (5.8) defines a possible direction of search for the simplex method. Our implementation of the simplex method described in Chapter 3 essentially computes all possible edge moves away from the current point x. In the *revised* simplex method, we aim to avoid calculating those edge directions along which we are not interested in moving. Rather, we pick *one* direction that we know a priori will yield descent in the objective and compute this direction efficiently.

5.2 The Revised Simplex Method

The revised simplex method is a succinct and efficiently implementable algebraic representation of the simplex method described in Chapter 3. Only a small part of the condensed tableau is actually calculated, namely, the tableau entries corresponding to the last column, the bottom row, the pivot column, and the labels. These entries are all we need to completely determine a pivot step, and the resulting economy of computation has proved the key to practical software implementations of the simplex method.

Instead of representing the whole tableau explicitly, we manipulate the basic and nonbasic variable sets B and N, which form a partition of $\{1, 2, \ldots, l\}$. Each step of the revised simplex method ensures that B always has the property that $\mathcal{A}_{.B}$ is a valid basis matrix. In addition, the corresponding simplex iterate x must be a basic feasible solution for $\mathcal{A}x = b$, $x \geq 0$, for which $x_N = 0$, where x_N is the subvector of x made up of the components in N. Together, these conditions can be stated concisely as follows:

$$x_B = \mathcal{A}_{.B}^{-1} b \geq 0. \tag{5.9}$$

In principle, the complete tableau can always be constructed from knowledge of B and N, and so by storing B and N we are in essence storing an implicit representation of the tableau. We show this by writing the constraint in (5.3) as follows:

$$\mathcal{A}_{.B} x_B + \mathcal{A}_{.N} x_N = b \iff x_B = \mathcal{A}_{.B}^{-1} (b - \mathcal{A}_{.N} x_N).$$

The canonical form is then equivalent to

$$\begin{aligned} \min \quad & z = p_B' x_B + p_N' x_N \\ \text{subject to} \quad & x_B = \mathcal{A}_{.B}^{-1}(b - \mathcal{A}_{.N} x_N), \\ & x_B, x_N \geq 0. \end{aligned}$$

Substituting for x_B in the objective, we obtain

$$\text{min} \qquad z = p_B' A_{.B}^{-1} b + (p_N' - p_B' A_{.B}^{-1} A_{.N}) x_N$$
$$\text{subject to} \qquad x_B = A_{.B}^{-1} (b - A_{.N} x_N),$$
$$x_B, x_N \geq 0.$$

This problem can now be written in tableau form as follows:

		x_N	1
x_B	$=$	$-A_{.B}^{-1} A_{.N}$	$A_{.B}^{-1} b$
z	$=$	$p_N' - p_B' A_{.B}^{-1} A_{.N}$	$p_B' A_{.B}^{-1} b$

If we compare this tableau to the general tableau (5.4) that we used in Chapter 3, we see that

$$H = -A_{.B}^{-1} A_{.N}, \qquad h = A_{.B}^{-1} b, \qquad c' = p_N' - p_B' A_{.B}^{-1} A_{.N}, \qquad \alpha = p_B' A_{.B}^{-1} b.$$

As mentioned above, we do not construct this complete tableau but only the part of it that is needed to decide on a direction that moves toward a new basic feasible solution and yields a decrease in the objective. Specifically, we need to know only the *last column*, the *bottom row*, and the *pivot column* of the tableau. We can obtain these vectors by carrying out the following calculations.

Algorithm 5.1 (Revised Simplex Method).

1. *Calculate $h = A_{.B}^{-1} b$.*

2. *Calculate $u' := p_B' A_{.B}^{-1}$ and $c' := p_N' - u' A_{.N}$, the vector of* reduced costs.

3. *If $c \geq 0$, stop; the current solution is optimal. Otherwise, choose a pivot column s such that $c_s < 0$.*

4. *Calculate the pivot column $d = A_{.B}^{-1} A_{.N(s)}$. Evaluate the minimum ratio*

$$\frac{h_r}{d_r} = \min \{h_i / d_i \mid d_i > 0\}$$

   to determine the pivot row r. If $d \leq 0$, then stop; the problem is unbounded. (Note the change of sign here. The vector d represents the negative of the pivot column that is found in the tableau.)

5. *Update B and N by swapping $B(r)$ (the rth component of the basic variable set) with $N(s)$ (the sth component of the nonbasic variable set).*

To carry out each iteration of the simplex method, we need to apply the inverse $A_{.B}^{-1}$ on three occasions, namely in Steps 1, 2, and 4. Instead of computing the inverse explicitly, we observe that the resulting vectors h, u and d can be calculated alternatively from solving 3 systems of equations involving $A_{.B}$, namely the following:

$$A_{.B} h = b \qquad \text{(Step 1: Compute last column } h),$$
$$A_{.B}' u = p_B \qquad \text{(Step 2: Compute basic dual variable } u),$$
$$A_{.B} d = A_{.N(s)} \qquad \text{(Step 4: Compute pivot column } d).$$

The notation $\mathcal{A}_{\cdot B}$, $\mathcal{A}_{\cdot N(s)}$, b, and p_B all refer to the data in the original problem (5.3). We use a single LU factorization of $\mathcal{A}_{\cdot B}$ to solve all three systems. (See Section 2.6 for details on this factorization procedure.) Following the outline given in Section 2.5 and using MATLAB notation, we solve the three systems as follows:

$$
\begin{aligned}
LU &= \mathcal{A}_{\cdot B} & &\text{(Compute the factorization),} \\
LUh &= b & &\Longleftrightarrow \ h = U\backslash(L\backslash b), \\
U'L'u &= p_B & &\Longleftrightarrow \ u = L'\backslash(U'\backslash p_B), \\
LUd &= \mathcal{A}_{\cdot N(s)} & &\Longleftrightarrow \ d = U\backslash(L\backslash \mathcal{A}_{\cdot N(s)}).
\end{aligned}
$$

In general, it is not obvious how to make an initial choice of basis B to satisfy the condition (5.9). We discuss this issue below in Section 5.2.2. For the special case in which the problem (5.3) was derived from a standard-form problem (5.1) with $b < 0$ via the transformations (5.2), we can make the initial choice B $= \{n+1, n+2, \ldots, n+m\}$, so that $\mathcal{A}_{\cdot B} = -I$ and $x_B = \mathcal{A}_{\cdot B}^{-1} b = -b \geq 0$. This was exactly the situation encountered at the start of Chapter 3, where the simplex method did not require Phase I to identify a starting point.

We now illustrate the revised simplex method on a simple example, assuming for now that a valid initial choice of B is known.

Example 5-2-1. Consider the following problem, for which it is known a priori that B $=$ [3 4 6] corresponds to a basic feasible solution:

$$
\begin{aligned}
\min \quad & z = p'x \\
\text{subject to} \quad & \mathcal{A}x = b, \quad x \geq 0,
\end{aligned}
$$

where

$$
\mathcal{A} = \begin{bmatrix} 1 & 6 & 0 & 1 & 0 & 0 \\ -1 & -4 & -1 & 0 & 1 & 0 \\ 2 & 14 & 0 & 0 & 0 & 1 \end{bmatrix}, \qquad b = \begin{bmatrix} 3 \\ -1 \\ 5 \end{bmatrix}, \qquad p = \begin{bmatrix} 2 \\ 9 \\ 3 \\ 0 \\ 0 \\ 0 \end{bmatrix}.
$$

The first steps in MATLAB are as follows:

```
>> load ex5-2-1
>> [m,l] = size(A);
>> B = [3 4 6];
>> N = setdiff(1:l,B);
>> [L,U] = lu(A(:,B));
>> x_B = U\(L\b)
```

$$
x_B = \begin{bmatrix} 1 \\ 3 \\ 5 \end{bmatrix}
$$

The vector x_B then contains the last column of the tableau. Since these are nonnegative, Phase I is unnecessary. We now proceed to calculate the values in the bottom row of the tableau. Since $\mathcal{A}_{\cdot B} = LU$ and $\mathcal{A}'_{\cdot B} u = p_B$, it follows that $u = (L')^{-1}(U')^{-1} p_B$. Hence, we can reuse the L and U factors of $\mathcal{A}_{\cdot B}$ calculated above (leading to important savings in practice) and proceed in MATLAB as follows:

```
>> u = L'\(U'\p(B));
>> c = p(N)-A(:,N)'*u
```

$$c = \begin{bmatrix} -1 \\ -3 \\ 3 \end{bmatrix}$$

Since c' makes up the bottom row of the tableau (except for the bottom-right element) and since it has negative entries, the current tableau is not optimal. Therefore, we select an index to enter the basis, corresponding to one of the negative entries in c, and calculate the column in the tableau that corresponds to this variable $x_N(s)$.

```
>> s = 2;
>> d = U\(L\A(:,N(s)))
```

$$d = \begin{bmatrix} 4 \\ 6 \\ 14 \end{bmatrix}$$

We now carry out the ratio test with the values of d and x_B calculated above and find that $r = 1$ is the index that leaves the basis. We swap r and s between the sets B and N and continue with the next iteration of the revised simplex method.

```
>> r=1; swap = B(r);
>> B(r) = N(s); N(s) = swap;
>> [L,U] = lu(A(:,B));
>> x_B = U\(L\b)
>> u = L'\(U'\p(B));
>> c = p(N)-A(:,N)'*u
```

$$c = \begin{bmatrix} -0.25 \\ 0.75 \\ 2.25 \end{bmatrix}$$

Now c has the value shown above. It indicates that we have not yet achieved optimality, as there is still a negative component. We identify the component $s = 1$ to enter the basis and a component $r = 1$ to leave, and we continue with the next simplex iteration.

```
>> s = 1; d = U\(L\A(:,N(s)));
>> r = 1; swap = B(r);
>> B(r) = N(s); N(s) = swap;
>> [L,U] = lu(A(:,B));
>> x_B = U\(L\b)
>> u = L'\(U'\p(B));
>> c = p(N)-A(:,N)'*u
```

$$c = \begin{bmatrix} 1 \\ 1 \\ 2 \end{bmatrix}$$

Since we now have $c \geq 0$, the current tableau is optimal. The corresponding solution is $x = (1, 0, 0, 2, 0, 3)'$, with $z = p'x = p'_B x_B = 2$. ∎

Exercise 5-2-2. In MATLAB, solve the following linear program using the revised simplex method as shown above. You will need to convert the data of this problem to canonical form and choose an initial B that corresponds to the slack variables that you add during this conversion.

$$
\begin{array}{lrcrcrcrcl}
\text{max} & -x_1 & + & 2x_2 & + & x_3 \\
\text{subject to} & x_1 & + & x_2 & + & x_3 & \leq & 3, \\
& -x_1 & + & x_2 & - & x_3 & \leq & 1, \\
& 2x_1 & + & x_2 & - & x_3 & \leq & 1, \\
& -x_1 & + & x_2 & & & \leq & 4, \\
& & & & x_1, x_2, x_3 & \geq & 0.
\end{array}
$$

You should work through each step of the method explicitly, calculating each of the intermediate vectors, as in the example above.

In fact, it is computationally more efficient to update the LU factorization at each step of the revised simplex method instead of recomputing the factorization anew. Details on such procedures can be found in Section 5.2.3; in the absence of these methods the overall complexity of the revised simplex method will be larger than that of an implementation using Jordan exchanges.

Numerical rounding error causes many problems for implementations of the simplex method due to the special nature of the number zero and the need to determine positivity or negativity of components in crucial steps of the procedure. Typical commercial codes contain at least three types of "zero tolerances," small positive numbers below which a floating-point number is deemed indistinguishable from zero for purposes of the algorithm. The first tolerance `zer_tol` is used in comparisons of numbers against zero. In testing the bottom row of the tableau for negativity, we use the MATLAB code

```
if isempty(find(c < -zer_tol))
```

rather than the simple test

```
if isempty(find(c < 0))
```

This allows some of the values of c to be slightly negative. Chvátal (1983) suggests a value of 10^{-5} as a typical value for this tolerance.

The second tolerance is a pivot tolerance `piv_tol` that is used to safeguard against very small pivot elements and hence avoid the problems shown in Example 2-6-1. Pivots are deemed acceptable only if the potential pivot element is greater than `piv_tol` in absolute value. A typical value for this tolerance is 10^{-8}.

The third zero tolerance is a slack tolerance `slack_tol` that is a measure of how much error we will allow in the slack variables, that is, the difference between $A_B x_B$ and b. The use of this tolerance will be outlined further in the sections on advanced pivot selection mechanisms and basis updates. A typical value for `slack_tol` is 10^{-6}.

A simple version of the resulting revised simplex method that uses these tolerances and LU decomposition for solving the systems of linear equations is given in `rsm.m`. We illustrate the use of this routine on the example given above.

MATLAB file rsm.m: Revised simplex

```
function [x_B,B,u] = rsm(A,b,p,B)
% syntax: [x_B,B,u] = rsm(A,b,p,B)
% A revised simplex routine for min p'x st Ax=b, x>=0.
% on input A is mxl, b is mxl, p is lxl
% B is lxm index vector denoting the basic columns.
% on output u is the dual solution

[m,l] = size(A);
zer_tol = 1.0e-5; piv_tol = 1.0e-8;
N = setdiff(1:l,B);

while (1)
  [L,U] = lu(A(:,B));
  x_B = U\(L\b);
  if any(x_B < -zer_tol)
    error('current point is infeasible'); end;

  u = L'\(U'\p(B));
  c = p(N)'-u'*A(:,N);

  if isempty(find(c < -zer_tol))
    return; end;

  [min_red_cost,s] = min(c);

  d = U\(L\A(:,N(s)));
  blocking = find(d >= piv_tol);
  if isempty(blocking)
    error('problem is unbounded'); end;

  [min_ratio,index_r] = min(x_B(blocking)./d(blocking));
  r = blocking(index_r);

  swap = B(r); B(r) = N(s); N(s) = swap;
end;
```

```
≫ load ex5-2-1
≫ B = [3 4 6];
≫ [x_B,B] = rsm(A,b,p,B)
```

$$x_B = \begin{bmatrix} 1 \\ 2 \\ 3 \end{bmatrix}, \qquad B = \begin{bmatrix} 1 & 4 & 6 \end{bmatrix}$$

Note that $x_1 = 1$, $x_4 = 2$, $x_6 = 3$, and the other components are zero.

```
≫ z = p(B)'*x_B                                         z = 2
```

Exercise 5-2-3. In MATLAB, solve the following linear program using the revised simplex method:

$$\begin{array}{rlrrrrl}
\max & z = 2x_1 + 4x_2 + x_3 + x_4 \\
\text{subject to} & x_1 & + & 3x_2 & & + & x_4 & \le & 4, \\
& 2x_1 & + & x_2 & & & & \le & 3, \\
& & & x_2 & + & 4x_3 & + & x_4 & \le & 3, \\
& & & & & x_1, x_2, x_3, x_4 & \ge & 0.
\end{array}$$

(An initial basic feasible solution and choice of B should again be obvious from the conversion to canonical form.)

The revised simplex method solves the canonical form linear program. The dual of the canonical form problem was found in Example 4-7-2. How do you determine the dual solution when using the revised simplex method? Note that the KKT conditions for the general formulation (4.14) and (4.15) imply that

$$x_B'(p - \mathcal{A}'u)_B = 0$$

at the optimal basis B. Thus

$$p_B = \mathcal{A}'_{.B}u$$

can be solved for the dual solution vector u. The calculation for the vector c shows that this solution is in fact dual feasible.

Exercise 5-2-4. What is the dual of the problem given in Example 5-2-1? Use the results of the exercise to exhibit a dual solution and prove that this solution is optimal.

We now show that the revised simplex procedure can be extended in a straightforward way to deal with upper and lower bounds on the components of x. This extended procedure can be used to construct an initial basic feasible solution, as we show subsequently in Section 5.2.2.

5.2.1 Upper and Lower Bounds

We consider now the linear program

$$\begin{array}{rl}
\min & z = p'x \\
\text{subject to} & \mathcal{A}x = b, \quad \bar{\ell} \le x \le \bar{u}.
\end{array} \tag{5.10}$$

This formulation can be applied directly to many practical situations in which the variables of the given problem are subject naturally to both lower and upper bounds. We allow $\bar{\ell}_i = -\infty$ and/or $\bar{u}_i = +\infty$ so that free variables and one-sided bounds can be recovered as special cases.

Example 5-2-5. The continuous knapsack problem is a simple example of a problem of this type. A knapsack that has a given fixed capacity b, say, should be filled with a mixture of goods that have greatest value. The problem can be written as

$$\begin{array}{ll} \max & p'x \\ \text{subject to} & a'x \leq b, \quad 0 \leq x \leq \bar{u}. \end{array}$$

Here a_i is the (unit) volume of good i, and p_i is its (unit) street value, while $\bar{u}_i$ represents the total amount of the good that is available. Note that a closely related problem was discussed in Exercise 5-1-2, and a sorting scheme allows its efficient solution. ■

A simple way to reformulate this problem so that we can apply the revised simplex method is to apply a change of variables ($x = y + \bar{\ell}$) to move the lower bounds to zero (assuming all the lower bounds are finite) and to treat the remaining upper bounds as general constraints. These general constraints can then be converted into equality constraints (for canonical form) by simply adding slack variables s. Thus the problem becomes

$$\begin{array}{ll} \min_{y,s} & z = p'y + p'\bar{\ell} \\ \text{subject to} & \begin{bmatrix} A & 0 \\ I & I \end{bmatrix} \begin{bmatrix} y \\ s \end{bmatrix} = \begin{bmatrix} b - A\bar{\ell} \\ \bar{u} - \bar{\ell} \end{bmatrix}, \quad (y, s) \geq 0. \end{array}$$

The problem with this approach is that the number of variables and the number of constraints increase fairly dramatically, and hence (as we show later in an example) the work per iteration of the simplex method increases dramatically.

To handle this formulation directly using the simplex method, we redefine the basis to be those components of x that may be away from their bounds. Nonbasic variables are held at one of their bounds and allowed to move away from this bound only when they are chosen to enter the basis. Conceptually, these definitions represent a straightforward extension from the canonical formulation, for which $\bar{\ell} = 0$ and $\bar{u} = +\infty$. Obviously, the rules for pivot column and row selection must be altered to fit these new definitions, but the changes are straightforward. We now describe the modified procedure explicitly.

Algorithm 5.2 (Revised Simplex Method for Bounded Variables).

1. *Put the problem into the canonical form with bound constraints (5.10) by adding appropriate slack variables.*

2. *Determine an initial basic feasible solution x and the corresponding index sets* B *and* N.

3. *Calculate $u' := p'_B A_{\cdot B}^{-1}$ and $c' := p'_N - u' A_{\cdot N}$.*

4. *A nonbasic variable $N(s)$ is eligible to enter the basis if*

(a) $x_{N(s)} = \bar{\ell}_{N(s)}$ and $c_s < 0$; or

(b) $x_{N(s)} = \bar{u}_{N(s)}$ and $c_s > 0$.

Choose a pivot column s corresponding to an eligible nonbasic variable. If none are eligible, then stop; the current point is optimal.

5. *Calculate* $d = A_{\cdot B}^{-1} A_{\cdot N(s)}$ *and define*

$$x_j(\lambda) := \begin{cases} x_j + \lambda sign(c_s)d_i & if\ j = B(i), \\ x_{N(s)} - \lambda sign(c_s) & if\ j = N(s), \\ x_j & otherwise. \end{cases}$$

Determine the pivot row r by increasing λ until one of the variables $x_j(\lambda)$ hits a bound. If no variable hits a bound, then **stop**; *the problem is unbounded.*

6. *Update* $x = x(\lambda)$ *and (if necessary)* B *and* N *by swapping* B(r) *and* N(s).

7. *Go to Step* 3.

In practice, we encode N as a signed integer vector, with $N(s) > 0$ indicating $x_{N(s)} = \bar{u}_{N(s)}$ and $N(s) < 0$ indicating $x_{-N(s)} = \bar{\ell}_{-N(s)}$. In Step 4, we declare s to be eligible if c_s and $N(s)$ have the same signs.

Example 5-2-6. Consider the problem

$$\begin{array}{llllllll} \min & 3x_1 & - 4x_2 & + x_3 & - 2x_4 & + 4x_5 \\ \text{subject to} & x_1 & & + 3x_3 & + x_4 & - x_5 & = & 7, \\ & & x_2 & - 4x_3 & + x_4 & - x_5 & = & 6, \\ & & 0 \leq x_1 \leq 5, \\ & & 0 \leq x_2 \leq 5, \\ & & 0 \leq x_3 \leq 5, \\ & & 0 \leq x_4 \leq 5, \\ & & 0 \leq x_5 \leq 2. \end{array}$$

To solve this problem in MATLAB, we can start at the point $x = (2, 1, 0, 5, 0)'$ with $B = [1\ \ 2]$ and $N = [-3\ \ 4\ \ -5]$. (The first and third components of N are negative as the corresponding elements of x are at their lower bounds.) We first show that the basis specified corresponds to the given basic feasible solution:

```
>> load ex5-2-6
>> B = [1 2];
>> N = [-3 4 -5];
>> [m,l] = size(A);
>> x = zeros(1,1);
>> x(3:5) = [0 5 0]';
>> [L,U] = lu(A(:,B));
>> x(B) = U\(L\(b-A(:,abs(N))*x(abs(N))))
```

$$x = \begin{bmatrix} 2 \\ 1 \\ 0 \\ 5 \\ 0 \end{bmatrix}$$

The next step is to compute the reduced cost vector c:

```
>> u = L'\(U'\p(B));
>> c = p(abs(N))-A(:,abs(N))'*u
```
$$c = \begin{bmatrix} -24 \\ -1 \\ 3 \end{bmatrix}$$

We see that neither x_4 nor x_5 is eligible to enter the basis. The reduced cost corresponding to x_4 is negative, and so the objective function will increase as x_4 is decreased away from its upper bound, while the reduced cost corresponding to x_5 is positive, indicating again that the objective function will increase as x_4 is increased away from its lower bound. The variable x_3 is, however, a valid choice for decreasing the objective, and so we set the index s to indicate our choice of pivot and compute the (negative of the) column from the tableau that corresponds to x_3:

```
>> s = 1;
>> d = U\(L\(A(:,abs(N(s)))))
```
$$d = \begin{bmatrix} 3 \\ -4 \end{bmatrix}$$

It follows that when we let $x_3 = \lambda > 0$ that $x_1 = 2 - 3\lambda$ and $x_2 = 1 + 4\lambda$. Hence x_1 is the blocking variable—it hits its lower bound of 0 when $\lambda = 2/3$. We now update the relevant components of x for this choice of λ.

```
>> lambda = 2/3;
>> x(3) = x(3) + lambda;
>> x(B) = x(B) - lambda*d
```
$$x = \begin{bmatrix} 0 \\ 3.6667 \\ 0.6667 \\ 5 \\ 0 \end{bmatrix}$$

The values of the basic variables x_B (that used to correspond to the last column h of the tableau) have been computed by updating the previous value of x_B using the components of the pivot column d, rather than by solving a new linear system.

We now update B and N to reflect the pivot that just occurred.

```
>> B(1) = 3; N(1) = -1;
```

We now have B = [3   2] and N = [−1   4   − 5]. We are now ready to start another iteration of the method by calculating c once again.

```
>> [L,U] = lu(A(:,B));
>> u = L'\(U'\p(B));
>> c = p(abs(N))-A(:,abs(N))'*u;
```
$$c = \begin{bmatrix} 8 \\ 7 \\ -5 \end{bmatrix}$$

Both the second and third elements of N yield acceptable pivots here—the second element x_4 because it is at its upper bound and has a positive reduced cost, and the third element x_5 because it is at its lower bound with a negative reduced cost. (A useful way to remember this rule is that if N and c have the same sign, the corresponding variable is eligible to enter the basis.) For variety, we choose the second element x_4 as the pivot and compute the vector d that captures the dependence of the basic variables on x_4:

```
>> s = 2;
>> d = U\(L\(A(:,abs(N(s)))))
```

$$d = \begin{bmatrix} 0.3333 \\ 2.3333 \end{bmatrix}$$

It follows that when we let $x_4 = 5 - \lambda$ then $x_3 = 2/3 + (1/3)\lambda$ and $x_2 = (11/3) + (7/3)\lambda$. Thus the blocking variable is x_2 (which reaches its upper bound of 5 for $\lambda = 4/7$). By taking this step and updating the relevant components of x, we obtain

```
>> lambda = 4/7;
>> x(4) = x(4) - lambda;
>> x(B) = x(B) + lambda*d
```

$$x = \begin{bmatrix} 0 \\ 5 \\ 0.8571 \\ 4.4286 \\ 0 \end{bmatrix}$$

Again, we update B and N to reflect the pivot that just occurred:

```
>> B(2) = 4; N(2) = 2;
```

yielding $B = [3 \quad 4]$ and $N = [-1 \quad 2 \quad -5]$. Proceeding with the next step, we compute the reduced cost vector c:

```
>> u = L'\(U'\p(B));
>> c = p(abs(N))-A(:,abs(N))'*u
```

$$c = \begin{bmatrix} 4 \\ -3 \\ 2 \end{bmatrix}$$

At this stage, we see that the above x vector is optimal for our problem, since the signs of the corresponding components of c and N all differ. The variable x_2 at its upper bound has a negative reduced cost, while the variables x_1 and x_5 at their lower bounds have positive reduced costs. Hence, we are done. ∎

The MATLAB code that implements the simplex method with upper and lower bounds can be found in the file `rsmbdd.m`. An important feature of `rsmbdd` is that we perform only two solutions of linear systems involving the matrix $\mathcal{A}_B$ and its transpose at each iteration (as outlined in the example above), rather than the three solutions suggested by our earlier description. We illustrate the use of `rsmbdd` on Example 5-2-6.

Example 5-2-7. Again, we use the starting basis $B = [1 \quad 2]$ and $N = [-3 \quad 4 \quad -5]$ that corresponds to the basic feasible solution $x = (2, 1, 0, 5, 0)'$.

```
>> load ex5-2-6
>> B = [1 2]; N = [-3 4 -5];
>> [x,B,N] = rsmbdd(A,b,p,lb,ub,B,N);
```

As we mentioned above, we can solve the same problem by first converting to canonical form by adding slack variables for each of the five upper-bound constraints. By extending the starting basis from above to include the slacks that are initially away from their bounds of 0 and applying the rsm code, we proceed as follows:

```
>> load ex5-2-6
>> [m,l] = size(A);
>> A = [A zeros(m,l); eye(l) eye(l)];
>> b = [b; ub]; p = [p; zeros(l,1)];
>> B = [1 2 4 6 7 8 10];
>> [x_B,B] = rsm(A,b,p,B);
>> x = zeros(2*l,1); x(B) = x_B;
```

On an earlier version of MATLAB that allows us to count floating-point operations, it was shown that the solution procedure using rsmbdd used only about one-tenth as many operations as the version using rsm. Clearly, at least in this instance, it is much more efficient to treat bounds directly in the algorithm than to introduce slacks and convert them to general constraints. ∎

Exercise 5-2-8. Use the revised simplex procedure with upper and lower bounds to solve

$$\min \quad 3x_1 + x_2 + x_3 - 2x_4 + x_5 - x_6 - x_7 + 4x_8$$

$$\text{subject to} \quad \begin{bmatrix} 1 & 0 & 3 & 1 & -5 & -2 & 4 & -6 \\ 0 & 1 & -2 & -1 & 4 & 1 & -3 & 5 \end{bmatrix} x = \begin{bmatrix} 7 \\ -3 \end{bmatrix},$$

$$
\begin{aligned}
0 \le\ &x_1 \le\ 8, \\
&x_2 \le\ 6, \\
0 \le\ &x_3 \le\ 4, \\
&x_4 \le\ 15, \\
&x_5 \le\ 2, \\
0 \le\ &x_6 \le\ 10, \\
0 \le\ &x_7 \le\ 10, \\
0 \le\ &x_8 \le\ 3.
\end{aligned}
$$

An initial basic feasible solution is $x' = (0, 5.6, 0, 15, 1.6, 0, 0, 0)$. Use these values to construct appropriate choices for B and N and then invoke rsmbdd to solve the problem.

5.2.2 Generating Basic Feasible Solutions

As we have noted above, the revised simplex technique must be started from an initial basic feasible solution. We now show how to generate such a solution by solving a Phase I problem, without the need for converting to standard form.

We suppose first that the problem (5.10) has no free variables; that is, each variable has either an upper or a lower bound (or both). We construct a Phase I problem by adding extra artificial variables (in a manner that generalizes the technique of Section 3.4), choosing the original set of variables to belong to the initial nonbasic set N and the artificial variables to make up the initial basis B. Specifically, we set each x_j, $j = 1, 2, \ldots, l$, to one of its bounds $\bar{\ell}_j$ or $\bar{u}_j$ (choosing a finite bound, of course) and define $N(j) = -j$ if $x_j = \bar{\ell}_j$ and $N(j) = j$ if $x_j = \bar{u}_j$. For this x, we then define d_i, $i = 1, 2, \ldots, m$, as follows:

$$d_i = \begin{cases} 1 & \text{if } \mathcal{A}_i.x - b_i \geq 0, \\ -1 & \text{otherwise.} \end{cases} \tag{5.11}$$

We now add artificial variables x_{l+i}, $i = 1, 2, \ldots, m$, and construct the following Phase I problem:

$$\min \quad z_0 = \sum_{i=1}^{m} x_{l+i}$$

$$\text{subject to} \quad \mathcal{A}_i. \begin{bmatrix} x_1 \\ \vdots \\ x_l \end{bmatrix} = b_i + d_i x_{l+i}, \quad i = 1, \ldots, m, \tag{5.12}$$

$$\begin{array}{ll} \bar{\ell}_j \leq x_j, \leq \bar{u}_j, & j = 1, 2, \ldots, l, \\ 0 \leq x_{l+i}, & i = 1, 2, \ldots, m. \end{array}$$

It is easy to see that by choosing $x_1, x_2, \ldots, x_l$ as above and by setting the artificial variables to

$$x_{l+i} = \left| \mathcal{A}_i. \begin{bmatrix} x_1 \\ \vdots \\ x_l \end{bmatrix} - b_i \right|, \quad i = 1, 2, \ldots, m,$$

we obtain an initial basic feasible solution for (5.12), with basis B $= \{l+1, l+2, \ldots, l+m\}$. Note too that the constraint matrix for this problem has full row rank, since the m columns corresponding to the artificial variables themselves have full rank m.

Starting from this basis, we can now solve the Phase I problem (5.12) and declare the original problem (5.10) infeasible if Phase I terminates with a positive objective value. Otherwise, we reset the upper bound on all the artificial variables x_{l+i}, $i = 1, 2, \ldots, m$, to zero (thereby forcing all these variables to stay at 0 for the remainder of the computation) and proceed directly to Phase II. The first l components of the solution to the Phase II problem constitute the solution of the original problem (5.10).

Example 5-2-9. We revisit Example 5-2-6, showing how to set up the Phase I problem. Since each of the variables has two finite bounds, we can choose to set each variable either at lower or upper bound. Arbitrarily, we set all to their lower bound except x_4 and define the set N accordingly. We then augment the problem data to account for the artificial variables and solve the Phase I linear program. In the following discussion, we use v to denote our guess at the values for the variable x.

```
≫ load ex5-2-6
≫ N = [-1 -2 -3 4 -5];
≫ v = lb; v(4) = ub(4);
≫ d = sign(A*v-b);
```

```
>> A = [A -diag(d)]; w = [zeros(5,1); ones(2,1)];
>> lb = [lb; zeros(2,1)]; ub = [ub; inf*ones(2,1)];
>> B = [6 7];
>> [x,B,N] = rsmbdd(A,b,w,lb,ub,B,N);
>> w'*x
```

Since $w'x = 0$, B and N correspond to a basic feasible solution. The nonartificial components of the Phase I solution, and the set B, are the same as the initial basic feasible solution used in Example 5-2-6. We now reset the upper bounds on the artificial variables to 0 and continue with Phase II of the simplex method.

```
>> ub(6:7) = zeros(2,1); p(6:7) = zeros(2,1);
>> [x,B,N] = rsmbdd(A,b,p,lb,ub,B,N);
>> p'*x    ∎
```

Free Variables.   We now consider the case in which free variables are present in the formulation. Since free variables have no finite bound, we cannot choose the initial x in the manner described above by setting each component to one of its bounds. Instead, we include these variables in the initial basic set B. Moreover, we keep them in the basis throughout the algorithm, exactly as we did in Scheme II of Chapter 3.

Let us denote by F the set of free variables, where $F \subset \{1, 2, \ldots, l\}$, and suppose that F contains f indices. We then choose a subset I of $\{1, 2, \ldots, m\}$ of the rows of $\mathcal{A}$, also with f elements, in such a way that the submatrix $\mathcal{A}_{IF}$ is nonsingular. (Assume for the present that this is possible.) For the nonfree variables x_j, $j \in \{1, 2, \ldots, l\}\backslash F$, we set x_j to one of its (finite) bounds, as above. Given the values x_j for $j \notin F$, we obtain initial values for the free variables x_j for $j \in F$ by solving the following system of equations:

$$\sum_{j \in F} \mathcal{A}_{Ij} x_j = b_I - \sum_{j \notin F} \mathcal{A}_{Ij} x_j.$$

To set up the Phase I problem, we define d_i exactly as in (5.11) but only for the indices $i \in \{1, 2, \ldots, m\}\backslash I$. The Phase I problem is as follows:

$$\min \qquad z_0 = \sum_{i \in \{1,2,\ldots,m\}\backslash I} x_{l+i}$$

$$\text{subject to} \qquad \mathcal{A}_{i\cdot} \begin{bmatrix} x_1 \\ \vdots \\ x_l \end{bmatrix} = b_i + d_i x_{l+i}, \quad i \in \{1, 2, \ldots, m\}\backslash I,$$

$$\mathcal{A}_{i\cdot} \begin{bmatrix} x_1 \\ \vdots \\ x_l \end{bmatrix} = b_i, \qquad i \in I,$$

$$\bar{\ell}_j \le x_j \le \bar{u}_j, \qquad j = 1, 2, \ldots, l,$$
$$0 \le x_{l+i}, \qquad i \in \{1, 2, \ldots, m\}\backslash I.$$

An initial basic feasible solution to this Phase I problem is obtained by setting $x_1, x_2, \ldots, x_l$ as described above and setting the artificial variables as follows:

$$x_{l+i} = \left| \mathcal{A}_{i\cdot} \begin{bmatrix} x_1 \\ \vdots \\ x_l \end{bmatrix} - b_i \right|, \qquad i \in \{1, 2, \ldots, m\} \backslash I.$$

The initial basis B is the union of the free variable set F with the set of artificial variable indices $l + i$ for $i \in \{1, 2, \ldots, m\} \backslash I$. (It is easy to check that this B has m elements.) Because of the nonsingularity of the submatrix $\mathcal{A}_{IF}$, the rows of $\mathcal{A}_{\cdot B}$ are guaranteed to be linearly independent.

Example 5-2-10. Consider the problem

$$
\begin{array}{rlcccccl}
\min & x_1 + x_2 + x_3 \\
\text{subject to} & 3x_1 & + & x_2 & & & - & x_4 & = & -5, \\
& x_1 & - & 2x_2 & + & x_3 & & & = & 1, \\
& & & x_2 \geq 0, & x_3 \geq 2, & x_4 \geq 0.
\end{array}
$$

Since x_1 is free in this example, we have $F = \{1\}$. We first set $v_2 = \bar{\ell}_2$, $v_3 = \bar{\ell}_3$, and $v_4 = \bar{\ell}_4$ and choose $I = \{2\}$; that is, we use the second equation to determine the value of $v_1 = 1 + 2v_2 - v_3 = -1$.

```
≫ load ex5-2-10
≫ v = lb;
≫ N = [-2 -3 -4];
≫ v(1) = b(2) - A(2,2:4)*v(2:4)
```

$$v = \begin{bmatrix} -1 \\ 0 \\ 2 \\ 0 \end{bmatrix}$$

For the remaining equation, we need to add an artificial variable for the Phase I problem and construct the basis B accordingly:

```
≫ d = sign(A(1,:)*v - b(1));
≫ A = [A; [-d; 0]];
≫ lb(5) = 0; ub(5) = inf;
≫ B = [1 5];
≫ w = [zeros(4,1); 1];
≫ [x,B,N] = rsmbdd(A,b,w,lb,ub,B,N)
```

$$x = \begin{bmatrix} -1.6667 \\ 0 \\ 2.6667 \\ 0 \\ 0 \end{bmatrix}$$

$$B = \begin{bmatrix} 1 & 3 \end{bmatrix}$$

$$N = \begin{bmatrix} -2 & -5 & -4 \end{bmatrix}$$

Since $w'x = 0$ at this point, we can now update the upper bounds on x_5 and proceed to Phase II.

```
>> ub(5) = 0; p(5) = 0;
>> [x,B,N] = rsmbdd(A,b,p,lb,ub,B,N)
```

This results in the same output as above, since the given vector x is in fact optimal for the problem. ∎

The index set I can be found in a stable fashion by performing an LU factorization of $\mathcal{A}_{\cdot F}$ with pivoting and taking I to be the indices of the rows of $\mathcal{A}_{\cdot F}$ that are chosen as the first f pivot rows by the factorization procedure. In the above example, this would result in the following steps:

```
>> F = [1]; f = length(f);                              i = [1]
>> [L,U,P] = lu(A(:,F));
>> [k,i] = find(P(1:f,:));
>> i
```

Thus, in this case, we would use the first equation instead of the second to determine the value of $v(1)$.

Finally, we consider the case in which the free variable submatrix $\mathcal{A}_{\cdot F}$ has linearly dependent columns, so that it is not possible to choose a subset of rows I such that $\mathcal{A}_{IF}$ is nonsingular. A modified Phase I procedure is derived in the following exercise.

Exercise 5-2-11. Suppose that the columns of $\mathcal{A}_{\cdot F}$ are linearly dependent. Assume that you can identify an invertible submatrix $\mathcal{A}_{IK}$ of $\mathcal{A}_{\cdot F}$ with the largest possible number of columns. Thus the set $J := F \setminus K$ represents the linearly dependent free variables. Now set $v_J = 0$ and solve the linear system $\sum_{j \in K} \mathcal{A}_{Ij} v_j = b_I - \sum_{j \notin K} \mathcal{A}_{Ij} v_j$ to determine values for v_j, $j \in K$.

(i) Using (5.11), construct a Phase I problem using the sets I, J, and K that fixes the variables in J at 0 and otherwise proceeds as before.

(ii) Describe how I and K can be calculated using an LU decomposition of $\mathcal{A}_{\cdot F}$ in MATLAB.

We are now able to use `rsmbdd` to solve linear programs stated in canonical form, with upper and lower bounds (5.10). A complete simplex code implementing the two-phase approach outlined above (including the cases in which free variables are present with dependent columns of $\mathcal{A}$) is given in `simplex.m`.

The function `simplex` can be tested, for example, on Example 3-4-1. We explicitly convert the general constraints from standard form to canonical form by adding slacks to the problem. When calling `simplex` with only three arguments, the code assumes that the lower bounds are zero and the upper bounds are $+\infty$.

```
>> load ex3-4-1
>> p = [p; zeros(5,1)];
>> x = simplex([A -eye(5)],b,p)
```

≫ p'*x

As a second example, we can solve Example 3-6-5, first adding slack variables for the first two constraints to convert to canonical form. In this example, however, x_3 is a free variable, and so we add explicit lower and upper bounds to the arguments of `simplex` to account for this:

```
≫ load ex3-6-5
≫ A = [A [-eye(2); zeros(1,2)]];
≫ p = [p; zeros(2,1)];
≫ lb = [0; 0; -inf; 0; 0]; ub = inf*ones(5,1);
≫ x = simplex(A,b,p,lb,ub);
```

Exercise 5-2-12. Use the two-phase simplex procedure to solve

$$\begin{array}{lrcrcrcrcrcrcl}
\min & z = 10x_1 & + & 12x_2 & + & 8x_3 & + & 10x_4 \\
\text{subject to} & 4x_1 & + & 5x_2 & + & 4x_3 & + & 5x_4 & + & x_5 & & & = & 1000, \\
& x_1 & + & x_2 & + & x_3 & + & x_4 & & & + & x_6 & = & 225,
\end{array}$$

$$0 \le x_1 \le 130, \quad 0 \le x_2 \le 110, \quad 0 \le x_3 \le 70,$$
$$0 \le x_4 \le 65, \quad 0 \le x_5, \quad\quad\; 0 \le x_6 \le 175.$$

Exercise 5-2-13. What is the radius of the largest ball that can be inscribed in the n-dimensional simplex

$$\left\{ x \in \mathbf{R}^n \mid x \ge 0, \; -\frac{e'x}{\sqrt{n}} + \frac{1}{\sqrt{n}} \ge 0 \right\},$$

where e is a vector of ones?

In the remainder of this chapter we outline some further enhancements that are used to improve the performance of the revised simplex method in practical codes for large problems.

5.2.3   Basis Updates

Each step of the simplex method requires the solution of equations involving $\mathcal{A}_{\cdot\text{B}}$ or $\mathcal{A}'_{\cdot\text{B}}$. As outlined previously, this can be done efficiently using a single LU factorization of the basis matrix $\mathcal{A}_{\cdot\text{B}}$, coupled with forward and backward substitution. Also, at each pivot step in the simplex method, we remove one column from the basis matrix and replace it with a new column. However, from the previous iteration, we already know an LU factorization of the original basis matrix $\mathcal{A}_{\cdot\text{B}}$, that is,

$$P\mathcal{A}_{\cdot\text{B}} = LU \tag{5.13}$$

(where P is a permutation matrix, L is unit lower triangular, and U is upper triangular). Rather than compute a new LU factorization from scratch after the one-column modification, we can *update* the current L and U factors at much lower cost. We now specify a factorization updating scheme that requires work of $O(m^2) + O(k^3)$ operations, where k is the number of column changes since the last factorization from scratch was performed. (By contrast, a full refactorization requires $O(m^3)$ operations.) Clearly, when k grows too large, it is better to refactor from scratch rather than to use this update scheme; practical simplex codes perform refactorizations periodically.

We first establish notation. Suppose that $\mathcal{A}_{.B}$ denotes the basis matrix for which the factorization (5.13) is known and that k subsequent column updates have produced the new basis matrix $\mathcal{A}_{.K}$. We can express the relationship as follows:

$$\mathcal{A}_{.K} = \mathcal{A}_{.B} + RS',$$

where each column of $R \in \mathbf{R}^{m \times k}$ is the difference between the entering and the leaving column of the basis and each column of $S \in \mathbf{R}^{m \times k}$ is a unit vector representing the location of the column being updated. For example, if $B = (4, 1, 6)$, with the first update replacing x_1 by x_7 and the second update replacing x_4 by x_3, then $K = (3, 7, 6)$ and R and S are given by

$$R = \begin{bmatrix} \mathcal{A}_{.7} - \mathcal{A}_{.1} & \mathcal{A}_{.3} - \mathcal{A}_{.4} \end{bmatrix}, \qquad S = \begin{bmatrix} 0 & 1 \\ 1 & 0 \\ 0 & 0 \end{bmatrix}.$$

In MATLAB, we can store S not as a full matrix but rather as a vector of length k, with `s(i)=j` indicating that column i of the matrix S contains the unit vector with 1 in the jth position and 0 elsewhere; this results in $s = [2\ 1]$ in the above example.

The following formula, known as the Sherman–Morrison–Woodbury formula, expresses the inverse of $\mathcal{A}_{.K}$ in terms of the inverse of $\mathcal{A}_{.B}$:

$$\mathcal{A}_{.K}^{-1} = (\mathcal{A}_{.B} + RS')^{-1} = \mathcal{A}_{.B}^{-1} - \mathcal{A}_{.B}^{-1} R (I + S' \mathcal{A}_{.B}^{-1} R)^{-1} S' \mathcal{A}_{.B}^{-1}. \tag{5.14}$$

(This formula can be verified by direct multiplication; multiply the right-hand side by $\mathcal{A}_{.B} + RS'$ and check that the identity I is recovered.)

We now show how to use this formula efficiently to calculate the vector $\mathcal{A}_{.K}^{-1} b$. A key idea is to store an update matrix, H, whose columns contain information about the kth column update, that is,

$$H_{.k} = \mathcal{A}_{.B}^{-1} R_{.k} = U^{-1} L^{-1} P R_{.k}.$$

This calculation requires a permutation of the kth column of R, together with triangular solves involving the factors L and U from (5.13), and requires $m^2 + O(m)$ operations. The relationship in (5.14) is thus simplified to

$$\mathcal{A}_{.K}^{-1} = \mathcal{A}_{.B}^{-1} - H(I + S'H)^{-1} S' \mathcal{A}_{.B}^{-1}.$$

The matrix $S'H$ can be obtained by selecting the components stored in the vector s from H, and instead of inverting the matrix $(I + S'H)$, we compute its LU factors as follows:

```
>> [L1,U1] = lu(eye(k) + H(s(1:k),1:k));
```

If we then compute $d = \mathcal{A}_{.B}^{-1} b$ using the factors from (5.13), at a cost of approximately $2m^2$ operations, it follows that

$$\mathcal{A}_{.K}^{-1} b = \mathcal{A}_{.B}^{-1} - H(I + S'H)^{-1} S' \mathcal{A}_{.B}^{-1} b = d - H U_1^{-1} L_1^{-1} S' d,$$

with L_1^{-1} and U_1^{-1} simply representing forward and backward substitution with these triangular matrices. In MATLAB, this corresponds to the following operation:

```
>> d = U\(L\(P*b));
>> d = d - H(:,1:k)*(U1\(L1\d(s(1:k))));
```

Since the second step requires about $2k^2 + 2mk$ operations, the work is dominated by the initial solve for d when $m \gg k$.

Example 5-2-14. Suppose

$$A = \begin{bmatrix} 1 & 3 & -3 \\ -4 & 3 & -2 \\ -1 & 0 & 1 \end{bmatrix}$$

is given in factored form as

$$P = \begin{bmatrix} 0 & 0 & 1 \\ 1 & 0 & 0 \\ 0 & 1 & 0 \end{bmatrix}, \quad L = \begin{bmatrix} 1 & 0 & 0 \\ -1 & 1 & 0 \\ 4 & 1 & 1 \end{bmatrix}, \quad U = \begin{bmatrix} -1 & 0 & 1 \\ 0 & 3 & -2 \\ 0 & 0 & -4 \end{bmatrix}.$$

Replace the first column of A with $a = (1, 5, -2)'$ to form $\hat{A}$ and then solve $\hat{A}x = b$ for $b = (1, 8, -3)'$ using the procedure outlined above. (Do not compute a new factorization.)

```
>> load ex5-2-14
>> H = U\(L\(P*(a-A(:,s))));
>> d = U\(L\(P*b));
>> s = 1;
>> d = d-H*(d(s)/(1+H(s,1)));
```

$$d = \begin{bmatrix} 1.6667 \\ 0.1111 \\ 0.3333 \end{bmatrix}$$

To check that d is in fact the solution, perform the following step:

```
>> norm([a A(:,2:3)]*d-b)
```

which should return a value close to 0.   ∎

Exercise 5-2-15. Replace the second column of A with $a = (1, 1, 1)'$ to form $\hat{A}$ and then solve $\hat{A}x = b$, where

$$A = \begin{bmatrix} 7 & 8 & 3 \\ 4 & 5 & 9 \\ 1 & 2 & 6 \end{bmatrix}, \quad b = \begin{bmatrix} 1 \\ 2 \\ 3 \end{bmatrix}.$$

Use the MATLAB routine `lu` to compute P, L, and U for A. Do not form or factor $\hat{A}$ explicitly, but use a similar procedure to that of Example 5-2-14.

The steps to calculate $c'A_{\cdot K}^{-1}$ are similar:

$$c'A_{\cdot K}^{-1} = (c' - c'H(I + S'H)^{-1}S')A_{\cdot B}^{-1} = (c' - d')A_{\cdot B}^{-1},$$

where $d' = c'H(I + S'H)^{-1}S'$. This corresponds to the following MATLAB code:

```
>> d = zeros(m,1); d(s(1:k)) = L1'\(U1'\(H(:,1:k)'*c));
>> u = P'*(L'\(U'\(c - d)));
```

The original factors from (5.13) are used to solve $(c' - d')\mathcal{A}_{\cdot\text{B}}^{-1}$, requiring $2m^2$ operations, which dominates the computation, assuming that the inner products involving the first $k - 1$ columns of H have been stored at previous iterations.

An implementation of this updating scheme can be found as part of the revised simplex routine rsmupd. This routine has the same calling sequence as rsm and can be used interchangeably with that routine. The maximum number of updates allowed before refactorization occurs is given as the variable max_upd, which by default is set to 10. The *Devex* scheme (Harris (1973)), described in Section 5.2.4, is also used in rsmupd to improve numerical accuracy on large problems.

There are other techniques for updating factorizations that achieve similar computational savings as the method outlined above. Although modern software uses an updating scheme for the *LU* factors of the basis matrix, other techniques generate a product form for the *inverse* of the basis matrix. A simple scheme that is appropriate in the dense setting can be found in Fletcher & Matthews (1984). (This approach does not perform as well as the code outlined above in MATLAB since it does not exploit the sparse matrix primitives that are incorporated in MATLAB.) Schemes that are better suited to sparse matrices are described in the following references: Bartels & Golub (1969), Forrest & Tomlin (1972), Gay (1978), Reid (1982). A full description of these techniques is beyond the scope of this book; interested readers are referred to Nazareth (1987) or Chvátal (1983, Chapter 24).

5.2.4   Advanced Pivot Selection Mechanisms

In large problems, it is often prohibitively expensive to compute the complete reduced cost vector c at every iteration. In fact, it is not necessary to do so, since we need to identify just one negative element of this vector in order to determine a valid pivot column for the simplex method. In *partial pricing* schemes, we calculate only a subvector of c (see Step 2 of the Revised Simplex Method) and select the pivot column from one of the negative elements of this subvector. One variant of partial pricing works with the same "window" of elements in c for several successive iterations and then moves to the next window for the next few iterations, cycling through the whole vector in this fashion. Another variant continues to calculate elements of c until it has found some specified fraction of m elements that are negative and chooses the most negative among these as the pivot column. Eventually, near the end of the simplex procedure, this scheme will require all the elements of c to be calculated.

Another technique for choosing the entering variable that is becoming popular in large scale codes is the *steepest edge*. These rules, which have both primal and dual variants, are described in detail by Forrest & Goldfarb (1992) and Goldfarb & Reid (1977). They essentially look for the entering variable such that the *decrease of the objective per unit distance moved* along the corresponding edge of the feasible polyhedron is as large as possible. (Contrast this choice with the choice of variable with the most negative reduced cost, which gives the largest decrease *per unit increase of the entering variable*.) Steepest-edge rules require additional computations to be performed for each candidate pivot and thus are more expensive to implement than simple rules such as finding the most negative reduced cost. However, on many problems, the total number of simplex pivots needed to find the solution is decreased considerably, and so the extra cost per iteration is worthwhile. Steepest-edge rules can be used in conjunction with partial pricing.

Turning now to the ratio test for computation of the steplength and the blocking variable, a modification known as Devex is often used (Harris (1973)). The obvious implementation of the ratio test from the routine `rsm.m` proceeds as follows:

```
blocking = find(d >= piv_tol);
if isempty(blocking)
error('problem is unbounded'); end;

[min_ratio,index_r] = min(x_B(blocking)./d(blocking));
r = blocking(index_r);
```

However, to ascertain whether the basic variables are within their bounds, the test

```
if any(x_B < -zer_tol)
```

is used. It seems unreasonable that the constraints are treated as hard constraints in the ratio test but are allowed to be slightly violated in other tests. Furthermore, when the coefficients of x_B are close to zero, the ratio test may well make a poor choice in determining the variable to leave the basis, since any numerical rounding error in the degenerate values will dominate such a test.

A more robust implementation of the ratio test is as follows:

```
blocking = find(d >= piv_tol);
if isempty(blocking)
error('problem is unbounded'); end;

elbow = x_B(blocking)+zer_tol;
min_ratio = min(elbow./d(blocking));

eligible = find(x_B(blocking)./d(blocking) <= min_ratio);
[max_piv,index_r] = max(d(blocking(eligible)));
r = blocking(eligible(index_r));
min_ratio = x_B(r)/d(r);
```

This implementation essentially moves the lower bound from zero to `-zer_tol` in order to effectively deal with rounding error. Other techniques based on this idea can be found in Gill et al. (1989).

Exercise 5-2-16. Modify the routine `rsm` to incorporate the Devex pivot rule. Test out the routine on Example 5-2-1.

5.3 Network Flow Problems

In Section 1.3.6 we described a minimum-cost network flow problem, in which we minimize the cost of moving a commodity along the arcs of a network to meet given demand patterns

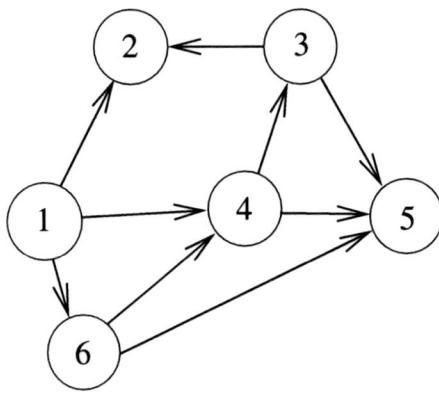

Figure 5.1. *Example network.*

at the nodes. By taking advantage of the special structure of this linear program, we can devise specialized, highly efficient versions of the simplex method that can be interpreted graphically in terms of the node-arc graph of the network. In this section, we discuss this problem in more detail, along with several other types of problems associated with networks that are useful in many applications.

A network is defined as a graph consisting of nodes and arcs. The node set $\mathcal{N}$ typically has a labeling scheme associated with it, which for the purposes of this chapter will be assumed to be $\{1, 2, \ldots, m\}$. In applications, the nodes may be cities, airports, or stages in a manufacturing process. Schematically, nodes are normally represented by labeled circles; see Figure 5.1.

Each arc connects one node (the origin) to another node (the destination). If the origin is i and the destination is j, we label the arc as (i, j), which is an element of the Cartesian product set $\mathcal{N} \times \mathcal{N}$. The orientation of each arc is significant; that is, $(1, 2)$ is not the same as $(2, 1)$. In applications, the nodes may represent roads, flight paths, or precedence relationships. An arc may be represented as in Figure 5.1 by a directed arrow joining two nodes. Undirected arcs can be replaced by two directed arcs between the two nodes in question, one pointing in each direction. The full set of arcs is denoted by $\mathcal{A}$.

5.3.1 Minimum-Cost Network Flow

Recapping the terminology of Section 1.3.6, the minimum-cost network flow problem associates three pieces of data with each arc $(i, j) \in \mathcal{A}$:

  c_{ij} is the cost per unit flow of the commodity along the arc (i, j);

  l_{ij} is the lower bound on the flow along (i, j);

  u_{ij} is the upper bound on the flow along (i, j), also known as the capacity of arc (i, j).

A *divergence* b_i associated with each node $i \in \mathcal{N}$ defines how much of the commodity is produced ($b_i > 0$) or consumed ($b_i < 0$) at node i. If $b_i > 0$, the node is called a supply node or *source*; if $b_i < 0$, it is a demand node or *sink*; if $b_i = 0$, node i is a *transshipment node*. Typically, all the data objects c_{ij}, l_{ij}, u_{ij}, and b_i are assumed to be integral.

The variables of the problem are the flows x_{ij} on each of the arcs $(i, j) \in \mathcal{A}$. As in Section 1.3.6, the resulting minimum-cost network flow problem is

$$\min_{x} \quad z = \sum_{(i,j)\in\mathcal{A}} c_{ij}x_{ij}$$

$$\text{subject to} \quad \sum_{j:(i,j)\in\mathcal{A}} x_{ij} - \sum_{j:(j,i)\in\mathcal{A}} x_{ji} = b_i \quad \text{for all nodes } i \in \mathcal{N},$$

$$l_{ij} \leq x_{ij} \leq u_{ij} \quad \text{for all arcs } (i, j) \in \mathcal{A},$$

which can be written in matrix form as follows:

$$\begin{aligned} \min \quad & c'x \\ \text{subject to} \quad & \mathcal{I}x = b, \quad l \leq x \leq u. \end{aligned} \tag{5.15}$$

Here the *node-arc incidence matrix* $\mathcal{I}$ is an $|\mathcal{N}| \times |\mathcal{A}|$ matrix, the rows being indexed by nodes and the columns being indexed by arcs. Every column of $\mathcal{I}$ corresponds to an arc $(i, j) \in \mathcal{A}$ and contains two nonzero entries: a $+1$ in row i and a -1 in row j. For the example network of Figure 5.1, the matrix $\mathcal{I}$ is given by

$$\mathcal{I} = \begin{bmatrix} 1 & 1 & 1 & & & & & & \\ -1 & & & & -1 & & & & \\ & & & 1 & 1 & -1 & & & \\ & & -1 & & & 1 & 1 & -1 & \\ & & & & & -1 & & -1 & -1 \\ & & -1 & & & & & 1 & 1 \end{bmatrix},$$

where we have taken the arcs in the order

$$\{(1, 2), (1, 4), (1, 6), (3, 2), (3, 5), (4, 3), (4, 5), (6, 4), (6, 5)\}.$$

For efficient implementations, it is crucial not to store or factor the complete matrix $\mathcal{I}$ but rather to use special schemes that exploit the special structure of this matrix.

Since $e'\mathcal{I} = 0$, we require $e'b = 0$ for a feasible point to exist. This relationship ensures that supply (the total amount of commodity produced over all the nodes) equals demand (the total amount consumed). We will show later how this assumption can be relaxed, but first we demonstrate that a number of typical operations research problems can be formulated as minimum-cost flow problems.

5.3.2 Shortest-Path Problem

The problem of finding the *shortest path* between two given nodes in the network is easily formulated as a minimum-cost network flow problem by setting the costs on the arcs to be the distances between the nodes that the arc connects. If knowing the shortest path between nodes s and t is required, we set the divergences as follows:

$$b_s = 1, \qquad b_t = -1, \qquad b_i = 0 \qquad \forall i \in \mathcal{N} \setminus \{s, t\}.$$

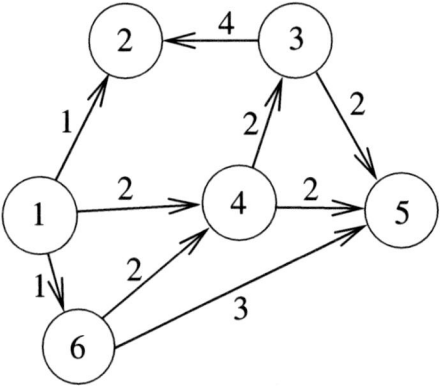

Figure 5.2. *Network with edge lengths.*

The lower bounds l_{ij} on the flows should be set to zero, while the upper bounds u_{ij} can be infinite.

In the example of Figure 5.2, where the distances c_{ij} are noted next to each arc, a shortest path from node $s = 1$ to node $t = 5$ is 1, 6, 5, with a cost of 4. Another solution (with, of course, the same distance) is 1, 4, 5. If the simplex method is applied to this formulation of the shortest-path problem, and the solution is unique, the solution can be recognized easily as consisting of those arcs on which the optimal flows are 1.

If we wish to know the shortest path from s to all the other nodes $i \in \mathcal{N}$, we define the network flow problem in the same way except that the divergences are different:

$$b_s = |\mathcal{N}| - 1, \qquad b_i = -1 \qquad \forall i \in \mathcal{N} \setminus \{s\}.$$

Having obtained the solution, we can recognize the shortest path from node s to a given node i by starting at i and backtracking. We choose the final leg of the path to be one of the arcs flowing into i for which the optimal flow is nonzero and backtrack to the origin node for this arc. We repeat the process to find the second-last leg of the path and continue until we reach the source node s.

5.3.3 Max-Flow Problem

Given a network and two special nodes s and t, the *max-flow* problem is to determine the maximum amount of flow that can be sent from s to t. Of course, for the problem to be meaningful, some of the arcs in the network must have finite capacities u_{ij}.

This problem can be formulated as a minimum-cost network flow problem by adding an arc (t, s) to the network with infinite capacity, zero lower bound, and a cost of -1. The divergences b_i at all the nodes are set to zero; the costs c_{ij} and lower bounds l_{ij} on all the original arcs are also set to zero. The added arc ensures that all the flow that is pushed from s to t (generally along multiple routes) is returned to s again and generates a profit (negative cost) corresponding to the flow on this arc.

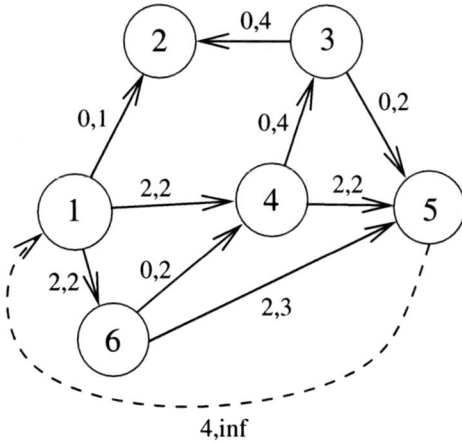

Figure 5.3. *Max-Flow: Edges labeled by (flow, upper bound).*

In the following example we take $s = 1$ and $t = 5$. The labels on the arcs correspond to x_{ij}, u_{ij}. The maximum flow from node 1 to node 5 is 4: 2 units flow along the path 1, 4, 5 and 2 units along the path 1, 6, 5. The total flow of 4 returns to node 1 along the dummy arc (5, 1). Note that the incidence matrix $\mathcal{I}$ for the network shown in Figure 5.3 is

$$
\mathcal{I} = \begin{bmatrix}
1 & 1 & 1 & & & & & & & -1 \\
-1 & & & -1 & & & & & & \\
 & & & 1 & 1 & -1 & & & & \\
 & -1 & & & 1 & 1 & -1 & & \\
 & & & & -1 & & -1 & -1 & 1 \\
 & & -1 & & & & & 1 & 1 \\
\end{bmatrix},
$$

where the final column corresponds to the dummy arc that has been added. This reformulation is an example of a circulation problem, which is a general term referring to a problem in which all divergences are zero.

We can also define max-flow problems with multiple sources, in which we wish to find the maximum amount of flow originating at any of the given sources that can be sent to the specified destination. To formulate as a minimum-cost network flow problem, we add a "super-source" as shown in Figure 5.4 and define arcs from the super-source to the original sources with infinite capacities and zero costs. We also add the arc from the sink to the super-source in the manner described above. Max-flow problems with multiple sinks (or multiple sources *and* multiple sinks) can be formulated similarly. See Figure 5.4.

5.3.4 Transportation Problem

In the transportation problem, we partition the nodes into two sets (sources and sinks) so that every arc is directed from one of the sources to one of the destinations. As is usual in network flow, we seek a flow that minimizes the total cost of moving the commodity along

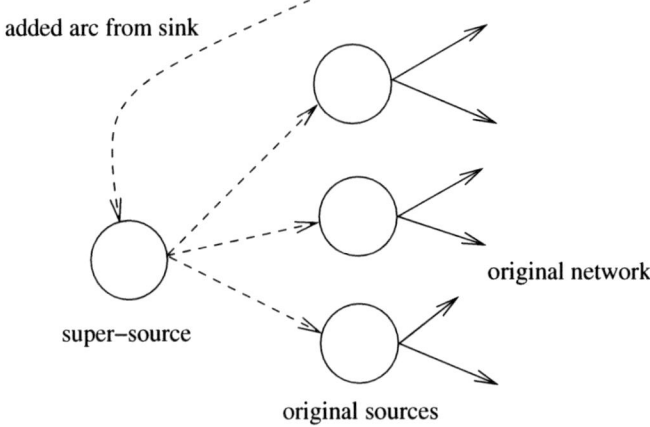

Figure 5.4. *Replacing multiple sources by a super-source.*

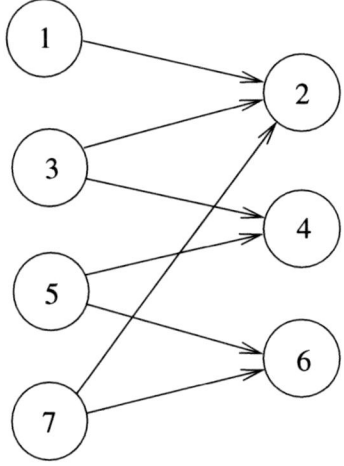

Figure 5.5. *Transportation network.*

the arcs, so as to satisfy all demands at the sinks while not exceeding the supply at each source. A typical application is to gasoline transportation, where each source is a refinery and each sink is a market.

We denote the sources by $\mathcal{N}_1$ and the sinks by $\mathcal{N}_2$, so that $\mathcal{N}_1 \cup \mathcal{N}_2$ forms a partition of $\mathcal{N}$. Each arc $(i, j) \in \mathcal{A}$ has $i \in \mathcal{N}_1$ and $j \in \mathcal{N}_2$. (A graph with this property is termed a *bipartite graph.*) A sample problem is shown in Figure 5.5, where $\mathcal{N}_1 = \{1, 3, 5, 7\}$ and $\mathcal{N}_2 = \{2, 4, 6\}$.

Given a supply s_i of the commodity at node $i \in \mathcal{N}_1$, the total flow out of node i cannot exceed this quantity. Similarly, given a demand d_j at node $j \in \mathcal{N}_2$, the total flow into node

j must be at least equal to this amount. Given a cost c_{ij} per unit flow along arc (i, j), we can formulate the problem as follows:

$$\min_{x_{ij}} \sum_{(i,j)\in\mathcal{A}\subset\mathcal{N}_1\times\mathcal{N}_2} c_{ij}x_{ij}$$

$$\text{subject to} \quad \sum_{j\in\mathcal{N}_2} x_{ij} \leq s_i \qquad \forall i \in \mathcal{N}_1,$$

$$\sum_{i\in\mathcal{N}_1} x_{ij} \geq d_j \qquad \forall j \in \mathcal{N}_2,$$

$$x_{ij} \geq 0 \qquad \forall i \in \mathcal{N}_1 \text{ and } \forall j \in \mathcal{N}_2.$$

Exercise 5-3-1. Show that a necessary condition for the existence of a feasible x_{ij}, $i \in \mathcal{N}_1$, $j \in \mathcal{N}_2$, is that

$$\sum_{j\in\mathcal{N}_2} d_j \leq \sum_{i\in\mathcal{N}_1} s_i.$$

If the graph is complete (that is, there are arcs between every $i \in \mathcal{N}_1$ and every $j \in \mathcal{N}_2$), show that this condition is also sufficient.

The problem as formulated above does not conform to the minimum-cost format (5.15), since the general constraints are inequalities, rather than equalities representing conservation of flow. However, by applying a conversion procedure that adds "dummy" nodes to the network, we can obtain a problem in the familiar form (5.15), which can then be solved by using software designed specifically for problems in this format. (We leave the details as an exercise.)

5.3.5   Assignment Problem

In the assignment problem we have two sets of nodes $\mathcal{N}_1$ and $\mathcal{N}_2$ of equal size. Given a specified cost c_{ij} for pairing a node $i \in \mathcal{N}_1$ with a node $j \in \mathcal{N}_2$, we wish to pair off each node in $\mathcal{N}_1$ with a partner in $\mathcal{N}_2$ (making a one-one correspondence between the two sets) so as to minimize the total cost of pairing. We can formulate this problem as a minimum-cost flow problem by defining the divergences as follows:

$$b_i = 1 \qquad \forall i \in \mathcal{N}_1, \qquad b_i = -1 \qquad \forall i \in \mathcal{N}_2,$$

while the lower bounds and capacities are defined as follows:

$$l_{ij} = 0, \qquad u_{ij} = 1 \qquad \forall (i, j) \in \mathcal{A} \subset \mathcal{N}_1 \times \mathcal{N}_2.$$

5.3.6   Network Simplex Method

It can be shown that the basis matrix $\mathcal{A}_{.\text{B}}$ corresponds to a (rooted) tree when $\mathcal{A}$ is the node-arc incidence matrix of a network (Ahuja, Magnanti & Orlin (1993)). Using this observation, the solution of linear systems involving the basis matrix corresponds to a tree walk, while the ratio test corresponds to examining the cycle formed in this tree by the addition of an (incoming) arc. Changing the basis then corresponds to rehanging the tree from a potentially different root. Implementation of these operations can be done much more efficiently than the steps we have outlined in other parts of this chapter for canonical form problems (without network structure). This topic is beyond the scope of the book, but details can be found in (Ahuja, Magnanti & Orlin (1993)).

A node-arc incidence matrix is an example of a totally unimodular matrix, that is, a matrix for which the determinant of every square submatrix is equal to 0, 1, or -1. When $\mathcal{A}$ is totally unimodular and $\tilde{b}$, b, $\tilde{d}$, and d are integer vectors, it can be shown that if the set $\{x \in \mathbf{R}^n \mid \tilde{b} \leq \mathcal{A}x \leq b, \tilde{d} \leq x \leq d\}$ is not empty, then all its extreme points are integer vectors (see Nemhauser & Wolsey (1988) for further details). Since the simplex method moves from one extreme point to another, this result guarantees that the network simplex method will produce solutions x that are integer vectors.

It is well known that restricting some or all of the variables in a problem to take on only integer values can make a general linear program a significantly harder problem to solve (more precisely, while linear programs are known to be polynomially solvable, mixed integer programs are NP-hard), and so in a certain sense this result about the integrality of solutions to network linear programs is rather remarkable. Further information on this and other results can be found in the texts by Ahuja, Magnanti & Orlin (1993), Nemhauser & Wolsey (1988), and Schrijver (1986).

Chapter 6

Sensitivity and Parametric Linear Programming

In this chapter, we examine how the solution of a linear program (and its optimal objective value) are affected when changes are made to the data of the problem. This topic is interesting for several reasons. First, given that the problem data is often uncertain, modelers may wish to know how the solution will be affected (slightly, drastically, or not at all) if they assign a slightly different value to a particular element of the constraint right-hand side or the cost vector. Second, from a computational point of view, a user of linear programming software would hope that having computed a solution (and an optimal basis) for one linear program, they will have a "warm start" in computing the solution of a second problem in which the data is only slightly different. Rather than restarting the simplex method from scratch for the modified problem we hope to be able to start with the optimal basis for the original problem and perhaps perform a few simplex steps to find the optimal basis for the modified problem.

In Section 6.1, we look at *small* changes to the constraint right-hand side and the cost vector and examine their effect on the optimal objective value. This topic is commonly called *sensitivity analysis*. We discuss the question of adding variables or constraints to the problem in Section 6.2. Sections 6.3 and 6.4 discuss *parametric programming*, which concerns more extensive changes to the data that are parametrized by a single variable. Again, we consider only changes to the cost vector (Section 6.3) and to the right-hand side (Section 6.4). (Changes to the coefficient matrix are considerably more difficult to analyze.)

6.1 Sensitivity Analysis

In this section we refer to the standard form of linear programming, which is

$$
\begin{aligned}
&\min_{x} && z = p'x \\
&\text{subject to} && Ax \geq b, \quad x \geq 0
\end{aligned}
\tag{6.1}
$$

(where $A \in \mathbf{R}^{m \times n}$), as well as the canonical form, which is

$$
\begin{aligned}
&\min && z = p'x \\
&\text{subject to} && \mathcal{A}x = b, \quad x \geq 0
\end{aligned}
\tag{6.2}
$$

(where $A \in \mathbf{R}^{m \times l}$). A standard form problem can be converted into a canonical form problem by adding slack variables, in which case $\mathcal{A} = [A \ -I]$, but this form of $\mathcal{A}$ is not always available.

From Chapter 5, we recall that a tableau for the canonical form problem is fully described by specifying the set B of basic variables, provided that the basis matrix $\mathcal{A}_{.\mathrm{B}}$ is invertible. With dual labels, the tableau for a given basis B is

$$
\begin{array}{rrr|c|c}
 & & & u_{\hat{\mathrm{B}}} = & w = \\
 & & & x_{\mathrm{N}} & 1 \\
\hline
-u_{\hat{\mathrm{N}}} & x_{\mathrm{B}} & = & -\mathcal{A}_{.\mathrm{B}}^{-1}\mathcal{A}_{.\mathrm{N}} & \mathcal{A}_{.\mathrm{B}}^{-1}b \\
\hline
1 & z & = & p_{\mathrm{N}}' - p_{\mathrm{B}}'\mathcal{A}_{.\mathrm{B}}^{-1}\mathcal{A}_{.\mathrm{N}} & p_{\mathrm{B}}'\mathcal{A}_{.\mathrm{B}}^{-1}b
\end{array}
\tag{6.3}
$$

Recall that the bottom row of the tableau is the vector of *reduced costs*, which we denote by c. The tableau (6.3) is optimal for (6.2) if and only if the reduced costs and the final column are all nonnegative; that is,

$$
c' := p_{\mathrm{N}}' - p_{\mathrm{B}}'\mathcal{A}_{.\mathrm{B}}^{-1}\mathcal{A}_{.\mathrm{N}} \geq 0, \qquad \mathcal{A}_{.\mathrm{B}}^{-1}b \geq 0.
$$

To be specific, $\mathcal{A}_{.\mathrm{B}}^{-1}b \geq 0$ connotes primal feasibility (dual optimality), while $c \geq 0$ connotes primal optimality (dual feasibility).

Note that a change in b affects only the last column of the tableau, while a change in p affects only the final row. If the problem data is changed in such a way that the last row and last column of the tableau both remain nonnegative, then the tableau remains optimal for the perturbed data. Specifically, if b is changed to $\tilde{b}$, then the tableau remains optimal if $\mathcal{A}_{.\mathrm{B}}^{-1}\tilde{b} \geq 0$. If p is changed to $\tilde{p}$ and we still have $\tilde{p}_{\mathrm{N}}' - \tilde{p}_{\mathrm{B}}'\mathcal{A}_{.\mathrm{B}}^{-1}\mathcal{A}_{.\mathrm{N}} \geq 0$, again the tableau remains optimal without any change to the basis B.

Example 6-1-1. We work with the following problem (and several variants) throughout this section:

$$
\begin{array}{rlcccccc}
\min & x_1 + 1.5x_2 + 3x_3 \\
\text{subject to} & x_1 & + & x_2 & + & 2x_3 & \geq & 6, \\
 & x_1 & + & 2x_2 & + & x_3 & \geq & 10, \\
 & & & x_1, x_2, x_3 & & & \geq & 0.
\end{array}
\tag{6.4}
$$

By adding slack variables x_4 and x_5 we arrive at the canonical-form problem (6.2) with

$$
A = \begin{bmatrix} 1 & 1 & 2 & -1 & 0 \\ 1 & 2 & 1 & 0 & -1 \end{bmatrix}, \qquad b = \begin{bmatrix} 6 \\ 10 \end{bmatrix}, \qquad p = \begin{bmatrix} 1 \\ 1.5 \\ 3 \\ 0 \\ 0 \end{bmatrix}.
$$

Consider the basis B = $\{1, 2\}$, for which we have

$$
\mathcal{A}_{.\mathrm{B}} = \begin{bmatrix} 1 & 1 \\ 1 & 2 \end{bmatrix}, \qquad \mathcal{A}_{.\mathrm{N}} = \begin{bmatrix} 2 & -1 & 0 \\ 1 & 0 & -1 \end{bmatrix}, \qquad \mathcal{A}_{.\mathrm{B}}^{-1} = \begin{bmatrix} 2 & -1 \\ -1 & 1 \end{bmatrix},
$$

and the last column and bottom row of the tableau are, respectively,

$$A_{\cdot B}^{-1} b = \begin{bmatrix} 2 & -1 \\ -1 & 1 \end{bmatrix} \begin{bmatrix} 6 \\ 10 \end{bmatrix} = \begin{bmatrix} 2 \\ 4 \end{bmatrix},$$

$$p'_N - p'_B A_{\cdot B}^{-1} A_{\cdot N} = \begin{bmatrix} 3 & 0 & 0 \end{bmatrix} - \begin{bmatrix} 1 & 1.5 \end{bmatrix} \begin{bmatrix} 2 & -1 \\ -1 & 1 \end{bmatrix} \begin{bmatrix} 2 & -1 & 0 \\ 1 & 0 & -1 \end{bmatrix}$$

$$= \begin{bmatrix} 3 & 0 & 0 \end{bmatrix} - \begin{bmatrix} 1.5 & -0.5 & -0.5 \end{bmatrix} = \begin{bmatrix} 1.5 & 0.5 & 0.5 \end{bmatrix}.$$

Thus the basis $\{1, 2\}$ is optimal. ∎

Suppose now we wish to solve a problem that is the same except for a different right-hand side $\tilde{b}$. Rather than starting the simplex method from scratch to find the solution of the modified problem, we can first check to see if the basis B is still optimal for the new right-hand side. Since only the last column of the tableau is affected by the change from b to $\tilde{b}$, optimality will be maintained if $A_{\cdot B}^{-1} \tilde{b} \geq 0$.

As an example, suppose that the right-hand side in (6.4) is replaced by

$$\tilde{b} = \begin{bmatrix} 7 \\ 9 \end{bmatrix}.$$

We have

$$A_{\cdot B}^{-1} \tilde{b} = \begin{bmatrix} 2 & -1 \\ -1 & 1 \end{bmatrix} \begin{bmatrix} 7 \\ 9 \end{bmatrix} = \begin{bmatrix} 5 \\ 2 \end{bmatrix} \geq 0,$$

and so the basis $B = \{1, 2\}$ remains optimal for $\tilde{b}$. The optimal x and the optimal objective value have changed; we can read the new values from the modified tableau. More generally, suppose that the first component on the right-hand side is replaced by $6 + \epsilon$, where ϵ is some scalar. The basic components of the solution become

$$A_{\cdot B}^{-1} \begin{bmatrix} 6 + \epsilon \\ 10 \end{bmatrix} = \begin{bmatrix} 2 + 2\epsilon \\ 4 - \epsilon \end{bmatrix}.$$

The basis $B = \{1, 2\}$ remains optimal when this vector is nonnegative, which is true for all values of ϵ in the range $[-1, 4]$.

The change in the optimal objective due to a perturbation in the right-hand side b can also be derived by examining the dual solution of the original problem, as we now show. Let us denote the optimal primal solution for the right-hand side b by $x(b)$ and the optimal objective by $z(b)$. The standard-form problem (6.1) has the dual

$$\begin{aligned} \max \quad & b'u \\ \text{subject to} \quad & A'u \leq p, \quad u \geq 0. \end{aligned}$$

In the optimal tableau, the reduced costs c represent the optimal values of the nonzero components of u, while the condition $A_{\cdot B}^{-1} b \geq 0$ establishes optimality of these variables. When b is changed to $\tilde{b}$, the reduced costs are not affected. Hence, the *same* u is feasible for the modified dual problem

$$\begin{aligned} \max \quad & \tilde{b}'u \\ \text{subject to} \quad & A'u \leq p, \quad u \geq 0, \end{aligned}$$

and it will also be optimal, provided that $A_{\cdot B}^{-1}\tilde{b} \geq 0$. By duality theory, we have in this case that

$$z(b) = p'x(b) = b'u,$$
$$z(\tilde{b}) = p'x(\tilde{b}) = \tilde{b}'u,$$

so that the change in optimal objective is as follows:

$$p'x(\tilde{b}) - p'x(b) = z(\tilde{b}) - z(b) = (\tilde{b} - b)'u = \sum_{i=1}^{m}(\tilde{b} - b)_i u_i.$$

This formula indicates that a change of ϵ in the ith component of b induces a change of $u_i \epsilon$ in the objective, where ϵ is small enough to ensure that the basis B is optimal for both original and modified problems. In fact, we can write

$$\frac{\partial z(b)}{\partial b_i} = u_i$$

to indicate the sensitivity of $z(b)$ to b_i in this case. Because u_i is the price to be paid per unit change in constraint i, the term *shadow prices* is often used for the dual variables.

We now examine perturbations to the coefficients of the objective function (that is, the elements of the cost vector). When p is perturbed to $\tilde{p}$, the tableau remains optimal if the reduced cost vector (the last row of the tableau) remains nonnegative, that is, if $\tilde{p}_N' - \tilde{p}_B' A_{\cdot B}^{-1} A_{\cdot N} \geq 0$. When just a single element p_j is perturbed to $p_j + \delta$, we can consider two cases. First, when $j = N(s)$, the reduced cost corresponding to element j becomes $(p_j + \delta) - p_B' A_{\cdot B}^{-1} A_{\cdot j} = c_s + \delta$, where $j = N(s)$ (that is, the s element of the reduced cost vector corresponds to the j element of p). Hence, the basis remains optimal if $c_s + \delta \geq 0$, since none of the other reduced costs are affected by this perturbation. In the example (6.4), when p_3 is perturbed to $p_3 + \delta$, the corresponding reduced cost is $1.5 + \delta$. Hence, the basis remains optimal for all δ in the range $[-1.5, \infty)$.

In the second case, when the perturbation element j belongs to the basis B, the analysis becomes slightly more complicated because *all* the reduced costs may be affected. Suppose that $j = B(r)$; that is, the r element of the basis B is the index j. When p_j is perturbed to $p_j + \delta$, the bottom row of the tableau becomes

$$p_N' - (p_{B(1)}, \ldots, p_{B(r)} + \delta, \ldots, p_{B(m)}) A_{\cdot B}^{-1} A_{\cdot N} = c' - \delta(A_{\cdot B}^{-1})_r. A_{\cdot N},$$

and the tableau remains optimal if this vector is still nonnegative. To determine whether this is the case, we need to calculate the rth row of the inverse basis matrix. In the example (6.4), if $p_1 = 1$ is changed to $1 + \delta$, the tableau remains optimal if the following condition is satisfied by δ:

$$\begin{bmatrix} 1.5 & 0.5 & 0.5 \end{bmatrix} - \delta \begin{bmatrix} 2 & -1 \end{bmatrix} \begin{bmatrix} 2 & -1 & 0 \\ 1 & 0 & -1 \end{bmatrix} = \begin{bmatrix} 1.5 - 3\delta & 0.5 + 2\delta & 0.5 - \delta \end{bmatrix} \geq 0.$$

This inequality holds for all δ in the range $[-1/4, 1/2]$.

When the perturbations to the elements of p do not produce a change in optimal basis, the primal solution x does not change. This is because the nonzero values of the optimal x

appear in the last column of the tableau, which is not affected by changes in p. It is easy to see that in this case, the optimal objective depends on p as follows:

$$\frac{\partial z(p)}{\partial p_i} = x_i.$$

We can think of the optimal values of x_i as shadow prices on the dual constraints.

Similar arguments to those above can be used to ascertain whether changes to the values of A_{ij} affect the optimal value and solution of the original problem. The issues are more complicated, however, since all parts of the tableau can be affected (including both the last column and last row), and we will not discuss them here.

Two other changes to the problem that might affect the optimal value of the linear program or the optimal solution vector are addition of a new variable and addition of a new constraint. In the next section, we use an analysis like that above to determine the effect of such changes to the linear program.

Exercise 6-1-2. Consider the following linear program:

$$
\begin{array}{rrrrrrl}
\min & -x_1 & - & 4x_2 & - & x_3 & \\
\text{subject to} & 2x_1 & + & 2x_2 & + & x_3 & = 4, \\
& x_1 & & & - & x_3 & = 1, \\
& & & x_1, x_2, x_3 & & \geq & 0.
\end{array}
$$

1. Verify that an optimal basis for this problem is $B = \{1, 2\}$, and calculate the quantities $A_{\cdot B}$, $A_{\cdot B}^{-1}$, $A_{\cdot N}$, p_B, p_N, and x_B for this basis, together with the reduced cost vector $c = p_N' - p_B' A_{\cdot B}^{-1} A_{\cdot N}$.

2. Suppose that the right-hand side 1 of the second constraint is replaced by $1 + \epsilon$. Calculate the range of ϵ for which the basis B remains optimal, and give the solution x for each value of ϵ in this range.

3. Suppose that the coefficient of x_2 in the objective is replaced by $-4 + \delta$. Find the range of δ for which the basis B remains optimal.

6.2 Adding New Variables or Constraints

Suppose we have solved a problem with an optimal basis B and we wish to add an extra nonnegative variable with constraint matrix column $f \in \mathbf{R}^m$ and objective coefficient $\pi \in \mathbf{R}$; that is, we now have

$$
\begin{array}{rl}
\min & p'x + \pi x_{l+1} \\
\text{subject to} & Ax + f x_{l+1} = b, \quad x, x_{l+1} \geq 0.
\end{array}
$$

To check whether adding this column affects the basis, we just calculate the reduced cost corresponding to this new column. If the entry is nonnegative, that is,

$$\pi - p_B' A_{\cdot B}^{-1} f \geq 0,$$

then the basis B for the original problem is still optimal for the augmented problem. Otherwise, we need to perform further simplex iterations to recover a solution. We can initiate the revised simplex method by choosing the new index $l + 1$ to be the index to enter the basis B. As in Step 4 of Algorithm 5.1, we calculate the pivot column $d = A_{\cdot B}^{-1} f$ and perform the ratio test to determine which index leaves B. We then update B and N, and start a new iteration of Algorithm 5.1 (applied to the augmented problem), until a solution is found.

Example 6-2-1. If we add the column

$$f = \begin{bmatrix} 4 \\ -1 \end{bmatrix}, \qquad \pi = 2$$

to the problem (6.4), then we have

$$\pi - p_B' A_{\cdot B}^{-1} f = 2 - \begin{bmatrix} 1 & 1.5 \end{bmatrix} \begin{bmatrix} 2 & -1 \\ -1 & 1 \end{bmatrix} \begin{bmatrix} 4 \\ -1 \end{bmatrix} = 0.5 \geq 0,$$

and so the basis remains optimal. ∎

Now consider the case in which a new inequality constraint $a'x \geq \beta$ is added to a canonical-form problem, that is,

$$\begin{aligned} \min \quad & p'x \\ \text{subject to} \quad & Ax = b, \quad a'x \geq \beta, \quad x \geq 0. \end{aligned}$$

We define a slack variable x_{l+1} corresponding to this constraint. Our strategy is to check whether the basis $\tilde{B} := B \cup \{l + 1\}$ is optimal for this modified problem (where B is the optimal basis for the original problem). Accordingly, we define the augmented basis matrix as

$$\tilde{A}_{\cdot \tilde{B}} = \begin{bmatrix} A_{\cdot B} & 0 \\ a_B' & -1 \end{bmatrix},$$

whose inverse is

$$\tilde{A}_{\cdot \tilde{B}}^{-1} = \begin{bmatrix} A_{\cdot B}^{-1} & 0 \\ a_B' A_{\cdot B}^{-1} & -1 \end{bmatrix}.$$

The tableau for the augmented problem is obtained by adding an extra row (corresponding to the slack variable for the new constraint) to the original optimal tableau as follows:

(6.5)

		x_N	1
x_B	=	$-A_{\cdot B}^{-1} A_{\cdot N}$	$A_{\cdot B}^{-1} b$
x_{l+1}	=	$-a_B' A_{\cdot B}^{-1} A_{\cdot N} + a_N'$	$a_B' A_{\cdot B}^{-1} b - \beta$
z	=	$p_N' - p_B' A_{\cdot B}^{-1} A_{\cdot N}$	$p_B' A_{\cdot B}^{-1} b$

Note that the bottom row is unchanged from the optimal tableau for the original problem, because

$$p_N' - p_B' \tilde{A}_{\cdot \tilde{B}}^{-1} \begin{bmatrix} A_{\cdot N} \\ a_N' \end{bmatrix} = p_N' - \begin{bmatrix} p_B' & 0 \end{bmatrix} \begin{bmatrix} A_{\cdot B}^{-1} & 0 \\ a_B' A_{\cdot B}^{-1} & -1 \end{bmatrix} \begin{bmatrix} A_{\cdot N} \\ a_N' \end{bmatrix}$$

$$= p_N' - p_B' A_{\cdot B}^{-1} A_{\cdot N}.$$

Since these reduced costs are still nonnegative, and since $A_{\cdot B}^{-1} b \geq 0$, optimality of the tableau (6.5) is assured if $a_B' A_{\cdot B}^{-1} b - \beta \geq 0$. If this condition is *not* satisfied, we can recover optimality (or determine dual unboundedness/primal infeasibility) by applying some pivots of the dual simplex method to (6.5). Our first move would be to choose the row corresponding to x_{l+1} as the pivot row and then use the ratio test to select the pivot column.

Example 6-2-2. Suppose we add the constraint

$$x_1 + x_2 \geq 5$$

to (6.4). We have

$$a_{_B}' \mathcal{A}_{\cdot_B}^{-1} b - \beta = \begin{bmatrix} 1 & 1 \end{bmatrix} \begin{bmatrix} 2 \\ 4 \end{bmatrix} - 5 = 1,$$

and so the augmented tableau is optimal. (This fact is fairly obvious, since the basic feasible solution $x_1 = 2$, $x_2 = 4$ of the original problem satisfies the added constraint and hence remains optimal.) If instead the constraint added was

$$x_1 + x_2 \geq 7,$$

then

$$a_{_B}' \mathcal{A}_{\cdot_B}^{-1} b - \beta = \begin{bmatrix} 1 & 1 \end{bmatrix} \begin{bmatrix} 2 \\ 4 \end{bmatrix} - 7 = -1,$$

and so the problem is no longer feasible. We proceed with the dual simplex method by calculating the remaining elements in this row of the augmented tableau (6.5):

$$-a_{_B}' \mathcal{A}_{\cdot_B}^{-1} \mathcal{A}_{\cdot_N} + a_{_N}' = -\begin{bmatrix} 1 & 1 \end{bmatrix} \begin{bmatrix} 2 & -1 \\ -1 & 1 \end{bmatrix} \begin{bmatrix} 2 & -1 & 0 \\ 1 & 0 & -1 \end{bmatrix} = \begin{bmatrix} -2 & 1 & 0 \end{bmatrix}.$$

The dual simplex ratio test (recall that the bottom row $c' = (1.5, 0.5, 0.5)$) determines variable x_4 as the entering variable. The resulting pivot gives a new basis of $\tilde{B} = \{1, 2, 4\}$, which happens to be optimal. ■

Suppose that instead of adding an inequality constraint, we wish to add the equation

$$a' x = \beta$$

to the problem. Our strategy here is to augment the tableau as in (6.5) and then immediately perform a pivot on the added row to make the newly added slack x_{l+1} nonbasic. (Recall that this is how we handle equality constraints in Scheme II of Chapter 3.) After this pivot, we remove the column corresponding to x_{l+1} from the tableau and proceed with further simplex pivots to recover optimality. Note that the Scheme II pivot may introduce nonpositive entries in both the last row and last column of the tableau, so that it might be necessary to apply a Phase I technique to recover a primal feasible tableau before proceeding.

Example 6-2-3. If we add the constraint

$$x_1 + x_2 = 5$$

to the standard example, then as calculated above, we have

$$a_{_B}' \mathcal{A}_{\cdot_B}^{-1} b - \beta = \begin{bmatrix} 1 & 1 \end{bmatrix} \begin{bmatrix} 2 \\ 4 \end{bmatrix} - 5 = 1.$$

Since for feasibility of the equation this value must be zero, we perform a pivot to remove x_6 from the basis. As above, the corresponding row of the tableau is calculated as

$$-a_{_B}' \mathcal{A}_{\cdot_B}^{-1} \mathcal{A}_{\cdot_N} + a_{_N}' = \begin{bmatrix} -2 & 1 & 0 \end{bmatrix},$$

and hence either the first or second nonbasic variable (x_3 or x_4) can be chosen to enter the basis. By choosing x_3 and performing the Jordan exchange of x_3 and x_6, we obtain an optimal tableau after the column corresponding to x_6 is deleted, yielding the optimal basis $B = \{1, 2, 3\}$. If we choose x_4, we obtain a dual feasible tableau after the pivot and deletion of the x_6 column. One iteration of dual simplex then yields the optimal basis. ∎

Exercise 6-2-4. Solve the canonical-form linear program (6.2) with the following data:

$$p = \begin{bmatrix} 1 \\ -1 \\ 0 \\ 0 \end{bmatrix}, \qquad \mathcal{A} = \begin{bmatrix} -1 & 2 & 1 & 0 \\ 0 & 1 & 0 & 1 \end{bmatrix}, \qquad b = \begin{bmatrix} 2 \\ 4 \end{bmatrix}.$$

(You may want to try the basis $B = \{2, 4\}$.) Use the techniques of this section to perform the following.

1. Change b to $\tilde{b} = (1, 1)'$. Is the same basis still optimal? Is the same solution optimal? Justify.

2. Go back to the original data and add the extra nonnegative variable $x_5 \geq 0$, with cost coefficient $\pi = -1$ and column $f = (3, 1)'$, so that the new constraints are $\mathcal{A}x + fx_5 = b$. What are the optimal solution and optimal value now?

3. Add the constraint $x_1 + x_2 \geq 0.5$ to the original data and determine the solution and the optimal value. Write down the dual of this extended problem. What is the corresponding dual solution and its value? (Hint: you may wish to use the KKT conditions for the general formulation given in (4.14), (4.15).)

We now turn to *parametric programming*, in which possibly large changes to the data result in changes to the optimal basis.

6.3   Parametric Optimization of the Objective Function

We first give a simple geometric example of parametric programming, in which the cost vector is a linear function of a scalar parameter t. Consider

$$\begin{array}{rl} \min_{x_1, x_2} & z(t) := \begin{bmatrix} -1 - t \\ -1 + t \end{bmatrix}' \begin{bmatrix} x_1 \\ x_2 \end{bmatrix} = -x_1 - x_2 + t(-x_1 + x_2), \quad (-\infty < t < \infty) \\ \text{subject to} & -x_1 \;-\; x_2 \;\geq\; -4, \\ & \qquad\;\; -\; x_2 \;\geq\; -2, \\ & -x_1 \qquad\qquad \geq\; -3, \\ & \qquad x_1, x_2 \;\geq\;\; 0. \end{array}$$

Figure 6.1 depicts the feasible region of the parametric linear program together with optimal value contours of the objective function for various values of the parameter $t \in (-\infty, \infty)$. For example, when $t = -1$, we have $z(-1) = -2x_2$, and so the contour indicated by the dotted line running along the top edge of the feasible region is the contour along which the optimal objective value $z(-1) = -4$ is attained. For $t \in (-\infty, 1)$, the

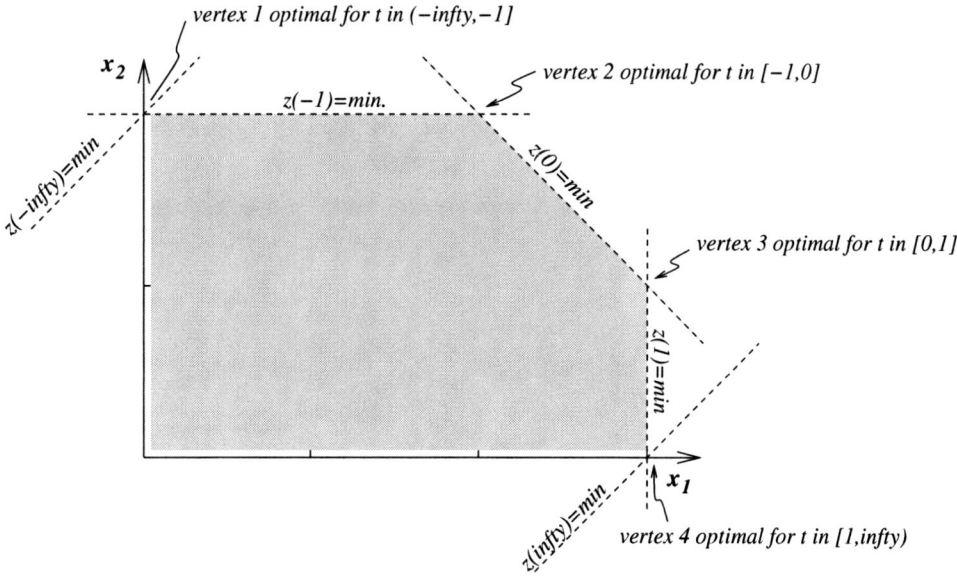

Figure 6.1. *Parametric linear program in* $\mathbf{R}^2$ *with parameter t: Optimal value contours and optimal vertices for* $t \in (-\infty, \infty)$. *(Feasible region is shaded.)*

vertex at $(x_1, x_2) = (0, 2)$ is optimal, as the best objective contour touches the feasible region at only this point (the limiting contour as $t \to -\infty$ is illustrated).

In general, we write the parametrized cost vector as $p + tq$, where p are fixed costs and q are variable costs. In a financial application, the ith element of p might represent the expected rate of inflation, so that if the corresponding element q_i is nonzero, we can consider the effects of a range of inflation (and deflation!) rates by allowing t to vary within a certain range. The cost-parametrized version of standard form is then

$$z(t) = \min_x (p + tq)'x \quad \text{subject to} \quad Ax \geq b, \quad x \geq 0. \tag{6.6}$$

We seek (i) the optimal vertex or vertices, and (ii) the optimal objective value $z(t)$ for each t in the given range.

To show how the tableau form of the simplex method can be extended to perform parametric linear programming, we consider again the example given at the start of the chapter and add the variable cost vector q defined by

$$q := \begin{bmatrix} -3 \\ -4 \\ -5 \end{bmatrix}.$$

The parametric linear program is then

$$\begin{array}{rrrrrrrrr}
\text{min} & x_1 + 1.5x_2 + 3x_3 + t(-3x_1 - 4x_2 - 5x_3) & & & & & \\
\text{subject to} & x_1 & + & x_2 & + & 2x_3 & \geq & 6, \\
& x_1 & + & 2x_2 & + & x_3 & \geq & 10, \\
& & & x_1, x_2, x_3 & & & \geq & 0.
\end{array}$$

We first construct the standard tableau with the cost vector p and then add an extra row containing q and labeled by z_0.

```
≫ load ex6-1-1

≫ T = totbl(A,b,p);

≫ T = addrow(T,[q' 0],'z0',4);
```

	x_1	x_2	x_3	1
$x_4 =$	1	1	2	-6
$x_5 =$	1	2	1	-10
$z =$	1	1.5	3	0
$z_0 =$	-3	-4	-5	0

In general, we must perform Phase I pivots to determine a feasible basis. However, in this case, recall from page 152 that $B = \{1, 2\}$ is a feasible basis, and so we carry out two pivots needed to establish this basis:

```
≫ T = ljx(T,1,1);

≫ T = ljx(T,2,2);
```

	x_4	x_5	x_3	1
$x_1 =$	2	-1	-3	2
$x_2 =$	-1	1	1	4
$z =$	0.5	0.5	1.5	8
$z_0 =$	-2	-1	0	-22

As $t \to -\infty$, the reduced cost row of the parametric tableau is $z + tz_0$, which is dominated by z_0. Since all the entries of the z_0 row are negative, all the reduced costs in the parametric tableaus become positive as t approaches $-\infty$, and so this tableau happens to be optimal for all t sufficiently negative. By examining the parametric objective $z + tz_0$ one component at a time, we find that it is nonnegative whenever

$$0.5 - 2t \geq 0, \qquad 0.5 - t \geq 0, \qquad 1.5 \geq 0,$$

that is, whenever $t \leq 1/4$. The corresponding optimal basis is $B = \{1, 2\}$, and therefore $x_B = [2, 4]'$ for all such values of t. By examining the elements at the bottom right of the tableau, we see that the objective function value is $z(t) = 8 - 22t$. Note that the solution vector x is the same for all $t \leq 1/4$, while the optimal objective value is a linear function of t.

As t increases through $1/4$, the reduced cost for x_4 becomes negative. We handle this event by doing what we usually do when we identify a negative cost—choose its column as the pivot column and use the ratio test to select a pivot row. We therefore pivot on the $(2, 1)$ element as follows:

```
≫ T = ljx(T,2,1);
```

	x_2	x_5	x_3	1
$x_1 =$	-2	1	-1	10
$x_4 =$	-1	1	1	4
$z =$	-0.5	1	2	10
$z_0 =$	2	-3	-2	-30

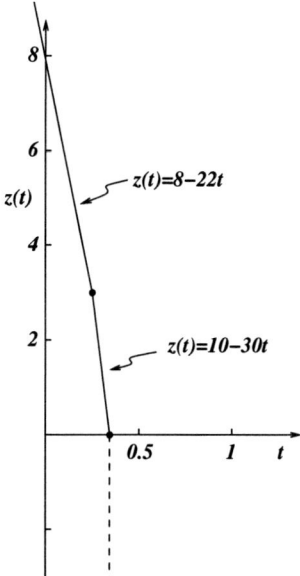

Figure 6.2. *Minimum parametric objective value as a function of t.*

In this tableau, we find that the reduced costs are positive, provided that all of the following three conditions are satisfied:

$$-0.5 + 2t \geq 0, \qquad 1 - 3t \geq 0, \qquad 2 - 2t \geq 0.$$

(As before, we obtain these conditions by examining the components of the reduced cost vector one at a time.) All three conditions are satisfied, provided that t satisfies $1/4 \leq t \leq 1/3$. The corresponding basis is $B = \{1, 4\}$, the solution is $x_B = [10, 4]'$, and the objective function value is $z(t) = 10 - 30t$.

As t increases through $1/3$, the reduced cost for x_5 becomes negative. We therefore seek to pivot on column 2 of the tableau, but the ratio test reveals no suitable pivot row. We conclude that the problem is unbounded for $t > 1/3$.

At this point, we have determined $z(t)$ for all real values of t, and it is possible to plot the objective function as a function of t. The resulting plot is shown in Figure 6.2. It can be seen that this function is *piecewise linear* and *concave*. In fact, we can easily prove a theorem to this effect. A fuller description of convex and concave functions can be found in Appendix A.

Theorem 6.3.1. *Let X be a nonempty set in $\mathbf{R}^n$, and let z(t) be defined by*

$$z(t) := \inf_{x \in X} (p + tq)'x.$$

Then $z \colon \mathbf{R} \to \mathbf{R} \cup \{-\infty\}$ is a concave function.

Proof. Let $\lambda \in (0, 1)$, and let t_1, $t_2 \in \mathbf{R}$. We need to show that $z((1 - \lambda)t_1 + \lambda t_2) \geq (1 - \lambda)z(t_1) + \lambda z(t_2)$. If $z(t_1) = -\infty$ or $z(t_2) = -\infty$, there is nothing to prove, and so suppose they are both finite. Then

$$
\begin{aligned}
z((1 - \lambda)t_1 + \lambda t_2) &= \inf_{x \in X} (p + ((1 - \lambda)t_1 + \lambda t_2)q)'x \\
&= \inf_{x \in X} ((1 - \lambda)p + \lambda p + ((1 - \lambda)t_1 + \lambda t_2)q)'x \\
&\geq \inf_{x \in X} (1 - \lambda)(p + t_1 q)'x + \inf_{x \in X} \lambda(p + t_2 q)'x \\
&= (1 - \lambda)z(t_1) + \lambda z(t_2).
\end{aligned}
$$

The inequality above follows from the fact that

$$
\inf_{x \in X} (f(x) + g(x)) \geq \inf_{x \in X} f(x) + \inf_{x \in X} g(x).
$$

The last equality in the sequence above follows from $\lambda \in (0, 1)$. $\square$

A simple application of this result shows that if $z(t)$ switches from finite to $-\infty$ as t increases through some value, then z will remain at $-\infty$ for all larger values of t. Similarly, if z becomes $-\infty$ as t is decreased through some value, it will remain at $-\infty$ for all smaller values of t.

A general approach for solving linear programs with parametric cost functions is summarized below.

1. Choose some starting value of t. Solve the linear program, using the simplex method, with the cost vector fixed at $p + tq$. If the problem is infeasible, then it is infeasible for all values of t (since the constraints do not depend on t). If the problem is unbounded, choose a different t and restart. Otherwise, determine the range (t_L, t_U) on which the basis is optimal, and determine the solution and the parametrized objective value $z(t)$ on this range.

2. If $t_U < \infty$, determine which component(s) of the reduced cost vector change their sign as t increases through t_U. Perform the pivots needed to restore optimality or establish unboundedness. If unbounded, set $z(t) = -\infty$ for all values $t > t_U$ and go to Step 3. Otherwise, determine the new range (t_L, t_U) on which this basis is optimal, and determine the solution and the parametrized objective value $z(t)$ on this range. Repeat this step as necessary until $t_U = \infty$.

3. Return to the lower limit t_L identified at the initial value of t.

4. If $t_L > -\infty$, determine which components of the reduced cost vector become negative as t decreases through t_L. Perform the pivots needed to restore optimality or establish unboundedness. If unbounded, set $z(t) = -\infty$ for all $t < t_L$ and stop. Otherwise, determine the new range (t_L, t_U) on which this basis is optimal, and determine the solution and the parametrized objective value $z(t)$ on this range. Repeat this step as necessary until $t_L = -\infty$.

Note that if the problem is feasible but unbounded for *all* t (so that we cannot identify a t for which a solution exists in Step 1), then we have by Theorem 4.4.2 (case (ii)) that the dual of (6.6) is infeasible for all t.

In Step 2, each time we move to a new interval (t_L, t_U), the new lower bound t_L equals the old upper bound t_U. A similar observation can be made about Step 4.

We can now use the procedure above to prove the piecewise linearity of $z(t)$.

Corollary 6.3.2. *Let X and $z(t)$ be as defined in Theorem 6.3.1. If X is also polyhedral, then z is a piecewise-linear concave function.*

Proof. We know already from Theorem 6.3.1 that z is concave. When $z(t)$ is finite, we have from Theorem 3.6.1 that the simplex method will identify a (vertex) solution of the problem. Since X has only a finite number of vertices, the procedure above can identify only a finite number of vertex solutions. Furthermore, if a vertex is optimal at t_1 and t_2, where $t_1 < t_2$, then that vertex is also optimal at any point in the interval (t_1, t_2) (this follows easily from the linearity of the objective). Hence the range $(-\infty, \infty)$ can be partitioned into a finite number of subintervals separated by "breakpoints" $t_1, t_2, t_3, \ldots, t_M$ at which the solution either switches from $-\infty$ to finite or switches from one vertex to another.

Suppose that x_i is a vertex that solves the problem over one of these subintervals $[t_i, t_{i+1}]$. We then have $z(t) = p'x_i + tq'x_i$ for $t \in [t_i, t_{i+1}]$, and so $z(t)$ is linear on this interval. If $z(t)$ is finite on the next subinterval in the sequence, then $z(t)$ is continuous across the breakpoint t_{i+1}, because both x_i and x_{i+1} solve the problem (with the same objective value) when $t = t_{i+1}$. $\quad\square$

Exercise 6-3-1. Form the appropriate matrices A, p, q, solve the following problem for $0 \le t < \infty$, and plot the optimal objective value $z(t)$ as a function of t:

$$
\begin{aligned}
\min \quad & (1-t)x_1 - x_2 \\
\text{subject to} \quad & x_1 \; - \; 2x_2 \; + \; 2 \; \ge \; 0, \\
& \quad\;\; - \;\; x_2 \; + \; 4 \; \ge \; 0, \\
& x_1, x_2 \qquad\qquad\;\; \ge \; 0.
\end{aligned}
$$

Exercise 6-3-2. Consider the linear programming problem

$$
\begin{aligned}
\min \quad & -3x_1 + (4 - \theta)x_2 - x_3 + 15x_4 \\
\text{subject to} \quad & x_1 \; + \; 2x_2 \; + \; x_3 \; + \; 2x_4 \; = \; 2, \\
& -2x_1 \qquad\quad - \; x_3 \; + \; 5x_4 \; = \; -3, \\
& x_1, x_2, x_3, x_4 \; \ge \; 0.
\end{aligned}
$$

Determine the optimal x and the optimal objective value for each real value of θ. Show your work. (Suggestion: use the initial value $\theta = 0$ and the initial basis $\{1, 3\}$.)

Exercise 6-3-3. Solve the following problem for all t:

$$
\begin{aligned}
\min \quad & -8x_1 + 10x_2 + t(x_2 - x_1) \\
\text{subject to} \quad & x_1 \; + \; 2x_2 \; \ge \; 4, \\
& x_1, x_2 \; \ge \; 0.
\end{aligned}
$$

Use Phase I to obtain a feasible initial tableau.

6.4 Parametric Optimization of the Right-Hand Side

We consider now the case in which the right-hand side depends linearly on a parameter t, so that it has the form $b + th$ for some fixed vectors b and h. We can apply duality to the results of the previous section to obtain theoretical results. The computational technique for tracking the solution as t is varied is also related to the technique of the previous section through duality.

The problem we consider is as follows:

$$z(t) = \min_x p'x \quad \text{subject to} \quad Ax \geq b + th, \quad x \geq 0, \tag{6.7}$$

for each t in some specified range. The dual of this problem is

$$w(t) = \max_u \left\{ (b + th)'u \mid A'u \leq p, \quad u \geq 0 \right\}$$
$$= -\min_u \left\{ -(b + th)'u \mid -A'u \geq -p, \quad u \geq 0 \right\}.$$

The following result is an immediate corollary of Corollary 6.3.2.

Corollary 6.4.1. *The function* $z \colon \mathbf{R} \to \mathbf{R} \cup \{\infty\}$ *defined by*

$$z(t) := \inf \left\{ p'x \mid Ax \geq b + th, \quad x \geq 0 \right\}$$

is a piecewise-linear convex function.

We can prove this result by taking the dual and invoking the method of the previous section. The details are left as an exercise.

If the problem (6.7) is unbounded for some value of t, then it is unbounded or infeasible for *all* values of t. We see this fact by writing the dual of (6.7), which is

$$\max_u (b + th)'u \quad \text{subject to} \quad A'u \leq p, \quad u \geq 0. \tag{6.8}$$

The dual constraints do not depend on t, so that the problem is dual infeasible for some t if and only if it is dual infeasible for all t. Since, by strong duality, dual infeasibility means that the primal is either infeasible or unbounded, our claim is proved.

We now outline a computational technique based on the dual simplex method for solving problems of the form (6.7). Again we work with a variant of Example 6-1-1, in which we introduce the variable component $h = (1, -2)'$ in the right-hand side. The problem is then as follows:

$$
\begin{array}{llrcrcrclcr}
\min & x_1 & + & 1.5x_2 & + & 3x_3 & & & & & \\
\text{subject to} & x_1 & + & x_2 & + & 2x_3 & \geq & 6 & + & t, \\
& x_1 & + & 2x_2 & + & x_3 & \geq & 10 & - & 2t, \\
& & & x_1, x_2, x_3 & \geq & 0. & & & &
\end{array}
$$

Suppose that we want to find the solutions for all $t \in (-\infty, \infty)$. We set up the tableau as usual, except that we add an extra column containing $-h$ and labeled with t. (Note that $-h$ is used in the tableau because h is moved from the right-hand side.)

```
>> load ex6-1-1
>> T = totbl(A,b,p);
>> T = addcol(T,[-h; 0],'t',5);
```

		x_1	x_2	x_3	1	t
x_4	=	1	1	2	-6	-1
x_5	=	1	2	1	-10	2
z	=	1	1.5	3	0	0

We start by considering what happens as $t \to -\infty$, applying either the primal or dual simplex method to obtain a tableau that is optimal for all t sufficiently negative. In this particular case, since the cost vector is already nonnegative, it makes sense to apply the dual simplex method. For all sufficiently negative t, the combination of the last two columns in the tableau will be negative in the second component, and so we choose the second row of the tableau as the pivot row. The ratio test for dual simplex leads us to take the second column as the pivot column, and we obtain

```
>> T = ljx(T,2,2);
```

		x_1	x_5	x_3	1	t
x_4	=	0.5	0.5	1.5	-1	-2
x_2	=	-0.5	0.5	-0.5	5	-1
z	=	0.25	0.75	2.25	7.5	-1.5

It happens that this tableau is optimal for all sufficiently negative t. We can find the precise range of validity by noting that the basis is B = {4, 2} and the basic part of the solution is

$$x_B = \begin{bmatrix} -1 - 2t \\ 5 - t \end{bmatrix}.$$

All components of x_B are nonnegative if $t \le -1/2$ and so we conclude that this basis is optimal for $t \in (-\infty, -1/2]$. By inspecting the elements in the bottom right of the tableau, we see that the optimal objective value is $z(t) = 7.5 - 1.5t$. Note that, unlike the case of the parametric cost vector, the solution changes as we move across this interval. However, as we could see by adding dual labels to the tableau and reading off the values of these variables, the optimal *dual* vertex does not change across this interval.

Consider now what happens as t increases through $-1/2$. The first component of the final tableau column becomes negative, and so we seek to pivot on the first row. We identify the $(1, 1)$ element as a pivot element and obtain the following:

```
>> T = ljx(T,1,1);
```

		x_4	x_5	x_3	1	t
x_1	=	2	-1	-3	2	4
x_2	=	-1	1	1	4	-3
z	=	0.5	0.5	1.5	8	-0.5

This basis B = {1, 2} is optimal if

$$2 + 4t \ge 0, \qquad 4 - 3t \ge 0,$$

that is, when $t \in [-1/2, 4/3]$, with objective value $z(t) = 8 - 0.5t$. As t increases through $4/3$, then x_2 becomes negative. The resulting dual simplex pivot on element $(2, 2)$ yields the following:

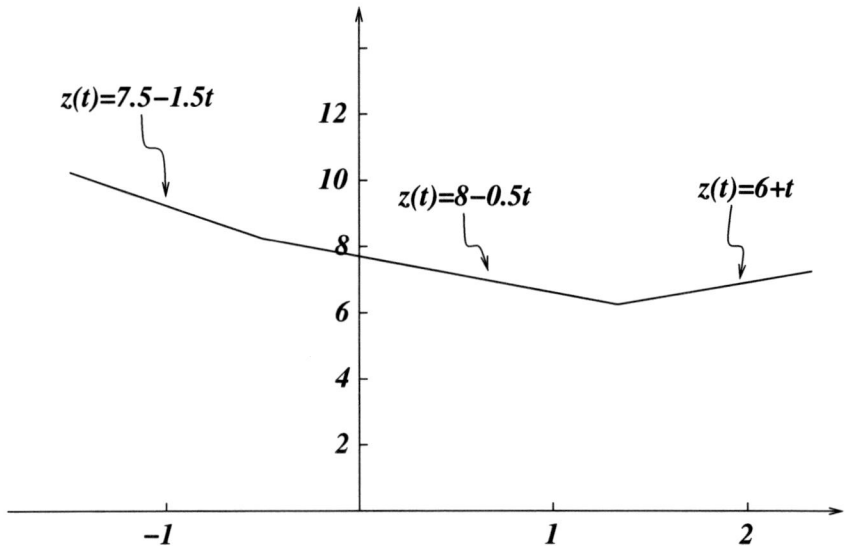

Figure 6.3. *Minimum parametric objective value for parametrized right-hand side.*

$\gg$ T = ljx(T,2,2);

		x_4	x_2	x_3	1	t
x_1	=	1	-1	-2	6	1
x_5	=	1	1	-1	-4	3
z	=	1	0.5	1	6	1

This basis $B = \{1, 5\}$ is feasible if

$$6 + t \geq 0, \qquad -4 + 3t \geq 0,$$

that is, $t \in [4/3, \infty)$, with objective value $z(t) = 6 + t$.

A plot of the objective function values $z(t)$ is given in Figure 6.3.

A computational procedure similar to that of the previous section can be devised to find the solution and the optimal objective $z(t)$ across the full range of t values.

1. Choose some starting value of t. Solve the linear program, using the simplex method, with the right-hand side fixed at $b + th$. If the problem is infeasible, choose a different value of t and restart. If the problem is unbounded, then it is unbounded for all t (see below). Otherwise, determine the range (t_L, t_U) on which the basis is optimal, and determine the solution $x(t)$ and the parametrized objective value $z(t)$ on this range.

2. If $t_U < \infty$, determine which component(s) of the basic part of the solution become negative as t increases through t_U. Perform the dual simplex pivots needed to restore optimality or establish infeasibility. If infeasible, set $z(t) = +\infty$ for all values $t > t_U$ and go to Step 3. Otherwise, determine the new range (t_L, t_U) on which this basis is optimal, and determine the solution and the parametrized objective value $z(t)$ on this range. Repeat this step as necessary until $t_U = \infty$.

3. Return to the lower limit t_L identified at the initial value of t.

4. If $t_L > -\infty$, determine which components of the basic part of the solution become negative as t decreases through t_L. Perform the pivots needed to restore optimality or establish infeasibility. If infeasible, set $z(t) = +\infty$ for all $t < t_L$ and stop. Otherwise, determine the new range (t_L, t_U) on which this basis is optimal, and determine the solution and the parametrized objective value $z(t)$ on this range. Repeat this step as necessary until $t_L = -\infty$.

To justify the statement in Step 1 that if the problem is unbounded for some t, it is unbounded for all t, we make use again of strong duality. If (6.7) is feasible but unbounded for some t, then by Theorem 4.4.2(ii) the dual of (6.7) is infeasible for this t. But since the dual constraints do not depend on t, the dual is infeasible for *all* t. Hence, by using Theorem 4.4.2(ii) again, we conclude that (6.7) is unbounded for all t.

In Step 1, it is possible to find a value of t for which the problem (6.7) is feasible by means of a Phase I problem, rather than trial and error. We introduce the artificial variable x_0 and solve the following problem:

$$\min_{x,t,x_0} x_0 \quad \text{subject to} \quad Ax + x_0 e \ge b + th, \quad x \ge 0, \quad x_0 \ge 0, \tag{6.9}$$

where $e = (1, 1, \ldots, 1)'$. Note that the parameter t is a *free variable* in this problem. If we have $x_0 = 0$ at the solution of this problem, we can choose the value of t from its optimum as the starting value in Step 1 and proceed. If, on the other hand, $x_0 > 0$ at the solution, then the original problem (6.7) is infeasible for all t.

Exercise 6-4-1. Consider the parametric linear program (6.7) with the following data:

$$p = \begin{bmatrix} 4 \\ 2 \end{bmatrix}, \quad A = \begin{bmatrix} 1 & -2 \\ -2 & 1 \end{bmatrix}, \quad b = \begin{bmatrix} -2 \\ -1 \end{bmatrix}, \quad h = \begin{bmatrix} -1 \\ 1 \end{bmatrix}.$$

Determine $z(t)$ for all t. (Include the ranges of t, if any, on which the problem is infeasible ($z(t) = \infty$) or unbounded ($z(t) = -\infty$).)

Exercise 6-4-2. Let $z(t)$ be the solution of

$$
\begin{array}{rlrcl}
\min & 2x_1 + x_2 - 4 & & & \\
\text{subject to} & x_1 + x_2 & \ge & 6 & - \ 2t, \\
& x_1 - x_2 & \ge & 2 & - \ t, \\
& x_1 & \ge & 0. &
\end{array}
$$

(Note that x_2 is a free variable.) Find $z(t)$ for all values of t. What properties does $z(t)$ have (as a function of t)? Is the solution vector $x(t)$ uniquely determined for each value of t? Justify your claim.

Exercise 6-4-3. Consider the linear programming problem

$$\max_u b'u \quad \text{subject to} \quad A'u \le c, \quad u \ge 0,$$

where

$$A' = \begin{bmatrix} 2 & 1 & 1 \\ 1 & 4 & 0 \end{bmatrix}, \quad c = \begin{bmatrix} 3 \\ 4 \end{bmatrix}, \quad b = \begin{bmatrix} 2 \\ 1 \\ 3 \end{bmatrix}.$$

Let

$$\Delta c = \begin{bmatrix} -1 \\ 4 \end{bmatrix} \quad \text{and} \quad \Delta b = \begin{bmatrix} 1 \\ -8 \\ 4 \end{bmatrix}.$$

Let $f(\theta)$ be the optimal value of the linear program when b is replaced by $b + \theta \Delta b$, and let $g(\theta)$ be the optimal value of the linear program when c is replaced by $c + \theta \Delta c$. Evaluate $f(\theta)$ and $g(\theta)$ as functions of θ. (Hint: the original problem can be thought of as the dual to a standard-form problem.)

Exercise 6-4-4. Consider the problem

$$
\begin{array}{rrcrcl}
\min & 3x_1 & - & 2x_2 & & \\
\text{subject to} & x_1 & - & x_2 & \geq & 7 + t, \\
& -x_1 & & & \geq & -5 + t, \\
& & x_1, x_2 & \geq & 0. &
\end{array}
$$

1. Set up and solve the problem (6.9) to find a value of t for which this problem is feasible.

2. Starting from this value of t, find the solutions of this problem for all values of t.

Chapter 7

Quadratic Programming and Complementarity Problems

In this chapter, we consider some extensions of linear programming to other mathematical programming problems. We first look at nonlinear problems, deriving optimality conditions based solely on calculus arguments and convexity properties. A reader may wish to review the material in Appendix A before proceeding. We then focus on quadratic programming (Section 7.2), an optimization problem that encompasses linear programming as a special case and that is being used in an increasingly wide range of interesting applications, such as least-squares approximation, classification, portfolio optimization, and optimal control. We investigate the optimality conditions of quadratic programs in Section 7.2.2 and look at the basic duality relationships in Section 7.2.3. To solve quadratic programs, one can apply Lemke's method, discussed in Section 7.4. Lemke's method was designed originally for linear complementarity problems (LCPs) (Section 7.3), which include convex quadratic programming as a special case. LCPs are studied in detail in the classic text of Cottle, Pang & Stone (1992); they are one of the most fundamental and elegant constructs of mathematical programming. They arise not only from optimality conditions for linear and quadratic programming but also from game theory applications and equilibrium problems such as those made famous by J. Nash and others.

7.1 Nonlinear Programs: Optimality Conditions

In this section, we consider the following optimization problem:

$$\min f(x) \quad \text{subject to} \quad x \in S, \tag{7.1}$$

where S is a convex subset of $\mathbf{R}^n$ and $f : \mathbf{R}^n \to \mathbf{R}$ is a smooth, possibly nonlinear function. We normally term S the *feasible region* of (7.1). (As shown in Appendix A, the feasible region of a linear program is in fact a convex set.)

We say that $\bar{x}$ is a *local solution* of (7.1) if it is feasible and has a function value no larger than that of all other x in a small feasible neighborhood. Formally, there is some $\delta > 0$ such that

$$f(x) \geq f(\bar{x}), \quad \text{whenever } \|x - \bar{x}\| < \delta \quad \text{and} \quad x \in S,$$

where $\| \cdot \|$ is any norm on $\mathbf{R}^n$ (see Section A.6). If $f(\bar{x}) \leq f(x)$ for *all* feasible x, we say that $\bar{x}$ is a *global solution* of (7.1).

Example 7-1-1. The problem

$$\min(x_1 - 1)^2 - x_2^2 \quad \text{subject to} \quad x_1 \in \mathbf{R}, \ -1 \leq x_2 \leq 2,$$

has two local solutions, namely $(1, -1)'$ and $(1, 2)'$. The solution at $(1, 2)'$ is the global solution. ∎

We now determine optimality conditions for (7.1). These conditions rely on f being a continuously differentiable function and use Taylor's theorem (Section A.7) to provide a linear approximation to the function f. To prove that the conditions are sufficient for optimality, a further assumption of convexity of f is required. See Appendix A for further discussion of convex sets, convex functions, and linear approximations.

Proposition 7.1.1. *Let S be a convex set in $\mathbf{R}^n$ and f be a continuously differentiable function.*

(a) *If $\bar{x}$ is a local solution of (7.1), then $\bar{x}$ satisfies*

$$\bar{x} \in S \quad and \quad \nabla f(\bar{x})'(x - \bar{x}) \geq 0 \qquad \forall x \in S. \tag{7.2}$$

(b) *If $\bar{x}$ satisfies (7.2) and f is convex, then $\bar{x}$ is a global solution of (7.1).*

The condition (7.2) is known as a minimum principle.

Proof. For part (a), let $\delta > 0$ be the constant associated with local optimality. Choose $x \in S$ arbitrarily, and define

$$y(\lambda) = (1 - \lambda)\bar{x} + \lambda x = \bar{x} + \lambda(x - \bar{x})$$

for $\lambda \in [0, 1]$. By the convexity of S, $y(\lambda) \in S$, and so for all sufficiently small λ, we have

$$\|y(\lambda) - \bar{x}\| < \delta.$$

Therefore, by local optimality, we have

$$f(y(\lambda)) - f(\bar{x}) \geq 0$$

for all sufficiently small λ. By using this inequality together with the definition of $y(\lambda)$ and formula (A.10), we obtain

$$\lambda \nabla f(\bar{x})'(x - \bar{x}) + o(\lambda) \geq 0,$$

where $o(\lambda)/\lambda \to 0$ as $\lambda \to 0$; see Appendix A.6. Dividing both sides by $\lambda > 0$ and letting $\lambda \to 0$, we obtain

$$\nabla f(\bar{x})'(x - \bar{x}) \geq 0,$$

as required.

We turn now to part (b). It follows from the convexity of f and Proposition A.7.2 that

$$f(x) \geq f(\bar{x}) + \nabla f(\bar{x})'(x - \bar{x})$$

for all x. Since $\bar{x}$ satisfies (7.2), we deduce that $f(x) \geq f(\bar{x})$ for all $x \in S$, which is the required result. □

Example 7-1-2. For the example given above,

$$\nabla f(x) = \left[\begin{array}{c} 2(x_1 - 1) \\ -2x_2 \end{array} \right],$$

and hence the minimum principle states for the local solution at $(1, -1)'$ that

$$2(x_2 + 1) \geq 0 \qquad \forall x_2 \in [-1, 2]$$

and for the global solution at $(1, 2)$ that

$$-4(x_2 - 2) \geq 0 \qquad \forall x_2 \in [-1, 2].$$

Note that since f is not convex (its Hessian matrix is indefinite), Proposition 7.1.1(b) does not apply. ∎

We now consider two special choices of the feasible region S and examine the specialization of Proposition 7.1.1 to these cases.

Corollary 7.1.2. *Suppose the feasible region of the problem is the whole space, that is,* $S = \mathbf{R}^n$ *and* f *is continuously differentiable.*

(a) *If* $f(\bar{x}) \leq f(x)$ *for all* $x \in \mathbf{R}^n$, *then* $\nabla f(\bar{x}) = 0$.

(b) *If* f *is convex and* $\nabla f(\bar{x}) = 0$, *then* $f(\bar{x}) \leq f(x)$ *for all* $x \in \mathbf{R}^n$.

Proof. Since $\bar{x}$ is a global (and hence local) solution of (7.4), we have from Proposition 7.1.1(a) that $\nabla f(\bar{x})'(x - \bar{x}) \geq 0$ for all $x \in \mathbf{R}^n$. Setting $x = \bar{x} - \nabla f(\bar{x})$, we have $\nabla f(\bar{x})'(x - \bar{x}) = -\|\nabla f(\bar{x})\|_2^2 \geq 0$, which can be true only if $\nabla f(\bar{x}) = 0$. Hence we have proved (a). Part (b) follows immediately from Proposition A.5.1(b). □

Corollary 7.1.3. *Suppose the feasible region is the nonnegative orthant, that is,* $S = \{x \in \mathbf{R}^n \mid x \geq 0\}$ *and* f *is continuously differentiable.*

(a) *If* $\bar{x}$ *is a global solution of (7.1) (that is,* $f(\bar{x}) \leq f(x)$ *for all* $x \in S$), *then* $\nabla f(\bar{x}) \geq 0$, $\bar{x} \geq 0$, *and* $\nabla f(\bar{x})'\bar{x} = 0$.

(b) *If* f *is convex and* $\nabla f(\bar{x}) \geq 0$, $\bar{x} \geq 0$, *and* $\nabla f(\bar{x})'\bar{x} = 0$, *then* $\bar{x}$ *is a global solution of (7.1).*

Proof. For (a), since $\bar{x}$ is a global (and hence local) minimizer, we have from Proposition 7.1.1(a) that

$$\nabla f(\bar{x})'(x - \bar{x}) \geq 0 \qquad \forall x \geq 0. \tag{7.3}$$

By setting $x = 0$, we have $\nabla f(\bar{x})'\bar{x} \leq 0$, while by setting $x = 2\bar{x}$, we have $\nabla f(\bar{x})'\bar{x} \geq 0$. Hence, $\nabla f(\bar{x})'\bar{x} = 0$, as claimed.

To show that $\nabla f(\bar{x}) \geq 0$, suppose for contradiction that there exists a component i such that $[\nabla f(\bar{x})]_i < 0$. We define $x \geq 0$ as follows:

$$x_j = \bar{x}_j \quad (j \neq i), \qquad x_i = \bar{x}_i - [\nabla f(\bar{x})]_i > 0.$$

By substituting into (7.3), we obtain

$$0 \leq \nabla f(\bar{x})'(x - \bar{x}) = \sum_{j=1}^{n} [\nabla f(\bar{x})]_j (x_j - \bar{x}_j) = -[\nabla f(\bar{x})]_i^2 < 0,$$

which is a contradiction. Therefore, there is no component i with $[\nabla f(\bar{x})]_i < 0$, and so $\nabla f(\bar{x}) \geq 0$, as claimed.

For (b), the convexity of f and Proposition A.7.2 imply that

$$f(x) \geq f(\bar{x}) + \nabla f(\bar{x})'(x - \bar{x})$$

for all $x \geq 0$. The second term is nonnegative by assumption, so that $f(x) \geq f(\bar{x})$ for all $x \geq 0$. $\square$

Exercise 7-1-3. Apply Corollaries 7.1.2 and 7.1.3 to $f(x) = x^3$ and $f(x) = x^4$.

7.2   Quadratic Programming

In this section, we specialize the results of the previous section to the following quadratic programming problem:

$$\min f(x) := \tfrac{1}{2} x' Q x + p' x \quad \text{subject to} \quad x \in S, \tag{7.4}$$

where Q is a symmetric $n \times n$ matrix and S is a polyhedral subset of $\mathbf{R}^n$, that is, a set defined by a finite number of equality and inequality constraints. It is a *convex quadratic program* when, in addition, Q is a positive semidefinite matrix.

As indicated in Section A.5, any nonsymmetric Q can be replaced by $(1/2)(Q + Q')$ without changing $f(x)$, and so our assumption that Q is symmetric can be made without loss of generality. According to the definitions of Section A.3, the gradient and Hessian of f are as follows:

$$\nabla f(x) = Qx + p, \qquad \nabla^2 f(x) = Q.$$

(Note that these formulae hold only if Q is symmetric!)

Exercise 7-2-1. Find the symmetric matrix $Q \in \mathbf{R}^{3 \times 3}$ and the vector $p \in \mathbf{R}^3$ such that the following function can be written in the form of (7.4):

$$f(x_1, x_2, x_3) = x_1 - x_3 + 3x_1^2 - 2x_1 x_2 + x_2^2 - 2x_2 x_3 + 4x_1 x_3 + 4x_3^2.$$

7.2.1   Basic Existence Result

The following result gives conditions that ensure the existence of a global solution to a quadratic program and generalizes the existence result of linear programming.

Theorem 7.2.1 (Frank–Wolfe Existence Theorem). *Let S be a nonempty polyhedral set and f given by (7.4). Then either (7.4) has a global solution, or else there is a feasible half line in S along which f approaches $-\infty$ (that is, there are vectors x and $d \neq 0$ such that $x(\lambda) := x + \lambda d \in S$ for all $\lambda \geq 0$ and $f(x(\lambda)) \to -\infty$ as $\lambda \to \infty$). Hence, if f is bounded below on S, it attains its minimum value on S.*

Proof. See Frank & Wolfe (1956). ☐

Note that this result does not require Q to be positive semidefinite, nor S to be a bounded set. Note also that the result does not hold for more general nonlinear functions f, not even smooth convex functions. A simple counterexample is the problem

$$\min_{x \geq 0} e^{-x},$$

which has a strictly convex objective yet does not have a minimizer.

As a particular example of a quadratic program, we now consider the problem of projection onto a nonempty convex set. Simply stated, this problem takes a closed convex set S and a vector z and seeks the point $x \in$ S that is closest to z in the Euclidean sense. The obvious formulation is

$$\min \|x - z\|_2 \quad \text{subject to} \quad x \in \text{S}.$$

Since squaring the objective does not change the answer, we can reformulate this problem equivalently as the following quadratic program:

$$\min (x - z)'(x - z) \quad \text{subject to} \quad x \in \text{S}. \tag{7.5}$$

The quadratic objective function has $Q = 2I$, which is positive definite, and so by Proposition 7.1.1, the minimum principle (7.2) is both necessary and sufficient to characterize a solution of the problem which exists by invocation of the Frank–Wolfe theorem. The gradient of the objective function is $2(x - z)$ so that, from the minimum principle (7.2), we have that $\bar{x}$ solves the projection problem if and only if

$$(\bar{x} - z)'(y - \bar{x}) \geq 0 \quad \forall y \in \text{S}.$$

The point $\bar{x}$ is called the *projection of x onto* S.

7.2.2 KKT Conditions

In this section, we assume a particular representation of the polyhedral set S, that is,

$$\text{S} := \{x \mid Ax \geq b, x \geq 0\}. \tag{7.6}$$

(Note that S coincides with the feasible region for the standard-form linear program (1.2).) In this setting, (7.4) becomes

$$\min f(x) := \tfrac{1}{2} x' Q x + p' x \quad \text{subject to} \quad Ax \geq b, \quad x \geq 0. \tag{7.7}$$

The minimum principle of Section 7.1 is essential in establishing the classical KKT conditions for quadratic programming. These conditions are *necessary* for any quadratic program (7.7); that is, any local solution will satisfy them. They are *sufficient* for convex quadratic programs—if Q is positive semidefinite and x (along with dual variables u, typically called Lagrange multiplier vectors in this context) satisfies the KKT conditions, then x is a global solution of (7.7). The key to the proof is the duality theory of linear programming, which can be applied to a linearized problem arising from Proposition 7.1.1.

In the following equations, we use the notation "$\perp$" defined in (A.1) as a convenient shorthand for stating the complementarity condition; that is, the scalar product of two vectors of equal dimension is equal to zero.

Theorem 7.2.2 (KKT Conditions for Quadratic Programming). *If $\bar{x}$ is a local solution for (7.7), then there exists $\bar{u} \in \mathbf{R}^m$ such that*

$$0 \leq \bar{x} \perp Q\bar{x} - A'\bar{u} + p \geq 0, \tag{7.8a}$$

$$0 \leq \bar{u} \perp A\bar{x} - b \geq 0. \tag{7.8b}$$

Conversely, if a pair $(\bar{x}, \bar{u}) \in \mathbf{R}^n \times \mathbf{R}^m$ satisfies (7.8), and Q is positive semidefinite, then $\bar{x}$ is a global solution for (7.7).

Note that if $Q = 0$, then the KKT conditions above are precisely the KKT conditions (4.7), (4.8) for the linear programming problem in standard form.

Proof. Since $\bar{x}$ is a local solution of (7.7), it follows from Proposition 7.1.1 that $\bar{x}$ satisfies the minimum principle (7.2) and hence that $\bar{x}$ solves the linearized problem

$$\min_x \nabla f(\bar{x})'x \quad \text{subject to} \quad x \in S. \tag{7.9}$$

Note that (7.9) is a linear program because S is polyhedral (7.6). In fact, we can write (7.9) as follows:

$$\min_x (Q\bar{x} + p)'x \quad \text{subject to} \quad Ax \geq b, \quad x \geq 0. \tag{7.10}$$

Since $\bar{x} \in S$, we have

$$A\bar{x} \geq b, \qquad \bar{x} \geq 0.$$

By the strong duality theorem of linear programming (Theorem 4.4.2) applied to (7.10), there is some $\bar{u} \in \mathbf{R}^m$ which is feasible for the dual of (7.10), that is,

$$A'\bar{u} \leq Q\bar{x} + p, \qquad \bar{u} \geq 0.$$

Furthermore, the complementarity condition (4.8) for (7.10) also holds, so that

$$\bar{x}'(Q\bar{x} + p - A'\bar{u}) = 0 \quad \text{and} \quad \bar{u}'(A\bar{x} - b) = 0.$$

By combining all the relations above, we obtain (7.8). Hence, we have established the necessity of the KKT conditions.

Conversely, suppose that $\bar{x}$ and $\bar{u}$ satisfy (7.8). Then, by Theorem 4.5.1, it follows that $\bar{x}$ solves (7.10), and hence $\nabla f(\bar{x})'(x - \bar{x}) \geq 0$, for all $x \in S$. Thus, by Proposition 7.1.1(b), $\bar{x}$ is a global solution of (7.7). $\square$

Exercise 7-2-2. Consider the following linear program:

$$
\begin{array}{rrcrcr}
\min & 8x_1 & - & x_2 \\
\text{subject to} & x_1 & - & 2x_2 & \geq & -2, \\
& x_1 & - & x_2 & \geq & -7, \\
& & & x_1, x_2 & \geq & 0.
\end{array}
$$

1. Solve this problem.

2. Write down the dual of the given problem and the KKT conditions.

3. Find a dual solution u^* (by inspection of the KKT conditions).

4. Suppose that the objective function is replaced by the following quadratic:

$$\alpha x_1^2 + \beta x_2^2 + 8x_1 - x_2,$$

where α and β are nonnegative parameters. Write down the modified KKT conditions for the resulting problem.

5. How large can we make α and β before the solution of the quadratic problem becomes different from the solution of the original linear program?

Exercise 7-2-3. Consider the following linear program, obtained by omitting the quadratic term from (7.7):

$$\min_x \; p'x \quad \text{subject to} \quad Ax \geq b, \quad x \geq 0. \tag{7.11}$$

Are the following statements true or false? Prove your claim or give a counterexample.

1. If (7.11) is solvable, then so is the quadratic program (7.7) for *any* $n \times n$ positive semidefinite matrix Q. (Hint: use the Frank–Wolfe theorem (Theorem 7.2.1)).

2. The linear program (7.11) is solvable whenever the quadratic program (7.7) is solvable for *some* $n \times n$ positive semidefinite matrix Q.

Finally, we consider a generalized formulation of convex quadratic programming, containing both equality and inequality constraints and both nonnegative and free variables. This formulation can be viewed as an extension of the general linear program discussed in Section 4.7. We consider the following problem:

$$
\begin{aligned}
\min_{x,y} \quad & \tfrac{1}{2}\begin{bmatrix} x' & y' \end{bmatrix} \begin{bmatrix} Q & R \\ R' & T \end{bmatrix} \begin{bmatrix} x \\ y \end{bmatrix} + \begin{bmatrix} p' & q' \end{bmatrix} \begin{bmatrix} x \\ y \end{bmatrix} \\
\text{subject to} \quad & Bx + Cy \;\geq\; d, \\
& Ex + Fy \;=\; g, \\
& x \;\geq\; 0,
\end{aligned}
\tag{7.12}
$$

where the matrix

$$\begin{bmatrix} Q & R \\ R' & T \end{bmatrix}$$

is symmetric and positive semidefinite. The KKT conditions are as follows:

$$0 \leq \bar{x} \perp Q\bar{x} + R\bar{y} + p - B'\bar{u} - E'\bar{v} \geq 0, \tag{7.13a}$$

$$R'\bar{x} + T\bar{y} + q - C'\bar{u} - F'\bar{v} = 0, \tag{7.13b}$$

$$0 \leq \bar{u} \perp B\bar{x} + C\bar{y} - d \geq 0, \tag{7.13c}$$

$$E\bar{x} + F\bar{y} = g. \tag{7.13d}$$

7.2.3 Duality

Nonlinear programming, particularly in the convex case, has a rich duality theory that extends many of the results that were given in Chapter 4. We outline below a duality construction technique that we find especially pleasing, within the context of quadratic programming, and specifically the formulation (7.7). Much of this construction can be applied in a more general setting, but we leave details of this for further study.

The first step is to construct a *Lagrangian*, which is a weighted combination of the objective function and the constraints of the problem. The weights on the constraints are the dual variables and are more frequently referred to as *Lagrange multipliers* in this context. For the problem given as (7.7), the Lagrangian is

$$\mathcal{L}(x, u, v) = \tfrac{1}{2}x'Qx + p'x - u'(Ax - b) - v'x,$$

where u are the Lagrange multipliers on the constraint $Ax \geq b$ and v are the Lagrange multipliers on the constraint $x \geq 0$.

The dual problem for (7.7), typically attributed to Wolfe, is defined as follows:

$$\max_{x,u,v} \mathcal{L}(x, u, v) \quad \text{subject to} \quad \nabla_x \mathcal{L}(x, u, v) = 0, \quad u \geq 0, \quad v \geq 0. \tag{7.14}$$

As was the case in linear programming, the weak duality result is easy to prove.

Theorem 7.2.3 (Weak Duality: Quadratic Programming). *Let Q be positive semidefinite, and suppose that $\bar{x}$ is feasible for the primal problem (7.7) and $(\hat{x}, \hat{u}, \hat{v})$ is feasible for the dual problem (7.14). Then*

$$\tfrac{1}{2}\bar{x}'Q\bar{x} + p'\bar{x} \geq \mathcal{L}(\hat{x}, \hat{u}, \hat{v}).$$

Proof. By primal feasibility of $\bar{x}$ and dual feasibility of $(\hat{x}, \hat{u})$ we have

$$A\bar{x} \geq b, \qquad \bar{x} \geq 0, \qquad Q\hat{x} - A'\hat{u} + p - \hat{v} = 0, \quad \hat{u} \geq 0, \quad \hat{v} \geq 0.$$

Consequently, we have

$$\begin{aligned}
\tfrac{1}{2}\bar{x}'&Q\bar{x} + p'\bar{x} - \mathcal{L}(\hat{x}, \hat{u}, \hat{v}) \\
&= \tfrac{1}{2}(\bar{x} - \hat{x})'Q(\bar{x} - \hat{x}) - \hat{x}'Q\hat{x} + \hat{x}'Q\bar{x} + p'\bar{x} - p'\hat{x} + \hat{u}'(A\hat{x} - b) + \hat{v}'\hat{x} \\
&= \tfrac{1}{2}(\bar{x} - \hat{x})'Q(\bar{x} - \hat{x}) + \hat{x}'Q\bar{x} + p'\bar{x} - \hat{u}'b \\
&\geq \hat{x}'Q\bar{x} + p'\bar{x} - b'\hat{u} \\
&\geq \bar{x}'A'\hat{u} - b'\hat{u} \\
&\geq 0.
\end{aligned}$$

The first equality is simply an identity; the second equality follows from the substitution $\hat{v} = Q\hat{x} - A'\hat{u} + p$; the first inequality follows from the positive semidefiniteness of Q; and the second inequality follows from $\bar{x} \geq 0$ and $Q\hat{x} - A'\hat{u} + p \geq 0$. The final inequality follows from $\hat{u} \geq 0$ and $A\bar{x} - b \geq 0$. $\square$

The strong duality result follows from Theorem 7.2.2.

Theorem 7.2.4 (Strong Duality: Quadratic Programming). *Let Q be positive semidefinite. If $\bar{x}$ is a solution of (7.7), then there is a $\bar{u} \in \mathbf{R}^m$, $\bar{v} \in \mathbf{R}^n$ such that $(\bar{x}, \bar{u}, \bar{v})$ solves (7.14), and the extrema of the two problems are equal.*

Proof. Since $\bar{x}$ solves (7.7), Theorem 7.2.2 ensures the existence of $\bar{u}$ such that $(\bar{x}, \bar{u}, \bar{v})$ is dual feasible (where $\bar{v} = Q\bar{x} - A'\bar{u} + p$), because of conditions (7.8). Hence,

$$\tfrac{1}{2}\bar{x}'Q\bar{x} + p'\bar{x} \geq \mathcal{L}(\bar{x}, \bar{u}, \bar{v})$$
$$= \tfrac{1}{2}\bar{x}'Q\bar{x} + p'\bar{x} - \bar{u}'(A\bar{x} - b) - \bar{x}'(Q\bar{x} - A'\bar{u} + p)$$
$$= \tfrac{1}{2}\bar{x}'Q\bar{x} + p'\bar{x}.$$

The inequality follows from the weak duality theorem (Theorem 7.2.3); the first equality is a simple identity; and the second equality follows from the complementarity conditions in (7.8). We conclude that the dual objective function $\mathcal{L}(\bar{x}, \bar{u}, \bar{v})$ evaluated at the dual feasible point $(\bar{x}, \bar{u}, \bar{v})$ equals its upper bound $\tfrac{1}{2}\bar{x}'Q\bar{x} + p'\bar{x}$, and therefore $(\bar{x}, \bar{u}, \bar{v})$ must solve the problem (7.14). $\square$

Typically, when formulating a dual problem, the equality constraints are used to eliminate the variables v from the problem. Thus, (7.14) is equivalent to the following problem:

$$\max \ -\tfrac{1}{2}x'Qx + b'u \quad \text{subject to} \quad Qx - A'u + p \geq 0, \quad u \geq 0.$$

Note that if $Q = 0$, the problems (7.7) and the above reduce to the dual linear programs (4.2) and (4.3).

7.3   Linear Complementarity Problems

The linear complementarity problem (LCP) is not an optimization problem, in that we are not aiming to minimize an objective function. Rather, we seek a vector that satisfies a given set of relationships, specifically, linear equality constraints, and nonnegativity and complementarity conditions.

Just as a linear program in standard form is fully specified by the constraint matrix A, the right-hand side b, and the cost vector c, the LCP is specified by a matrix $M \in \mathbf{R}^{n \times n}$ and a vector $q \in \mathbf{R}^n$. We seek a vector $z \in \mathbf{R}^n$ such that the following conditions are satisfied:

$$\text{LCP}(M, q): \quad w = Mz + q, \quad z \geq 0, \quad w \geq 0, \quad z'w = 0, \tag{7.15}$$

or equivalently

$$w = Mz + q, \quad 0 \leq z \perp w \geq 0.$$

LCPs arise in many applications; we will see some examples later in this chapter. They are also closely related to linear and quadratic programming, because the KKT conditions for these problems actually make up an LCP. From the KKT conditions for quadratic programming (7.8), we have by taking

$$z = \begin{bmatrix} \bar{x} \\ \bar{u} \end{bmatrix}, \qquad M = \begin{bmatrix} Q & -A' \\ A & 0 \end{bmatrix}, \qquad q = \begin{bmatrix} p \\ -b \end{bmatrix} \tag{7.16}$$

that these conditions are identical to LCP(M, q) defined in (7.15). Linear programs give rise to LCPs in the same manner by setting $Q = 0$. In this case, M is skew symmetric, that is, $M = -M'$.

The relationship outlined above suggests that we can find the primal and dual solutions to linear and complementarity problems simultaneously by applying an algorithm for solving LCPs to their KKT conditions. In the next section, we outline such an algorithm.

Exercise 7-3-1. Consider the linear program

$$\min q'z \quad \text{subject to} \quad Mz + q \geq 0, \ z \geq 0,$$

where M satisfies the skew-symmetric property $M = -M'$, and suppose that z^* solves this problem. Write down the dual of this problem and give a solution of the dual in terms of z^*.

7.4   Lemke's Method

The first algorithm proposed to solve LCPs was the famous pivotal algorithm of Lemke (1965). To describe Lemke's method for solving LCP(M, q) we introduce some definitions.

Definition 7.4.1. Consider the vector pair $(z, w) \in \mathbf{R}^n \times \mathbf{R}^n$.

(a) (z, w) is *feasible* for LCP(M, q) if $w = Mz + q, z \geq 0, w \geq 0$.

(b) A component w_i is called the *complement* of z_i, and vice versa, for $i = 1, 2, \ldots, n$.

(c) (z, w) is *complementary* if $z \geq 0$, $w \geq 0$, and $z'w = 0$. (Note that a complementary pair must satisfy $z_i w_i = 0$ for $i = 1, 2, \ldots, n$.)

(d) (z, w) is *almost complementary* if $z \geq 0$, $w \geq 0$, and $z_i w_i = 0$ for $i = 1, 2, \ldots, n$ *except* for a single index j, $1 \leq j \leq n$.

For positive semidefinite M, Lemke's method generates a finite sequence of feasible, almost-complementary pairs that terminates at a complementary pair or an unbounded ray. Similarly to the simplex method, an initial pair must first be obtained, usually via a Phase I scheme. There are a variety of Phase I schemes tailored to LCPs with particular structures. We describe here the most widely applicable scheme, which requires only one pivot.

Algorithm 7.1 (Phase I: Generate a Feasible Almost-Complementary Tableau).

1. *If $q \geq 0$, STOP: $z = 0$ is a solution of LCP(M, q); that is, $(z, w) = (0, q)$ is a feasible complementary pair.*

2. *Otherwise, add the artificial variables z_0 and w_0 that are constrained to satisfy the following relationships:*

$$w = Mz + ez_0 + q, \qquad w_0 = z_0,$$

where e is the vector of ones in $\mathbf{R}^n$. Create the initial tableau

$$
\begin{array}{rl}
 & \begin{array}{ccc} z & z_0 & 1 \end{array} \\
\begin{array}{r} w = \\ w_0 = \end{array} & \begin{array}{|cc|c|} \hline M & e & q \\ \hline 0 & 1 & 0 \\ \hline \end{array}
\end{array}
$$

3. *Make this tableau feasible by carrying out a* Jordan exchange *on the z_0 column and the row corresponding to the most negative q_i. (This step corresponds to the "special pivot" in Phase I of the simplex method for linear programming.) Without removing the artificial variables from the tableau, proceed to Phase II.*

Phase II generates a sequence of almost-complementary vector pairs. It performs a pivot at each iteration, selecting the pivot row by means of a ratio test like that of the simplex method, whose purpose is to ensure that the components of z and w remain nonnegative throughout the procedure.

Algorithm 7.2 (Phase II: Generate a Feasible Complementary or Unbounded Tableau).
Start with a feasible almost-complementary pair (z, w) and the corresponding tableau in Jordan exchange form

$$
\begin{array}{ccc}
 & \begin{array}{ccc} w_{l_1} & z_{J_2} & 1 \end{array} \\
\begin{array}{c} z_{J_1} = \\ w_{l_2} = \end{array} & \boxed{\begin{array}{cc|c} H_{l_1 J_1} & H_{l_1 J_2} & h_{l_1} \\ H_{l_2 J_1} & H_{l_2 J_2} & h_{l_2} \end{array}}
\end{array}
$$

Take note of the variable that became nonbasic (i.e., became a column label) at the previous iteration. (At the first step, this is simply the component of w that was exchanged with z_0 during Phase I.)

1. *Pivot column selection: Choose the column s corresponding to the* complement *of the variable that became nonbasic at the previous pivot.*

2. *Pivot row selection: Choose the row r such that*

$$-h_r/H_{rs} = \min_i \left\{ -h_i/H_{is} \mid H_{is} < 0 \right\}.$$

If all $H_{is} \geq 0$, STOP: An unbounded ray has been found.

3. *Carry out a Jordan exchange on element H_{rs}. If (z, w) is complementary, STOP: (z, w) is a solution. Otherwise, go to Step 1.*

Step 1 maintains almost-complementarity by moving a component into the basis as soon as its complement is moved out. By doing so, we ensure that for all except one of the components, exactly one of z_i and w_i is basic while the other is nonbasic. Since nonbasic variables are assigned the value 0, this fact ensures that $w_i z_i = 0$ for all except one component—the almost-complementary property. (When the initial tableau of Phase II was derived from Phase I, it is the variables w_0 and z_0 that violate complementarity until an optimal tableau is found.) The ratio test in Step 2 follows from the same logic as in the simplex method. We wish to maintain nonnegativity of all the components in the last column, and so we allow the nonbasic variable in column s to increase away from zero only until it causes one of the basic variables to become zero. This basic variable is identified by the ratio test as the one to leave the basis on this iteration.

In practice, it is not necessary to insert the w_0 row into the tableau, since w_0 remains in the basis throughout and is always equal to z_0. In the examples below, we suppress the w_0 row.

Example 7-4-1. Consider LCP(M, q) with

$$M = \begin{bmatrix} 1 & 0 \\ -1 & 1 \end{bmatrix}, \qquad q = \begin{bmatrix} -2 \\ -1 \end{bmatrix}.$$

```
>> load ex7-4-1
>> T = lemketbl(M,q);
```

		z_1	z_2	1
w_1	=	1	0	-2
w_2	=	-1	1	-1

Since $q \not\geq 0$, we require a Phase I. First, add the z_0 column:

```
>> T = addcol(T,[1 1]','z0',3);
```

		z_1	z_2	z_0	1
w_1	=	1	0	1	-2
w_2	=	-1	1	1	-1

Then pivot z_0 with the row corresponding to the most negative entry in the last column (row 1):

```
>> T = ljx(T,1,3);
```

		z_1	z_2	w_1	1
z_0	=	-1	0	1	2
w_2	=	-2	1	1	1

This tableau represents an almost-complementary solution. We can now execute the steps of Lemke's method, as outlined in Phase II above. Since w_1 became nonbasic at the last pivot step, we choose its complement z_1 to become basic. Hence, the pivot column is $s = 1$, and the ratio test identifies the pivot row $r = 2$.

```
>> T = ljx(T,2,1);
```

		w_2	z_2	w_1	1
z_0	=	0.5	-0.5	0.5	1.5
z_1	=	-0.5	0.5	0.5	0.5

Since this is still almost complementary, we carry out another pivot. Since w_2 left the basis at the last pivot, its complement z_2 enters at this pivot, that is, $s = 2$. The ratio test determines z_0 as the leaving variable, and hence we perform a pivot on the element in position $(1, 2)$.

```
>> T = ljx(T,1,2);
```

		w_2	z_0	w_1	1
z_2	=	1	-2	1	3
z_1	=	0	-1	1	2

The resulting (z, w) pair is feasible and complementary, and the solution to the LCP is $z_1 = 2, z_2 = 3.$ ∎

We now give two examples on how to set up and solve (or determine unsolvability of) a convex quadratic program.

Example 7-4-2. Consider the quadratic program

$$\min \tfrac{1}{2}x_1^2 - x_1 x_2 + \tfrac{1}{2}x_2^2 + 4x_1 - x_2 \quad \text{subject to} \quad x_1 + x_2 - 2 \geq 0, \quad x_1, x_2 \geq 0.$$

This problem has the form (7.7) when we set

$$Q = \begin{bmatrix} 1 & -1 \\ -1 & 1 \end{bmatrix}, \qquad A = \begin{bmatrix} 1 & 1 \end{bmatrix}, \qquad p = \begin{bmatrix} 4 \\ -1 \end{bmatrix}, \qquad b = \begin{bmatrix} 2 \end{bmatrix}.$$

Following (7.16), we define an LCP from the KKT conditions for this problem by setting

$$M = \begin{bmatrix} 1 & -1 & -1 \\ -1 & 1 & -1 \\ 1 & 1 & 0 \end{bmatrix}, \qquad q = \begin{bmatrix} 4 \\ -1 \\ -2 \end{bmatrix}, \qquad \begin{bmatrix} z_1 \\ z_2 \\ z_3 \end{bmatrix} = \begin{bmatrix} x_1 \\ x_2 \\ u_1 \end{bmatrix}.$$

We now show how to solve the problem in MATLAB.

```
>> M = [1 -1 -1;-1 1 -1;1 1 0];
>> q = [4 -1 -2]';
>> T = lemketbl(M,q);
```

		z_1	z_2	z_3	1
w_1	=	1	-1	-1	4
w_2	=	-1	1	-1	-1
w_3	=	1	1	0	-2

```
>> T = addcol(T,[1 1 1]','z0',4);
```

		z_1	z_2	z_3	z_0	1
w_1	=	1	-1	-1	1	4
w_2	=	-1	1	-1	1	-1
w_3	=	1	1	0	1	-2

```
>> T = ljx(T,3,4);
```

		z_1	z_2	z_3	w_3	1
w_1	=	0	-2	-1	1	6
w_2	=	-2	0	-1	1	1
z_0	=	-1	-1	0	1	2

```
>> T = ljx(T,2,3);
```

		z_1	z_2	w_2	w_3	1
w_1	=	2	-2	1	0	5
z_3	=	-2	0	-1	1	1
z_0	=	-1	-1	0	1	2

```
>> T = ljx(T,3,2);
```

		z_1	z_0	w_2	w_3	1
w_1	=	4	2	1	-2	1
z_3	=	-2	0	-1	1	1
z_2	=	-1	-1	0	1	2

This final tableau is complementary, and so we have the following solution of the LCP: $z_0 = 0$, $z_1 = 0$, $z_2 = 2$, $z_3 = 1$. Since Q is positive semidefinite, it follows from Theorem 7.2.2 that $x_1 = z_1 = 0$, $x_2 = z_2 = 2$ is a global solution of the quadratic program, and the minimum objective value is $\frac{1}{2}(4) - 2 = 0$. ∎

The theory associated with Lemke's method hinges on whether or not the algorithm terminates at an unbounded ray. Under certain hypotheses on the matrix M, it is shown that either ray termination cannot occur or (if it does occur) there is no feasible pair (z, w). Two fundamental results are given in the following theorem.

Theorem 7.4.2.

(a) *If $M \in \mathbf{R}^{n \times n}$ is positive definite, then Lemke's algorithm terminates at the unique solution of $\mathrm{LCP}(M, q)$ for any $q \in \mathbf{R}^n$.*

(b) *If $M \in \mathbf{R}^{n \times n}$ is positive semidefinite, then for each $q \in \mathbf{R}^n$, Lemke's algorithm terminates at a solution of $\mathrm{LCP}(M, q)$ or at an unbounded ray. In the latter case, the set $\{z \mid Mz + q \geq 0, z \geq 0\}$ is empty; that is, there is no feasible pair.*

Proof. See Cottle & Dantzig (1968). □

When the LCP is derived from a quadratic program, the matrix M, defined by

$$M = \begin{bmatrix} Q & -A' \\ A & 0 \end{bmatrix},$$

is positive semidefinite if and only if Q is positive semidefinite, because

$$[x' \quad u']M \begin{bmatrix} x \\ u \end{bmatrix} = x'Qx - x'A'u + u'Ax = x'Qx.$$

(Note also that the matrix M above for the quadratic program is never positive definite, except in the special case in which A is null.) Therefore, it follows from part (b) of the theorem above that Lemke's algorithm either solves convex quadratic programs or else determines that the corresponding linear complementarity problem is infeasible. In the latter case, the quadratic program either is infeasible or unbounded below on the feasible region. Formally, we have the following result.

Corollary 7.4.3. *If Q is positive semidefinite, then Lemke's method solves (7.7) or else determines that (7.7) is infeasible or unbounded below on the feasible region.*

Example 7-4-3. Consider the following quadratic program:

$$\begin{array}{rl}
\min & x_1^2 + x_1 x_2 + 2x_2^2 + x_1 - x_2 \\
\text{subject to} & x_1 - 2x_2 - 2 \geq 0, \\
& -x_1 + x_2 + 1 \geq 0, \\
& x_1, x_2 \geq 0.
\end{array}$$

We can identify this problem with (7.7) when we set

$$Q = \begin{bmatrix} 2 & 1 \\ 1 & 4 \end{bmatrix}, \qquad A = \begin{bmatrix} 1 & -2 \\ -1 & 1 \end{bmatrix}, \qquad p = \begin{bmatrix} 1 \\ -1 \end{bmatrix}, \qquad b = \begin{bmatrix} 2 \\ -1 \end{bmatrix}.$$

Following (7.16), we can write the corresponding $LCP(M, q)$ as follows:

$$0 \leq \begin{bmatrix} x_1 \\ x_2 \\ u_1 \\ u_2 \end{bmatrix} \perp \begin{bmatrix} 2x_1 & +x_2 & -u_1 & +u_2 & +1 \\ x_1 & +4x_2 & +2u_1 & -u_2 & -1 \\ x_1 & -2x_2 & & & -2 \\ -x_1 & +x_2 & & & +1 \end{bmatrix} \geq 0.$$

We now show how to solve the LCP in MATLAB.

```
>> M = [2 1 -1 1;1 4 2 -1;
   1 -2 0 0;-1 1 0 0];
>> q = [1 -1 -2 1]';
>> T = lemketbl(M,q);
```

		z_1	z_2	z_3	z_4	1
w_1	=	2	1	−1	1	1
w_2	=	1	4	2	−1	−1
w_3	=	1	−2	0	0	−2
w_4	=	−1	1	0	0	1

```
>> T = addcol(T,[1 1 1 1]',
   'z0',5);
```

		z_1	z_2	z_3	z_4	z_0	1
w_1	=	2	1	−1	1	1	1
w_2	=	1	4	2	−1	1	−1
w_3	=	1	−2	0	0	1	−2
w_4	=	−1	1	0	0	1	1

```
>> T = ljx(T,3,5);
```

		z_1	z_2	z_3	z_4	w_3	1
w_1	=	1	3	−1	1	1	3
w_2	=	0	6	2	−1	1	1
z_0	=	−1	2	0	0	1	2
w_4	=	−2	3	0	0	1	3

```
>> T = ljx(T,1,3);
```

		z_1	z_2	w_1	z_4	w_3	1
z_3	=	1	3	−1	1	1	3
w_2	=	2	12	−2	1	3	7
z_0	=	−1	2	0	0	1	2
w_4	=	−2	3	0	0	1	3

```
>> T = ljx(T,4,1);
```

		w_4	z_2	w_1	z_4	w_3	1
z_3	=	−0.5	4.5	−1	1	1.5	4.5
w_2	=	−1	15	−2	1	4	10
z_0	=	0.5	0.5	0	0	0.5	0.5
z_1	=	−0.5	1.5	0	0	0.5	1.5

This tableau contains an unbounded ray when $z_4 \to \infty$, indicating that the LCP does not have a solution. This implies that the underlying quadratic program is either infeasible or unbounded. In fact, we can see that it is infeasible because by adding the two constraints we obtain $-x_2 - 1 \geq 0$, which is patently false. ∎

Exercise 7-4-4. Solve the following quadratic program by Lemke's method:

$$\begin{array}{ll}
\min & x_1^2 + x_2^2 + x_1 x_2 - x_1 - x_2 \\
\text{subject to} & x_1 + x_2 - 2 \geq 0, \\
& 2x_1 - 1 \geq 0, \\
& x_1, x_2 \geq 0.
\end{array}$$

Exercise 7-4-5. Solve the following quadratic program by Lemke's method:

$$\begin{array}{ll}
\min & x_1^2 + \frac{1}{2}x_2^2 + x_1 x_3 + \frac{1}{2}x_3^2 - x_1 \\
\text{subject to} & x_1 + x_2 + x_3 - 1 \geq 0, \\
& x_1, x_2, x_3 \geq 0.
\end{array}$$

Exercise 7-4-6. By using the minimum principle (7.2), show that if Q is positive definite, then the solution x of the quadratic program is unique. What about the Lagrange multipliers u? Relate your conclusions to the theorem and corollary above.

When the quadratic program is nonconvex, recall that the KKT conditions are only necessary, not sufficient. Hence, Lemke's method applied to these conditions may identify a solution of the LCP that may *not* be a solution of the underlying quadratic program. It may be a constrained local minimum, a saddle point, or even a maximum.

When convex quadratic programs have equality constraints or free variables, then we must instead solve the KKT conditions given as (7.13). It is straightforward to adapt the techniques we used for the simplex method to this setting as we now show.

Example 7-4-7. Consider the following quadratic program:

$$\begin{array}{ll}
\min & x_1^2 + x_1 x_2 + 2x_2^2 + x_1 - x_2 \\
\text{subject to} & x_1 - 2x_2 - 2 \geq 0, \\
& -x_1 + x_2 + 1 \geq 0,
\end{array}$$

which is the same as Example 7-4-3, except that the variables are no longer constrained to be nonnegative. The KKT conditions form the following complementarity problem:

```
>> M = [2 1 -1 1;1 4 2 -1;
   1 -2 0 0;-1 1 0 0];
>> q = [1 -1 -2 1]';
>> T = lemketbl(M,q);
```

		z_1	z_2	z_3	z_4	1
w_1	=	2	1	-1	1	1
w_2	=	1	4	2	-1	-1
w_3	=	1	-2	0	0	-2
w_4	=	-1	1	0	0	1

In this case the variables $z_1 = x_1$ and $z_2 = x_2$ are free, and consequently $w_1 = 0$ and $w_2 = 0$ must hold. We achieve this by pivoting w_1 and w_2 to the top of the tableau, while simultaneously moving z_1 and z_2 into the basis.

```
>> T = ljx(T,1,1);
>> T = ljx(T,2,2);
>> T = delcol(T,'w1');
>> T = delcol(T,'w2');
>> T = permrows(T,[3 4 1 2]);
```

		z_3	z_4	1
w_3	=	2.286	-1.571	-3.571
w_4	=	-1.571	1.143	2.143
z_1	=	0.857	-0.714	-0.714
z_2	=	-0.714	0.429	0.429

Ignoring the last two rows, we now apply Lemke's method to the (standard form) complementarity problem in w_3, w_4, z_3, and z_4, starting with the Phase I procedure:

```
≫ T = addcol(T,[1 1 0 0]',
    'z0',3);
≫ T = ljx(T,1,3);
```

		z_3	z_4	w_3	1
z_0	=	-2.286	1.571	1	-3.571
w_4	=	-3.857	2.714	1	5.714
z_1	=	0.857	-0.714	0	-0.714
z_2	=	-0.714	0.429	0	0.429

Two further complementary pivots generate the complementary solution.

```
≫ T = ljx(T,2,1);
≫ T = ljx(T,1,2);
```

		w_4	z_0	w_3	1
z_4	=	16	-27	11	5
z_3	=	11	-19	8	5
z_1	=	-2	3	-1	0
z_2	=	-1	2	-1	-1

The solution of the quadratic problem is thus $x_1 = 0$, $x_2 = -1$. ∎

7.5 Bimatrix Games

Some real-world problems can be better modeled with two objectives that are partially or totally opposed to each other. Consider, for example, the following game with two players, Player 1 and Player 2. The game consists of a large number of plays, and at each play Player 1 picks one of m choices and Player 2 picks one of n choices. These choices are called *pure strategies*. If, in a certain play, Player 1 chooses *pure strategy i* and Player 2 chooses *pure strategy j*, then Player 1 loses $\tilde{A}_{ij}$ dollars and Player 2 loses $\tilde{B}_{ij}$ dollars. (A positive value $\tilde{A}_{ij} > 0$ represents a loss to Player 1, while a negative value $\tilde{A}_{ij} < 0$ represents a gain; similarly for Player 2 and $\tilde{B}_{ij}$.) The matrices $\tilde{A}$ and $\tilde{B}$ are called *loss matrices*, and the game is fully determined by the matrix pair $(\tilde{A}, \tilde{B})$.

If $\tilde{A} + \tilde{B} = 0$, the game is known as a *zero-sum game*. As we show in Section 7.5.2, the game can then be described by a dual pair of linear programs. Otherwise, when $\tilde{A} + \tilde{B} \neq 0$, the game is known as a *bimatrix game* or, more long-windedly, a *two-person nonzero-sum game with a finite number of pure strategies*.

Often, in practice, it does not make sense for a player to choose a pure strategy, that is, to make the same play on every move. If the player does this, or otherwise follows a predictable pattern of plays, the opponent can strategize accordingly and presumably come out ahead. It is often better to follow a *mixed strategy*, in which the player chooses randomly from among the moves available to him, assigning a certain probability to each play. If Player 1 chooses play i with probability x_i (with $\sum_{i=1}^{m} x_i = 1$), while Player 2 chooses play j with probability y_j (with $\sum_{j=1}^{n} y_j = 1$), the vectors x and y define the mixed strategies for each player. Summing over all possible combinations of strategies for the two players, we find that the *expected loss* of Player 1 is $x'\tilde{A}y$. Similarly, the expected loss of Player 2 is $x'\tilde{B}y$.

The pair of mixed strategies $(\bar{x}, \bar{y})$ is called a *Nash equilibrium pair* (after John Nash, 1994 Nobel Laureate in Economics) of strategies if neither player benefits by unilaterally changing their own strategy while the other player holds their strategy fixed. That is,

$$\bar{x}'\tilde{A}\bar{y} \le x'\tilde{A}\bar{y} \qquad \forall x \ge 0, \qquad e'_m x = 1, \tag{7.17a}$$

$$\bar{x}'\tilde{B}\bar{y} \le \bar{x}'\tilde{B}y \qquad \forall y \ge 0, \qquad e'_n y = 1. \tag{7.17b}$$

Here e_m is a vector of length m whose elements are all 1; similarly for e_n.

If we add an arbitrary scalar $\alpha > 0$ to every element of $\tilde{A}$ and $\beta > 0$ to every entry of $\tilde{B}$, the values of $x'\tilde{A}y$ and $x'\tilde{B}y$ change by α and β, respectively. That is, defining A and B by

$$A = \tilde{A} + \alpha e_m e'_n, \qquad B = \tilde{B} + \beta e_m e'_n,$$

we have

$$x'\tilde{A}y = x'Ay - \alpha x'e_m e'_n y = x'Ay - \alpha, \qquad x'\tilde{B}y = x'By - \beta.$$

Therefore, $(\bar{x}, \bar{y})$ is an equilibrium pair for $(\tilde{A}, \tilde{B})$ if and only if it is an equilibrium pair for (A, B). Thus, we can choose α and β large enough to ensure that $A > 0$ and $B > 0$. It is convenient for the remainder of the analysis to assume that A and B have this property.

Example 7-5-1. An example of such a game is the prisoners' dilemma; see, for example, Murty (1976). Two well-known criminals are caught and during plea bargaining the District Attorney urges both criminals to confess and plead guilty. As encouragement, she states that if one confesses and the other does not, then the one who confesses will be acquitted and the other one will serve 10 years. If both confess, then they will each get 5 years. However, the criminals realize that the case against them is not strong, and so if they both choose not to confess, the worst the District Attorney can do is charge them with a minor violation for which they would each serve 1 year.

The loss matrix for this problem is the number of years the prisoner will spend in jail. The two pure strategies are "confess" (strategy 1) and "not confess" (strategy 2). The corresponding loss matrices are

$$A = \begin{bmatrix} 5 & 0 \\ 10 & 1 \end{bmatrix}, \qquad B = \begin{bmatrix} 5 & 10 \\ 0 & 1 \end{bmatrix}. \quad \blacksquare$$

7.5.1　Computing Nash Equilibria

The following lemma shows how to construct equilibrium pairs for bimatrix games by solving $\mathrm{LCP}(M, q)$ for a particular choice of M and q and *normalizing* its solution.

Lemma 7.5.1. *Suppose $A, B \in \mathbf{R}^{m \times n}$ are positive loss matrices representing a game (A, B), and suppose that $(s, t) \in \mathbf{R}^m \times \mathbf{R}^n$ solves $\mathrm{LCP}(M, q)$, where*

$$M := \begin{bmatrix} 0 & A \\ B' & 0 \end{bmatrix}, \qquad q := -e_{m+n} \in \mathbf{R}^{m+n}.$$

Then, defining $\bar{x} = s/(e'_m s)$ and $\bar{y} = t/(e'_n t)$, we have that $(\bar{x}, \bar{y})$ is an equilibrium pair of (A, B).

Proof. We write out the LCP conditions explicitly as follows:

$$0 \le At - e_m \perp s \ge 0,$$
$$0 \le B's - e_n \perp t \ge 0.$$

Since $At - e_m \ge 0$, we must have that $t \ne 0$, and similarly $s \ne 0$. Thus $\bar{x}$ and $\bar{y}$ are well defined. It is also clear from their definitions that $\bar{x} \ge 0$, $\bar{y} \ge 0$ and $e'_m \bar{x} = 1$, $e'_n \bar{y} = 1$. Therefore, $\bar{x}$ and $\bar{y}$ are mixed strategies.

By complementarity, we have that

$$\bar{x}'(At - e_m) = \frac{1}{e'_m s} s'(At - e_m) = 0,$$

so that $\bar{x}' At = \bar{x}' e_m = 1$. We use this relation to obtain the following:

$$A\bar{y} - (\bar{x}' A\bar{y})e_m = \frac{1}{e'_n t}(At - (\bar{x}' At)e_m) = \frac{1}{e'_n t}(At - e_m) \ge 0.$$

Given any strategy x, we have from $x \ge 0$ together with the expression above that

$$0 \le x'\left(A\bar{y} - e_m(\bar{x}' A\bar{y})\right) \quad \Rightarrow \quad x' A\bar{y} \ge (e'_m x)\bar{x}' A\bar{y} = \bar{x}' A\bar{y}.$$

Hence, the relationship (7.17a) is satisfied. We can prove (7.17b) similarly. We conclude that $(\bar{x}, \bar{y})$ is a Nash equilibrium pair, as claimed. $\square$

This result motivates the following procedure (the Lemke–Howson method) for solving bimatrix games.

1. Increase all the entries in each loss matrix by a constant amount to obtain A and B, whose entries are all greater than 0.

2. Solve LCP(M, q), where

$$M = \begin{bmatrix} 0 & A \\ B' & 0 \end{bmatrix}, \qquad q = \begin{bmatrix} -e_m \\ -e_n \end{bmatrix},$$

with the first two pivots especially designed to make the tableau feasible (see below), followed by Phase II of Lemke's method.

3. Set $\bar{x} = s/(e'_m s)$ and $\bar{y} = t/(e'_n t)$.

Theorem 7.5.2. *The Lemke–Howson method always terminates at a solution of the LCP and hence always finds a Nash equilibrium of the game* (A,B).

Proof. See Lemke & Howson (1964). $\square$

We now give some examples of bimatrix games.

Example 7-5-2 (The Matching Pennies Game). In this zero-sum ($\tilde{A} + \tilde{B} = 0$) game, two players simultaneously show each other a coin, deciding to themselves before they show it whether to put the heads side up or the tails side. If the coins match (both heads or both tails), then Player 2 wins both coins; otherwise Player 1 wins them both. The resulting loss matrices are

$$\tilde{A} = \begin{bmatrix} 1 & -1 \\ -1 & 1 \end{bmatrix}, \qquad \tilde{B} = \begin{bmatrix} -1 & 1 \\ 1 & -1 \end{bmatrix}.$$

We now apply Lemke's method to solve this problem. First, we find the most negative element in each matrix $\tilde{A}$ and $\tilde{B}$ and add a sufficiently large number to ensure that $A > 0$ and $B > 0$. In fact, the code below adds 2 to every element of these matrices to obtain

$$A = \begin{bmatrix} 3 & 1 \\ 1 & 3 \end{bmatrix}, \qquad B = \begin{bmatrix} 1 & 3 \\ 3 & 1 \end{bmatrix}.$$

We then set up the tableau.

```
≫ load ex7-5-2
≫ [m,n] = size(A);
≫ alpha = 1 - min(min(A)); A = A +
  alpha*ones(size(A));
≫ beta = 1 - min(min(B)); B = B +
  beta*ones(size(B));
≫ M = [zeros(m,m) A; B' zeros(n,n)];
≫ q = -ones(m+n,1);
≫ T = lemketbl(M,q);
```

		z_1	z_2	z_3	z_4	1
w_1	=	0	0	3	1	-1
w_2	=	0	0	1	3	-1
w_3	=	1	3	0	0	-1
w_4	=	3	1	0	0	-1

The Phase I procedure described earlier fails on bimatrix games; after the initial pivot to bring z_0 into the basis, the first pivot of Phase II encounters a column with all nonnegative elements, and so the ratio test fails. However, there is a specialized Phase I procedure for complementarity problems arising from bimatrix games, which we now describe. It takes the form of two pivots performed on the original tableau, with special rules for pivot row and column selection.

1. Pivot on column 1 and row r corresponding to the smallest positive element in column 1. (This pivot causes the bottom n rows of the tableau to become feasible.)

2. Pivot on the column corresponding to z_r (the complement of w_r) and on the row with the smallest positive element in that column. (This pivot causes the top m rows to become feasible and makes the tableau almost complementary.)

We then proceed with Phase II, the usual Lemke method.

For the problem at hand, the two Phase I pivots are as follows:

```
≫ T = ljx(T,3,1);
```

		w_3	z_2	z_3	z_4	1
w_1	=	0	0	3	1	-1
w_2	=	0	0	1	3	-1
z_1	=	1	-3	0	0	1
w_4	=	3	-8	0	0	2

≫ T = ljx(T,2,3);

$$
\begin{array}{r|cccc|c}
 & w_3 & z_2 & w_2 & z_4 & 1 \\
\hline
w_1 = & 0 & 0 & 3 & -8 & 2 \\
z_3 = & 0 & 0 & 1 & -3 & 1 \\
z_1 = & 1 & -3 & 0 & 0 & 1 \\
w_4 = & 3 & -8 & 0 & 0 & 2 \\
\end{array}
$$

Proceeding now to Phase II, we choose the first pivot column as the one corresponding to z_2, the complement of w_2, since z_2 was pivoted out of the basis on the previous step.

≫ T = ljx(T,4,2);

≫ T = ljx(T,1,4);

These steps yield the following tableau:

$$
\begin{array}{r|cccc|c}
 & w_3 & w_4 & w_2 & w_1 & 1 \\
\hline
z_4 = & 0 & 0 & 0.38 & -0.13 & 0.25 \\
z_3 = & 0 & 0 & -0.13 & 0.38 & 0.25 \\
z_1 = & -0.13 & 0.38 & 0 & 0 & 0.25 \\
z_2 = & 0.38 & -0.13 & 0 & 0 & 0.25 \\
\end{array}
$$

At this point, we have a complementary solution: For each component $i = 1, 2, 3, 4$, exactly one of the pair (z_i, w_i) is in the basis and the other is nonbasic. Partitioning z into s and t, we have $s = (1/4, 1/4)'$ and $t = (1/4, 1/4)'$, from which we obtain $\bar{x} = (1/2, 1/2)'$ and $\bar{y} = (1/2, 1/2)'$. This solution indicates that the best strategy for each player is simply to choose between heads and tails randomly, with an equal probability for each. The expected loss for Player 1 is $\bar{x}'\tilde{A}\bar{y} = 0$, while for Player 2 it is $\bar{x}'\tilde{B}\bar{y} = 0$. That is (not surprisingly) by following these strategies, both players can expect to break even in the long run.

It is easy to see why nonequilibrium strategies fail on this problem. Suppose, for instance, that Player 1 decides to choose randomly between heads and tails but assigns a higher probability to heads. Player 2 can then counter by showing heads all the time (i.e., following a pure strategy of "heads"). In the long run, the coins will match on the majority of plays, and Player 2 will win them both. ∎

We now present a variant on the matching pennies game, discussed (and answered incorrectly) in a column by Marilyn Vos Savant.[1]

Example 7-5-3. In this game, the two players again show pennies to each other. If both are heads, Player 2 receives \$3 from Player 1, and if both are tails, Player 2 receives \$1 from Player 1. On the other hand, if the pennies do not match, Player 2 pays \$2 to Player 1.

This is a zero-sum game that seems similar, at first glance, to Example 7-5-2. However, its Nash equilibrium solution turns out to give an edge to Player 1, as we will see. First, we set up the loss matrices. Row 1 of both $\tilde{A}$ and $\tilde{B}$ corresponds to Player 1 choosing heads, while row 2 is for Player 1 choosing tails. Similarly, columns 1 and 2 represent Player 2 choosing heads and tails, respectively. We then have

$$
\tilde{A} = \begin{bmatrix} 3 & -2 \\ -2 & 1 \end{bmatrix}, \qquad \tilde{B} = \begin{bmatrix} -3 & 2 \\ 2 & -1 \end{bmatrix}.
$$

---

[1] *Parade*, 31 March, 2002; discussed by Francis J. Vasko and Dennis D. Newhart, *SIAM News*, June 2003, p. 12.

By adding 3 to the elements of $\tilde{A}$ and 4 to the elements of $\tilde{B}$, we obtain the following positive matrices:

$$A = \begin{bmatrix} 6 & 1 \\ 1 & 4 \end{bmatrix}, \qquad B = \begin{bmatrix} 1 & 6 \\ 6 & 3 \end{bmatrix}.$$

We now set up and solve the problem using the same technique as in Example 7-5-2.

```
>> A=[3 -2; -2 1]; B=[-3 2; 2 -1];
>> [m,n] = size(A);
>> alpha = 1 - min(min(A)); A = A +
   alpha*ones(size(A));
>> beta = 1 - min(min(B)); B = B +
   beta*ones(size(B));
>> M = [zeros(m,m) A; B' zeros(n,n)];
>> q = -ones(m+n,1);
>> T = lemketbl(M,q);
```

		z_1	z_2	z_3	z_4	1
w_1	=	0	0	6	1	−1
w_2	=	0	0	1	4	−1
w_3	=	1	6	0	0	−1
w_4	=	6	3	0	0	−1

We now perform the two Phase I pivots (recalling that we need to use the special Phase I for bimatrix games, described above).

```
>> T = ljx(T,3,1);
>> T = ljx(T,2,3);
```

		w_3	z_2	w_2	z_4	1
w_1	=	0	0	6	−23	5
z_3	=	0	0	1	−4	1
z_1	=	1	−6	0	0	1
w_4	=	6	−33	0	0	5

Two further Lemke pivots (Phase II) are needed to solve the problem.

```
>> T = ljx(T,4,2);
>> T = ljx(T,1,4);
```

yielding the final tableau

		w_3	w_4	w_2	w_1	1
z_4	=	0	0	.2609	−.0435	.2174
z_3	=	0	0	−.0435	.1739	.1304
z_1	=	−.0909	.1818	0	0	.0909
z_2	=	.1818	−.0303	0	0	.1515

We obtain the solution $s = (.0909, .1515)'$ and $t = (.1304, .2174)'$, from which we recover (after adjusting for rounding errors) $\bar{x} = (3/8, 5/8)'$ and $\bar{y} = (3/8, 5/8)'$. The expected loss for Player 1 is $\bar{x}'\tilde{A}\bar{y} = -1/8$, while (because this is a zero-sum game) the expected loss for Player 2 is $1/8$. Hence, if both players follow this equilibrium strategy, Player 1 can expect to win an average of 12.5 cents per play from Player 2.

In fact, it can be shown that *whatever* strategy y is pursued by Player 2, Player 1 will win an average of 12.5 cents per play if she follows the strategy $\bar{x}$. To see this, let y be any mixed strategy of Player 2 (satisfying $y \geq 0$ and $y_1 + y_2 = 1$). We then have

$$\bar{x}'\tilde{A}y = \begin{bmatrix} 3/8 & 5/8 \end{bmatrix} \begin{bmatrix} 3 & -2 \\ -2 & 1 \end{bmatrix} \begin{bmatrix} y_1 \\ y_2 \end{bmatrix} = (-1/8)(y_1 + y_2) = -1/8. \quad \blacksquare$$

Example 7-5-4. Following the procedure described above, we solve the prisoners dilemma problem (Example 7-5-1) as follows:

```
≫ load ex7-5-1
≫ [m,n] = size(A);
≫ alpha = 1 - min(min(A)); A = A + alpha*ones(size(A));
≫ beta = 1 - min(min(B)); B = B + beta*ones(size(B));
≫ M = [zeros(m,m) A; B' zeros(n,n)];
≫ q = -ones(m+n,1);
≫ T = lemketbl(M,q);
```

The following two pivots make up the special Phase I procedure:

```
≫ T = ljx(T,3,1);
≫ T = ljx(T,1,3);
```

The resulting tableau is in fact complementary (not just almost complementary), and so no further pivots are required. The (unique) solution to this LCP(M, q) has $s = (1/6, 0)'$, $t = (1/6, 0)'$, leading to the (unique) equilibrium pair $\bar{x} = (1, 0)'$, $\bar{y} = (1, 0)'$. This solution indicates that the equilibrium strategy is for both to confess. Given that the other prisoner sticks to this strategy, there is no advantage for a prisoner to change to "not confess"—in fact, he will be worse off.

The nonequilibrium strategies for this simple problem are also interesting. If the prisoners collude and agree to "not confess," they will both do better than in the equilibrium strategy, serving 1 year each. However, this is risky for each prisoner. If one prisoner double-crosses and changes to "confess," he will escape a sentence, while his opposite number will serve 10 years. $\blacksquare$

Exercise 7-5-5. On a Friday evening, two newlyweds wish to go out together. The wife wishes to go to a musical concert, while the husband prefers to go to a baseball game. On a scale of 1 to 4, the loss matrix entry to each person is 4 if they go to separate events, 1 if they go together to his/her favored event, and 2 if they go together but to his/her less favored event.

1. Write down the resulting loss matrices for the game. (Let x represent the husband and y the wife, and let strategy 1 for each party represent "attend baseball game" while strategy 2 represents "attend concert.")

2. Using the bimatrix version of Lemke's algorithm, verify that $\bar{x} = (1, 0)'$, $\bar{y} = (1, 0)'$ is an equilibrium solution of the bimatrix game.

3. Verify by inspection of the loss matrices that $\tilde{x} = (0, 1)'$, $\tilde{y} = (0, 1)'$ is also an equilibrium pair.

4. Prove that another equilibrium pair is $\hat{x} = (0.6, 0.4)'$, $\hat{y} = (0.4, 0.6)'$.

7.5.2   Zero-Sum Games As Dual Linear Programs

We now show that in the case of a zero-sum game—that is, when $B = -A$—the equilibrium pair can be found via the primal-dual solution of a linear program, rather than the LCP technique of the previous section. (It does not matter in this section if A or B contains nonpositive entries, and so we can identify A with $\tilde{A}$ and B with $\tilde{B}$.)

Our equilibrium pair is a pair of mixed strategies $\bar{x}$ and $\bar{y}$ satisfying the following properties:

$$\bar{x} \geq 0, \quad e_m'\bar{x} = 1, \quad \text{and} \quad \bar{x}'A\bar{y} \leq x'A\bar{y} \quad \forall x \geq 0, \ e_m'x = 1, \quad (7.18a)$$
$$\bar{y} \geq 0, \quad e_n'\bar{y} = 1, \quad \text{and} \quad -\bar{x}'A\bar{y} \leq -\bar{x}'Ay \quad \forall y \geq 0, \ e_n'y = 1. \quad (7.18b)$$

We can therefore view $\bar{x}$ and $\bar{y}$ as the solutions of two linear programs:

$$\bar{x} = \underset{x}{\operatorname{argmin}} \ x'(A\bar{y}) \quad \text{subject to} \quad x \geq 0, \ e_m'x = 1; \quad (7.19a)$$

$$\bar{y} = \underset{y}{\operatorname{argmax}} \ y'(A'\bar{x}) \quad \text{subject to} \quad y \geq 0, \ e_n'y = 1. \quad (7.19b)$$

The KKT conditions for these linear programs are as follows, respectively:

$$0 \leq \bar{x} \perp A\bar{y} - \bar{\beta}e_m \geq 0, \qquad e_m'\bar{x} = 1;$$
$$0 \leq \bar{y} \perp A'\bar{x} - \bar{\alpha}e_n \leq 0, \qquad e_n'\bar{y} = 1.$$

Taken together, these are the optimality conditions of the following dual pair of linear programs:

$$
\begin{array}{ll}
\begin{aligned}
\max_{x,\alpha} \quad & -\alpha \\
\text{subject to} \quad & A'x \ - \ e_n\alpha \ \leq \ 0, \\
& -e_m'x \ \ \ \ \ \ \ = \ -1, \\
& \ \ \ \ \ \ x \ \geq \ 0,
\end{aligned}
&
\begin{aligned}
\min_{y,\beta} \quad & -\beta \\
\text{subject to} \quad & Ay \ - \ e_m\beta \ \geq \ 0, \\
& -e_n'y \ \ \ \ \ \ \ = \ -1, \\
& \ \ \ \ \ \ y \ \geq \ 0.
\end{aligned}
\end{array}
$$
$$(7.20)$$

Lemma 7.5.3. *The dual pair of linear programs (7.20) is always solvable.*

Proof. The result follows from strong duality, since both linear programs are clearly feasible (α large and positive, β large and negative).   □

We conclude that the equilibrium point for zero-sum games can be solved by finding the primal and dual solutions of one of these linear programs.

Example 7-5-6. For the matching pennies problem (Example 7-5-2), we can solve the dual problem as follows:

```
≫ A = [1 -1; -1 1];
≫ [m,n] = size(A);
≫ b = [zeros(m,1); -1];
≫ p = [zeros(n,1); -1];
≫ T = totbl([A -ones(m,1);
  -ones(1,n) 0],b,p);
≫ T = relabel(T,'x1','y1','x2','y2','x3','β');
```

		y_1	y_2	β	1
x_4	=	1	-1	-1	0
x_5	=	-1	1	-1	0
x_6	=	-1	-1	0	1
z	=	0	0	-1	0

Noting that x_6 is an artificial variable associated with an equation, we pivot and delete:

```
≫ T = ljx(T,3,1);
≫ T = delcol(T,'x6');
```

		y_2	β	1
x_4	=	-2	-1	1
x_5	=	2	-1	-1
y_1	=	-1	0	1
z	=	0	-1	0

Applying the following two pivots (luckily) generates an optimal tableau:

```
≫ T = ljx(T,1,1);
≫ T = ljx(T,2,2);
```

		x_4	x_5	1
y_2	=	-0.25	0.25	0.5
β	=	-0.5	-0.5	0
y_1	=	0.25	-0.25	0.5
z	=	0.5	0.5	0

Thus the optimal solution of the dual is $\bar{y} = (0.5, 0.5)$, and furthermore the optimal solution of the primal can be read of this tableau as $\bar{x} = (0.5, 0.5)$. This confirms the results given in Example 7-5-2. ■

Exercise 7-5-7. Two players play the game "rock, paper, scissors." If Player I calls rock and Player II calls scissors, then Player I takes $1 from Player II since "rock blunts scissors." If Player I calls scissors and Player II calls paper, then Player I again takes $1 from Player II, since "scissors cut paper." The same result occurs if Player I calls paper and Player II calls rock (since "paper covers rock"). If the above strategies are reversed, then Player I *pays* Player II $1. If they both call the same object, no money changes hands. Write down the loss matrices for each player. For zero-sum games, Player I's strategy can be determined from the linear programs in (7.20). Show that $x = (1/3, 1/3, 1/3)$ solves this problem with $\alpha = 0$, where A is the loss matrix of the rock, paper, scissors game. Is this solution unique? Exhibit an optimal solution of the dual of this problem. Use these solutions to construct a Nash equilibrium for the game.

Chapter 8

Interior-Point Methods

Interior-point methods follow a fundamentally different approach from the simplex method. The simplex approach moves from vertex to vertex, usually improving the objective function on each step. By contrast, the most successful interior-point approaches focus instead on the KKT conditions discussed in Section 4.5, searching for primal and dual variables that satisfy these conditions, and hence solve the primal and dual linear programs concurrently. Primal and dual variables that are required to be nonnegative at the solution are kept *strictly positive* at each interior-point iteration. That is, the iterates stay *interior* with respect to these constraints, though some of these variables will approach zero in the limit.

8.1 Motivation and Outline

We give specifics of the interior-point approach by referring to the *canonical* form of the linear programming problem (5.3), which we restate here as follows:

$$\begin{array}{ll} \min & z = p'x \\ \text{subject to} & Ax = b, \quad x \geq 0. \end{array} \tag{8.1}$$

Throughout this chapter, we make no assumptions on A other than that it has linearly independent rows. (We do not assume that A has the form (5.2), for instance.) By introducing Lagrange multipliers y for the constraints $Ax = b$ and applying the theory of Section 4.5, we can write the KKT conditions for this problem as follows:

$$Ax = b, \qquad x \geq 0, \qquad A'y \leq p, \qquad x'(p - A'y) = 0. \tag{8.2}$$

Following Theorem 4.5.1, there exist vectors $x \in \mathbf{R}^n$ and $y \in \mathbf{R}^m$ that satisfy (8.2) if and only if x is a solution of the primal linear program (8.1), while y solves the dual problem, which is

$$\begin{array}{ll} \max_{y} & b'y \\ \text{subject to} & A'y \leq p. \end{array}$$

195

If we introduce the slack vector $s \in \mathbf{R}^n$ defined by $s = p - \mathcal{A}'y$, we can rewrite the dual problem as follows:

$$\max_{y,s} \quad b'y$$
$$\text{subject to} \quad \mathcal{A}'y + s = p, \quad s \geq 0, \tag{8.3}$$

while the KKT conditions (8.2) can be recast as follows:

$$\mathcal{A}x = b, \tag{8.4a}$$
$$\mathcal{A}'y + s = p, \tag{8.4b}$$
$$x \geq 0, \quad s \geq 0, \quad x's = 0. \tag{8.4c}$$

The conditions (8.4c) imply that for each $i = 1, 2, \ldots, n$, one of the variables x_i or s_i is zero, and the other is nonnegative—possibly also zero. We can express (8.4c) alternatively as follows:

$$x \geq 0, \quad s \geq 0, \quad x_i s_i = 0, \quad i = 1, 2, \ldots, n. \tag{8.5}$$

Summarizing, if we define $e = (1, 1, \ldots, 1)'$ and let the diagonal matrices X and S be defined from the components of x and s,

$$X = \text{diag}(x_1, x_2, \ldots, x_n) = \begin{bmatrix} x_1 & & & \\ & x_2 & & \\ & & \ddots & \\ & & & x_n \end{bmatrix}, \tag{8.6a}$$

$$S = \text{diag}(s_1, s_2, \ldots, s_n) = \begin{bmatrix} s_1 & & & \\ & s_2 & & \\ & & \ddots & \\ & & & s_n \end{bmatrix}, \tag{8.6b}$$

then the KKT conditions can be written using X and S as follows:

$$\mathcal{A}x = b, \tag{8.7a}$$
$$\mathcal{A}'y + s = p, \tag{8.7b}$$
$$XSe = 0, \tag{8.7c}$$
$$x \geq 0, \quad s \geq 0. \tag{8.7d}$$

Interior-point methods of the "primal-dual" variety—the type that has been the most successful in practice—generate iterates (x, y, s) with the following properties:

(i) The inequalities (8.7d) are satisfied *strictly* at every iteration; that is, $x_i > 0$ and $s_i > 0$ for all $i = 1, 2, \ldots, n$.

(ii) The amount by which the equality conditions (8.7a), (8.7b) are violated decreases at each iteration.

(iii) The quantity μ defined by

$$\mu = \frac{x's}{n} = \frac{1}{n} \sum_{i=1}^{n} x_i s_i, \tag{8.8}$$

known as the *duality measure*, decreases at each iteration. Note that from (i), this quantity is strictly positive, while because of (8.4c), it will approach zero as the iterates approach a solution.

(iv) The pairwise products $x_i s_i$, $i = 1, 2, \ldots, n$, are kept roughly in balance. That is, although all these products approach zero as (x, y, s) approaches a primal-dual solution of (8.7), no single one of these quantities approaches zero much faster than the others. (From (8.8), we see that the duality measure μ is the average value of these pairwise products.)

In later sections, we specify how interior-point methods move from one iterate to the next. The iteration number appears as a superscript: The starting point is (x^0, y^0, s^0) and iterates are denoted by (x^k, y^k, s^k), $k = 0, 1, 2, \ldots$.

Exercise 8-1-1. Verify that if x is primal feasible for the canonical form and (y, s) is dual feasible, we have

$$p'x - b'y = x's = n\mu,$$

so that $x's$ properly measures the *duality gap* between the primal and dual objectives.

8.2   Newton's Method

Newton's method for well-determined systems of nonlinear equations (in which the number of equations equals the number of unknowns) plays an important role in the development of primal-dual interior-point methods. We give some background on Newton's method in this section.

Let $F(\cdot)$ be a vector function with N components that depends on N variables $z_1, z_2, \ldots, z_N$. An example with $N = 3$ is as follows:

$$F(z) = F(z_1, z_2, z_3) = \begin{bmatrix} z_1^2 + z_2^2 - \pi^2 \\ z_3 - \cos z_1 \\ z_3 - \sin z_2 \end{bmatrix}. \tag{8.9}$$

We seek a *solution* of the system $F(z) = 0$. The vector $z = (1.2926, 2.8634, 0.2747)$ is (to four digits of accuracy) a solution of $F(z) = 0$ for the function in (8.9).

Except for simple functions F, it is difficult to find a solution of $F(z) = 0$ by direct observation or simple calculation. An important special case occurs when F is a *linear* function of z; that is, $F(z) = Az - b$ for an $N \times N$ matrix A and a vector $b \in \mathbf{R}^N$. For such a function, the solution of $F(z) = 0$ satisfies $Az = b$. Therefore, if A is nonsingular, we can find the solution by factoring the matrix A (for example, by using the LU factorization discussed in Section 2.6) and solving triangular systems involving L and U.

When F is a scalar function of a scalar variable, it makes sense to approximate F by a simpler function that is close to F in the neighborhood of the latest estimate z of the solution. If we know the value of F and also its first derivative $F'(z)$, we can use the following approximation:

$$F(z + \Delta z) \approx F(z) + F'(z)\Delta z, \tag{8.10}$$

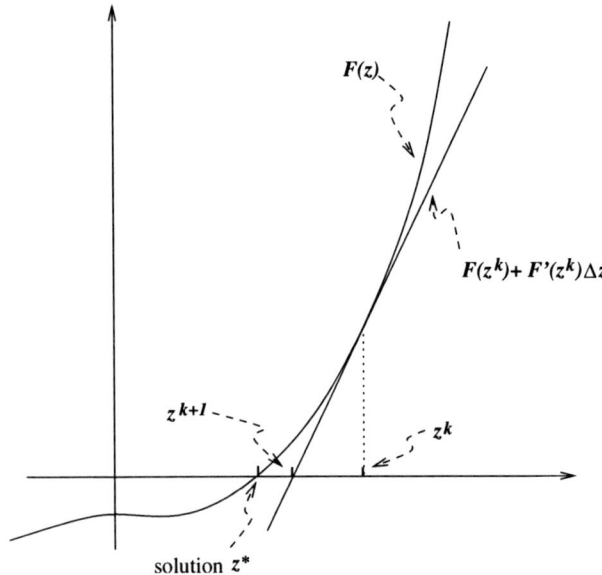

Figure 8.1. *Newton's method for $N = 1$, showing the linear approximation to F at z^k.*

where Δz represents a step away from the current guess z. This approximation is based on Taylor's theorem (stated in most calculus books; see also Section A.7) and is valid when Δz is not too large. The approximation (8.10) is referred to as the *first-order* or *linear approximation* to F about the point z. Newton's method chooses the step Δz to make this linear approximation equal to zero; that is, it sets

$$\Delta z = -\left[F'(z)\right]^{-1} F(z).$$

From (8.10), it follows that if we replace z by $z + \Delta z$, we will have $F(z) \approx 0$ at the new value of z.

Using this scheme, we can generate a sequence of guesses of the solution to $F(z) = 0$. We use a superscript to indicate the vectors in this sequence: $z^0, z^1, z^2, \ldots$. From each iterate $z^k, k = 0, 1, 2, \ldots$, we obtain the next iterate by performing the following calculations:

$$\Delta z^k = -\left[F'(z^k)\right]^{-1} F(z^k), \qquad z^{k+1} = z^k + \Delta z^k. \tag{8.11}$$

An illustration of one step of Newton's method is shown in Figure 8.1. Note that the graph of the linear approximation is, geometrically speaking, *tangent* to the graph of F at z^k and that the next iterate z^{k+1} is the point at which this tangent crosses the z-axis. Note also that z^{k+1} is considerably closer to the solution than is z^k in this example. We can continue this iterative process until the step Δz^k becomes short (indicating successful termination) or until problems are encountered (such as many iterations being taken without significant decrease of F).

For general functions $F : \mathbf{R}^N \to \mathbf{R}^N$, Newton's method can be used when each of the N components of F is at least a continuously differentiable function of z. (Note in the example (8.9) that each of the three components of F can be differentiated arbitrarily often with respect to each of the three components of z, and so this function is certainly smooth enough.) The *Jacobian* of F, denoted by $J(z)$, is the $N \times N$ matrix of first partial derivatives of F. The (i, j) element of $J(z)$ is the partial derivative of the ith component of the vector F with respect to the jth variable; that is,

$$[J(z)]_{ij} = \frac{\partial F_i(z)}{\partial z_j}, \qquad i = 1, 2, \ldots, N, \qquad j = 1, 2, \ldots, N. \qquad (8.12)$$

For the function (8.9) above, the Jacobian is

$$J(z) = \begin{bmatrix} 2z_1 & 2z_2 & 0 \\ \sin z_1 & 0 & 1 \\ 0 & -\cos z_2 & 1 \end{bmatrix}. \qquad (8.13)$$

When F is smooth, we can use the values of $F(z)$ and $J(z)$ to construct an approximation to the value of $F(z + \Delta z)$ for Δz small. Extending the use of Taylor's theorem to multivariate functions, we have the approximation

$$F(z + \Delta z) \approx F(z) + J(z)\Delta z. \qquad (8.14)$$

(Compare with (8.10) and see Section A.7 for discussion of Taylor's theorem.) As for the scalar case ($N = 1$), Newton's method chooses the step Δz^k from iterate z^k to make the linear approximation in (8.14) equal to zero, that is,

$$F(z^k) + J(z^k)\Delta z^k = 0.$$

In practice, we obtain Δz^k by solving the following $N \times N$ system of linear equations to find the step Δz^k:

$$J(z^k)\Delta z^k = -F(z^k). \qquad (8.15)$$

From (8.14), we have that $F(z^k + \Delta z^k) \approx 0$; that is, $z^k + \Delta z^k$ is an approximate solution of the true problem $F(z) = 0$. We then obtain the next iterate z^{k+1} by setting

$$z^{k+1} = z^k + \Delta z^k. \qquad (8.16)$$

Newton's method often converges rapidly once it reaches the vicinity of a solution of $F(z) = 0$. When F is smooth (Lipschitz continuously differentiable, to be precise) near the solution z^*, and when $J(z^*)$ is nonsingular, Newton's method actually converges *quadratically* to z^*. By this we mean that the error ϵ_k, defined by

$$\epsilon_k := \|z^k - z^*\|_2,$$

satisfies

$$\epsilon_{k+1} \leq C\epsilon_k^2 \qquad (8.17)$$

for some constant $C > 0$. The sequence of function norms $\|F(z^k)\|$ converges at a similar rate. In practice, the Newton algorithm can be terminated successfully when $\|\Delta z^k\|$ becomes small or unsuccessfully if it appears that convergence is not taking place.

MATLAB file newton.m: Newton's method for solving $F(z) = 0$

```
function [z] = newton(z0, F, J, eps, itmax)
%
% syntax: z = newton(z0, @myF, @myJ, eps, itmax)
%
% performs Newton's method from starting point z0, terminating
% when 2-norm of step is shorter than eps or when at itmax
% steps have been taken, whichever comes first. Call as follows:
%
% where z0 is the starting point, myF and myJ are the actual
% names of the function and Jacobian evaluation routines;
% method terminates when length of Newton step drops below eps
% or after at most itmax iterations (whichever comes first).

z=z0; iter = 0;
while iter<itmax
  Fval = feval(F,z); Jval = feval(J,z);
  fprintf(' iteration %3d, Fnorm=%9.4e\n', iter, norm(Fval));
  zstep = -Jval\Fval;
  z = z + zstep; iter = iter+1;
  if norm(zstep) < eps  % stop if the step is short
    break;
  end
end
return;
```

MATLAB file Ftrig.m: Evaluation of F from (8.9)

```
function Fval = Ftrig(z)
% evaluates the three-dimensional example at the vector x in
% R^3
Fval = [z(1)^2 + z(2)^2 - pi^2; z(3) - cos(z(1));
z(3) - sin(z(2))];
```

MATLAB file Jtrig.m: Evaluation of Jacobian of F from (8.9)

```
function Jval = Jtrig(z)
% evaluates the 3 x 3 Jacobian of the example function
Jval = [2*z(1) 2*z(2) 0; sin(z(1)) 0 1 ; 0 -cos(z(2)) 1];
```

A simple code implementing Newton's method is given in `newton.m`. Note that the argument list of `newton.m` includes the names of the MATLAB functions used to evaluate the function F and Jacobian J at a given point. We show the code for the function (8.9) and its Jacobian in `Ftrig.m` and `Jtrig.m`.

To invoke Newton's method for solving (8.9), starting from the point $z^0 = (1, 3, 0)'$, we can use the following code:

```
>> z0=[1; 3; 0]; eps = 1.e-12; itmax = 10;
>> z = newton(z0,@Ftrig,@Jtrig,eps,itmax);
```

The `newton` function produces the following output:

```
iteration    0, Fnorm=5.7345e-01
iteration    1, Fnorm=1.2284e-01
iteration    2, Fnorm=9.3889e-04
iteration    3, Fnorm=4.1466e-08
iteration    4, Fnorm=1.1102e-16
```

and returns with a value of z close to the solution indicated earlier. The sequence of norms $\|F(z^k)\|$ displayed by `newton` shows clear evidence of quadratic convergence.

Despite its excellent local convergence properties, Newton's method can behave erratically when the initial point z^0 is far from a solution. One way to improve the performance of the method is to introduce a line search. Instead of defining the new iterate by (8.16), we set

$$z^{k+1} = z^k + \alpha_k \Delta z^k, \tag{8.18}$$

where α_k is a positive scalar known as the *steplength*, chosen to make z^{k+1} a better approximate solution than z^k. We can for instance choose α_k to be the approximate minimizer of the function

$$\|F(z^k + \alpha_k \Delta z^k)\|_2$$

for the given z^k and Δz^k. This choice ensures that

$$\|F(z^0)\|_2 > \|F(z^1)\|_2 > \cdots \geq \|F(z^k)\|_2 > \|F(z^{k+1})\|_2 > \cdots$$

and often (but not always) results in the iterates making steady progress toward a solution of $F(z) = 0$.

Exercise 8-2-1. Use `newton` to solve (8.9) starting from the point $z^0 = (2, 2, 0.5)$. Tabulate the values of $\|F(z^k)\|_2$ for $k = 0, 1, 2, 3$.

8.3 Primal-Dual Methods

Primal-dual interior-point methods are founded in part on the observation that the equations (8.7a), (8.7b), (8.7c) form a system in which the number of equations equals the number of unknowns, and so Newton's method can be applied. However, the need to handle the nonnegativity conditions (8.7d) leads to significant (but interesting!) complications.

We define the function $F_0(x, y, s)$ of the primal-dual vector triple (x, y, s) as follows:

$$F_0(x, y, s) := \begin{bmatrix} Ax - b \\ A'y + s - p \\ XSe \end{bmatrix}. \tag{8.19}$$

Note that the vector triple (x, y, s) contains $2n + m$ components in all and that the vector function F_0 has $2n + m$ components. A vector triple (x^*, y^*, s^*) satisfies the KKT conditions (8.7) if and only if

$$F_0(x^*, y^*, s^*) = 0, \qquad x^* \geq 0, \qquad s^* \geq 0.$$

The Jacobian $J_0(x, y, s)$ of F_0 is the following block 3×3 matrix:

$$J_0(x, y, s) = \begin{bmatrix} A & 0 & 0 \\ 0 & A' & I \\ S & 0 & X \end{bmatrix}. \tag{8.20}$$

Starting from a point (x^0, y^0, s^0) with $x^0 > 0$ and $s^0 > 0$, primal-dual methods generate subsequent iterates by applying Newton's method to the function F_0—or a slightly modified version of this function—choosing the steplength α_k to ensure that the positivity conditions $x^k > 0$ and $s^k > 0$ are satisfied at all iterates (x^k, y^k, s^k).

We maintain positivity conditions $x^k > 0$ and $s^k > 0$ at all iterates for two reasons. First, vectors that solve the system $F_0(x, y, s) = 0$ and yet have negative components in x or s are of no interest in terms of solving the primal and dual problems (8.1) and (8.3). They are usually far from solutions of these linear programs, and we cannot recover such solutions easily from them. Second, when the matrix A has linearly independent rows, the Jacobian $J_0(x, y, z)$ is guaranteed to be nonsingular whenever the positivity conditions $x > 0$ and $s > 0$ hold, and so linear systems that have this matrix as the coefficient matrix are guaranteed to have a solution. We prove this assertion later in discussing implementation of primal-dual methods.

8.3.1   An Affine-Scaling Approach

The simplest primal-dual approach is to apply Newton's method directly to the function F_0, using a steplength α_k of less than one if this is necessary to maintain positivity of the components of x and s. Specifically, at the iterate (x^k, y^k, s^k), satisfying $x^k > 0$ and $s^k > 0$, we obtain a Newton direction by solving the system

$$J_0(x^k, y^k, s^k) \begin{bmatrix} \Delta x^k \\ \Delta y^k \\ \Delta s^k \end{bmatrix} = -F_0(x^k, y^k, s^k). \tag{8.21}$$

We replace F_0 and J_0 in the system (8.21) by their definitions (8.19) and (8.20) resulting in the system

$$\begin{bmatrix} A & 0 & 0 \\ 0 & A' & I \\ S^k & 0 & X^k \end{bmatrix} \begin{bmatrix} \Delta x^k \\ \Delta y^k \\ \Delta s^k \end{bmatrix} = - \begin{bmatrix} Ax^k - b \\ A'y^k + s^k - p \\ X^k S^k e \end{bmatrix}.$$

Using the definitions (8.6) of the diagonal matrices X^k and S^k, we can write out the last block row of this system componentwise as follows:

$$s_i^k \Delta x_i^k + x_i^k \Delta s_i^k = -x_i^k s_i^k, \qquad i = 1, 2, \ldots, n. \qquad (8.22)$$

We then define the new iterate $(x^{k+1}, y^{k+1}, s^{k+1})$ as follows:

$$(x^{k+1}, y^{k+1}, s^{k+1}) = (x^k, y^k, s^k) + \alpha_k (\Delta x^k, \Delta y^k, \Delta s^k) \qquad (8.23)$$

for some steplength α_k. We need to choose α_k to maintain positivity of the x and s components, that is,

$$x_i^k + \alpha_k \Delta x_i^k > 0, \qquad i = 1, 2, \ldots, n,$$
$$s_i^k + \alpha_k \Delta s_i^k > 0, \qquad i = 1, 2, \ldots, n.$$

It is not hard to see that the largest value of α_k that maintains these conditions is given by the following formula:

$$\alpha_{\max} = \min \left(\min_{i \,|\, \Delta x_i^k < 0} \frac{x_i^k}{-\Delta x_i^k}, \; \min_{i \,|\, \Delta s_i^k < 0} \frac{s_i^k}{-\Delta s_i^k} \right), \qquad (8.24)$$

which is similar to the ratio test used by the simplex method. We can step back from this maximum value, and prevent each x_i and s_i from being too close to zero, by defining α_k as follows:

$$\alpha_k = \min \left(1, \eta_k * \alpha_{\max} \right), \qquad (8.25)$$

where η_k is some factor close to, but less than, 1. (A typical value of η_k is .999.) The value in (8.25) is almost as long as it can possibly be while maintaining positivity and therefore makes as much progress as it reasonably can make in the Newton direction.

This approach is referred to as *primal-dual affine scaling*. While it can be shown to converge to a primal-dual solution, for an appropriate steplength selection strategy (more complicated than the simple formula (8.25)), it often requires many iterations. Usually, until the iterates reach the vicinity of a solution, the x and s components of the step— Δx^k and Δs^k—move too sharply toward the boundary of the sets defined by $x \geq 0$ and $s \geq 0$. Therefore, the steplength α_k needs to be set to a small value to avoid violating these conditions.

More successful approaches use modifications of the search direction that do not move so sharply toward the boundary. That is, the Δx^k and Δs^k components are reoriented so as to allow a longer steplength α_k to be used. These modifications are defined by applying Newton's method not to F_0 but to a perturbed version of this system, as we see in the next section.

Exercise 8-3-1. Show that for steps $(\Delta x^k, \Delta y^k, \Delta s^k)$ generated by the affine-scaling method, we have that for each $i = 1, 2, \ldots, n$, either $\Delta x_i^k < 0$ or $\Delta s_i^k < 0$ (or both). (Hint: use (8.22) and the fact that $x^k > 0$ and $s^k > 0$.)

Exercise 8-3-2. Verify that $\alpha_{\max}$ defined by (8.24) has the property that $x^k + \alpha \Delta x^k > 0$ and $s^k + \alpha \Delta s^k > 0$ for all $\alpha \in [0, \alpha_{\max})$, while for $\alpha \geq \alpha_{\max}$ we have that at least one component of $(x^k + \alpha \Delta x^k, s^k + \alpha \Delta s^k)$ is nonpositive.

MATLAB file steplength.m: Finds a steplength that maintains positivity of x and s

---

```
function [alpha, alphax, alphas]
  = steplength(x, s, Dx, Ds, eta)
%
% syntax: [alpha, alphax, alphas]
% = steplength(x, s, Dx, Ds, eta)
%
% given current iterate (x,s) and steps (Dx,Ds), compute
% steplengths that ensure that x + alphax*Dx>0 and
% s + alphas*Ds>0, and alpha = min(alphax,alphas). eta
% indicates the maximum fraction of step to the boundary
% (typical value: eta=.999)

    alphax = -1/min(min(Dx./x),-1);
    alphax = min(1, eta * alphax);
    alphas = -1/min(min(Ds./s),-1);
    alphas = min(1, eta * alphas);
    alpha = min(alphax, alphas);
```

Exercise 8-3-3. Show that the value

$$\alpha_{x,\max} = \min\left(\min_{i\,|\,\Delta x_i < 0} \frac{x_i}{-\Delta x_i}, 1\right) \tag{8.26}$$

also satisfies the definition

$$\alpha_{x,\max} = -\left[\min\left(\min_{i=1,2,\dots,n} \frac{\Delta x_i}{x_i}, -1\right)\right]^{-1}. \tag{8.27}$$

(The latter form is sometimes more convenient for computation and is used in the MATLAB routine `steplength.m`.)

8.3.2   Path-Following Methods

Path-following methods take a less "greedy" approach to satisfying the complementarity conditions than do affine-scaling methods. Rather than aiming with every iteration of Newton's method to satisfy these conditions $XSe = 0$ (the third component of the system $F_0(x, y, s) = 0$), they use Newton iterations to aim at points at which the n pairwise products $x_i s_i$, $i = 1, 2, \dots, n$, are reduced from their present value (though not all the way to zero) and are more "balanced." These algorithms are based on the concept of the central path, which we now introduce.

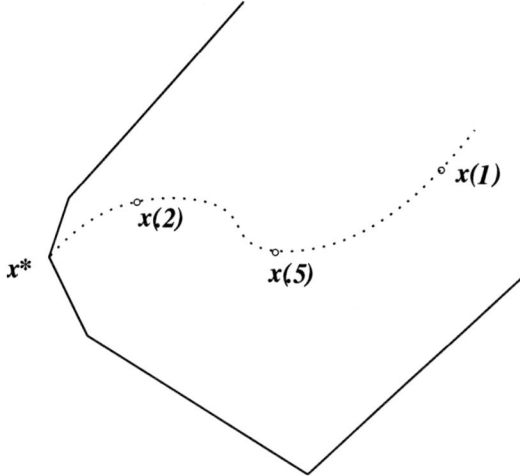

Figure 8.2. *Central path projected into space of primal variables, showing* $x(\tau)$
for the three values $\tau = 1$, $\tau = .5$, *and* $\tau = .2$.

We start by defining the vector triple $(x(\tau), y(\tau), s(\tau))$ to satisfy the following system
of equations for a given positive parameter τ:

$$Ax = b, \tag{8.28a}$$

$$A'y + s = p, \tag{8.28b}$$

$$XSe = \tau e, \tag{8.28c}$$

$$x > 0, \qquad s > 0. \tag{8.28d}$$

Note that these conditions represent a perturbation of the KKT conditions (8.7), in which
the pairwise products $x_i s_i$ are all required to take on the same positive value τ rather than
being zero. Provided that there is a vector triple (x, y, s) such that x is feasible for (8.1) with
$x > 0$ and (y, s) is feasible for (8.3) with $s > 0$, the system (8.28) has a *unique* solution
$(x(\tau), y(\tau), s(\tau))$ for any $\tau > 0$.

We define the *central path* C to be the set of points satisfying (8.28) for any $\tau > 0$;
that is,

$$C := \{(x(\tau), y(\tau), s(\tau)) \mid \tau > 0\}. \tag{8.29}$$

The projection of C into the space of x variables for a linear program in $\mathbf{R}^2$ is shown
in Figure 8.2. Note that as $\tau \downarrow 0$, $x(\tau)$ approaches the solution x^*. (When the problem
(8.1) does not have a unique solution, the path $x(\tau)$ approaches a unique point in the set of
solutions as $\tau \downarrow 0$.) In the space of dual variables, $(y(\tau), s(\tau))$ also approaches a solution
of (8.3) as $\tau \downarrow 0$. Note that for any $(x(\tau), y(\tau), s(\tau)) \in C$, the duality measure μ is equal
to τ, that is,

$$\mu = \frac{x(\tau)'s(\tau)}{n} = \tau.$$

Each step generated by a path-following method is a Newton step toward a point on the central path $\mathcal{C}$. For a given $\tau \geq 0$, we define the following modification of the function F_0 by taking the first three conditions in the system (8.28):

$$F_\tau(x, y, s) = \begin{bmatrix} \mathcal{A}x - b \\ \mathcal{A}'y + s - p \\ XSe - \tau e \end{bmatrix}. \tag{8.30}$$

The unique solution of $F_\tau(x, y, s) = 0$ for which $x > 0$ and $s > 0$ is $(x(\tau), y(\tau), s(\tau))$. Regarding τ as a fixed parameter, we have that the Jacobian of F_τ is

$$J_\tau(x, y, s) = \begin{bmatrix} \mathcal{A} & 0 & 0 \\ 0 & \mathcal{A}' & I \\ S & 0 & X \end{bmatrix}.$$

In fact, $J_\tau(x, y, s)$ is the same as the Jacobian $J_0(x, y, s)$ defined in (8.20) for all τ.

Long-step path-following methods generate each step by applying Newton's method to $F_{\tau_k}(x, y, s) = 0$, where $\tau_k = \sigma_k \mu_k$, $\mu_k = (x^k)'(s^k)/n$ is the duality measure at the current iterate (x^k, y^k, s^k) and σ_k is a parameter in the range $(0, 1)$. Note that the solution $(x(\tau_k), y(\tau_k), s(\tau_k))$ of $F_{\tau_k}(x, y, s) = 0$ has duality measure τ_k, which is smaller than the duality measure μ_k at iteration k. By setting $\sigma_k < 1$, we are aiming at a point that is not only on the central path but also more advanced toward the solution than the current point (x^k, y^k, s^k). As in the affine-scaling approach, we scale the Newton direction by a steplength α_k, chosen to ensure that the next iterate $(x^{k+1}, y^{k+1}, s^{k+1})$ also satisfies the positivity conditions $x^{k+1} > 0$ and $s^{k+1} > 0$.

The parameter σ_k is known as a *centering parameter*. When σ_k is close to 1, the Newton step for $F_\tau(x, y, s)$ tends to produce a point that is more "central" than the current iterate (in the sense that the n pairwise products $x_i s_i$, $i = 1, 2, \ldots, n$, all have similar values), but the new point does not make much progress toward the primal-dual solution. When σ_k is close to zero, the Newton step moves more aggressively toward the solution set, in the manner of the affine-scaling step of Subsection 8.3.1, but typically does not produce much of a balancing effect on the pairwise products $x_i s_i$, $i = 1, 2, \ldots, n$.

Path-following methods take just *one* Newton step for each value of τ_k. At each iteration, we reset $\tau_k = \sigma_k \mu_k$ for the chosen value of σ_k, which produces a generally decreasing sequence of values for τ_k, $k = 0, 1, 2, \ldots$. In a sense, the Newton iterates are chasing a moving target—the target is moving along the central path toward the primal-dual solution.

We can now state a simple algorithmic framework for a path-following method, using the ingredients described above.

Path-Following Algorithm
Choose σ_{min} and σ_{max} such that $0 < \sigma_{min} < \sigma_{max} < 1$;
Choose initial point (x^0, y^0, s^0) with $x^0 > 0$ and $s^0 > 0$;
for $k = 0, 1, 2, \ldots$
    Choose $\sigma_k \in [\sigma_{min}, \sigma_{max}]$;
    Solve for $(\Delta x^k, \Delta y^k, \Delta s^k)$:

$$\begin{bmatrix} \mathcal{A} & 0 & 0 \\ 0 & \mathcal{A}' & I \\ S^k & 0 & X^k \end{bmatrix} \begin{bmatrix} \Delta x^k \\ \Delta y^k \\ \Delta s^k \end{bmatrix} = - \begin{bmatrix} \mathcal{A}x^k - b \\ \mathcal{A}'y^k + s^k - p \\ X^k S^k e - \sigma_k \mu_k e \end{bmatrix}, \qquad (8.31)$$

where $\mu_k = (x^k)'s^k/n$, $X^k = \text{diag}(x_1^k, x_2^k, \ldots, x_n^k)$, etc.;
Choose $\alpha_{\max}$ to be the largest positive value of α such that

$$(x^k, s^k) + \alpha(\Delta x^k, \Delta s^k) \geq 0;$$

Set $\alpha_k = \min(1, \eta_k \alpha_{\max})$ for some $\eta_k \in (0, 1)$;
Set $(x^{k+1}, y^{k+1}, s^{k+1}) = (x^k, y^k, s^k) + \alpha_k(\Delta x^k, \Delta y^k, \Delta s^k)$;
end (for).

Reasonable values for the constants in this algorithm are $\sigma_{\min} = 10^{-2}$ and $\sigma_{\max} = .75$.

Note that we have considerable freedom in the choice of centering parameters σ_k and steplength scaling parameters η_k at each iteration. We would like to choose these parameters so as to make good progress in reducing the duality measure to zero on the current step (which requires σ_k close to zero) while at the same time maintaining steplengths α_k close to 1 (which is more likely to happen when σ_k is closer to 1). In general, it is a good idea to use a smaller, more aggressive value of σ_k on the current iteration when the steplength α_{k-1} was close to 1 on the previous iteration or when the affine-scaling direction computed at the current iteration allows a steplength close to 1 before violating the positivity conditions $x > 0$ and $s > 0$. We comment further on this point in Section 8.3.4.

It has proven effective in practice to choose η_k closer to 1 as we approach the solution but in any case to ensure that this parameter does not drop below a threshold quite close to 1, for example, 0.99. Again, see Section 8.3.4 for further discussion of this issue.

Under some additional assumptions and slightly different conditions on the choice of α_k, the algorithm above can be shown to have a *polynomial complexity* property. Specifically, given a small tolerance ϵ, a starting point (x^0, y^0, s^0) that satisfies the conditions (8.4a) and (8.4b) and is close to the central path in a certain sense, and a strategy for selecting α_k that keeps all iterates within a certain neighborhood of the central path, the method converges to a point (x, y, s) for which $x's \leq \epsilon$ in $O\left(n \log\left((x^0)'(s^0)/\epsilon\right)\right)$ iterations.

Exercise 8-3-4. Show that if (x^k, y^k, s^k) is feasible with respect to the conditions (8.4a) and (8.4b), then the next iterate $(x^{k+1}, y^{k+1}, s^{k+1})$ also satisfies these conditions. (It follows that in fact *all* subsequent iterates satisfy these feasibility conditions.)

Exercise 8-3-5. Show that if (x^k, y^k, s^k) is feasible with respect to (8.4a) and (8.4b), the following conditions hold:

$$(\Delta x^k)' \Delta s^k = 0,$$

$$(s^k)' \Delta x^k + (x^k)' \Delta s^k = (-1 + \sigma_k)(x^k)'s^k,$$

$$(x^k + \alpha \Delta x^k)'(s^k + \alpha \Delta s^k) = (1 - \alpha(1 - \sigma_k))(x^k)'s^k \qquad \forall \alpha > 0.$$

8.3.3   Solution of the Linear System at Each Interior-Point Iteration

The major computational operation at each step of any primal-dual interior-point algorithm, including the Path-Following Algorithm above, is to solve a linear system like (8.31), which we state here without the superscripts and in a more general form:

$$\begin{bmatrix} \mathcal{A} & 0 & 0 \\ 0 & \mathcal{A}' & I \\ S & 0 & X \end{bmatrix} \begin{bmatrix} \Delta x \\ \Delta y \\ \Delta s \end{bmatrix} = \begin{bmatrix} r_b \\ r_p \\ r_{xs} \end{bmatrix}. \tag{8.32}$$

Here S and X are the diagonal matrices with positive diagonal elements defined in (8.6), and r_b, r_p, and r_{xs} represent the three components of the right-hand side. In some variants of the algorithm (see next section), we may need to solve a system of this form more than once at each iteration, with the same coefficient matrix but different right-hand side components r_b, r_p, r_{xs}.

It is crucial to solve this system efficiently since, for practical linear programs, it is often very large. Moreover, the constraint matrix $\mathcal{A}$ (and therefore the overall matrix in (8.32)) is often quite sparse, so that sophisticated sparse linear algebra software is needed. Fortunately, MATLAB incorporates some good sparse solvers, and so it is possible to implement a fairly efficient linear programming solver based on interior-point methods purely in MATLAB—a feat that is not possible for the simplex method.

Although it is possible to apply a sparse solver directly to (8.32), we usually can do much better by exploiting the structure of this system and performing some preliminary block eliminations. By writing the third block of equations as

$$S\Delta x + X\Delta s = r_{xs},$$

we can express Δs in terms of Δx as follows:

$$\Delta s = -X^{-1}S\Delta x + X^{-1}r_{xs}. \tag{8.33}$$

By substituting into the second equation $\mathcal{A}'\Delta y + \Delta s = r_p$ of (8.32) and exchanging the first and second block rows, we obtain the following equivalent form:

$$\begin{bmatrix} -X^{-1}S & \mathcal{A}' \\ \mathcal{A} & 0 \end{bmatrix} \begin{bmatrix} \Delta x \\ \Delta y \end{bmatrix} = \begin{bmatrix} r_p - X^{-1}r_{xs} \\ r_b \end{bmatrix}. \tag{8.34}$$

The matrix in this system is *symmetric indefinite*, and there are sparse linear solvers that can exploit this property. However, we can reduce it to an even more compact and convenient form. Noting that $X^{-1}S$ is again a diagonal matrix with positive diagonal elements, whose inverse is $S^{-1}X$, we can use the first equation in (8.34) to eliminate Δx as follows:

$$\Delta x = -S^{-1}X\left[r_p - X^{-1}r_{xs} - \mathcal{A}'\Delta y\right]. \tag{8.35}$$

By substituting into the second equation in (8.34), we obtain the following system involving only Δy:

$$\mathcal{A}(S^{-1}X)\mathcal{A}'\Delta y = r_b + \mathcal{A}S^{-1}X\left[r_p - X^{-1}r_{xs}\right]. \tag{8.36}$$

This form is sometimes known as the *normal-equations* form. The matrix $\mathcal{A}(S^{-1}X)\mathcal{A}'$ is a symmetric positive definite matrix, provided that $\mathcal{A}$ has linearly independent rows. Note that this matrix needs to be recalculated at each iteration, as the diagonal matrices X and S change from one step to the next. Most software uses a sparse Cholesky factorization procedure to obtain a lower triangular matrix L and a permutation matrix P such that

$$LL' = P\mathcal{A}(S^{-1}X)\mathcal{A}'P'. \tag{8.37}$$

We can then perform triangular substitution (forward, then backward) to obtain the solution Δy of (8.36), then recover Δx using the formula (8.35), and finally recover Δs using the formula (8.33).

The permutation matrix P can be computed before the first iteration of the algorithm. It is chosen to ensure that the factor L stays reasonably sparse during the Cholesky factorization. (Techniques for choosing P appropriately have been the subject of a great deal of study in sparse matrix computations.) Since the nonzero structure of $\mathcal{A}(S^{-1}X)\mathcal{A}'$ does not depend on the contents of $S^{-1}X$, the permutation P computed at the first iteration can be used at all subsequent iterations.

The code `pathfollow.m` contains a simple implementation of an algorithm similar to the Path-Following Algorithm above. Use the command `help pathfollow` in MATLAB to see details of the calling sequence for this routine. Note that this solver uses the MATLAB sparse Cholesky routine `cholinc`. Study the `pathfollow.m` code closely to see how the choices of the parameters σ_k and η_k are made by this implementation and to understand how the implementation otherwise varies slightly from the Path-Following Algorithm specified above.

Exercise 8-3-6. Test the routine `pathfollow()` by solving the problem

$$\begin{array}{lrcrcrcrcl} \min & -x_1 & - & 5x_2 \\ \text{subject to} & x_1 & + & x_2 & + & x_3 & & & = & 5, \\ & x_1 & + & 3x_2 & & & + & x_4 & = & 7, \\ & & & & & x_1, x_2, x_3, x_4 & & \geq & 0. \end{array}$$

Note: You need to call the routine `pathfollow` with a matrix defined in the MATLAB sparse format. MATLAB uses the command `sparse()` to create a matrix in sparse format from a given dense matrix. For the matrix given here, you can define `Aden = [1 1 1 0; 1 3 0 1];` and then set `A = sparse(Aden)`. The data structure A can then be passed to `pathfollow()`.

The code `pathfollowTest.m` shows some sample code for calling the routine `pathfollow()` with random data. The dimensions m and n are specified, and the sparse matrix A is chosen at random to have a specified density. The remaining data p and b are chosen to ensure that the primal and dual problems are feasible. Note that `pathfollow()` prints one line of information at each interior-point iteration, showing the value of duality measure μ_k and a measure of the infeasibility in the conditions (8.4a) and (8.4b).

8.3.4   Practical Primal-Dual Methods

Practical implementations of interior-point method underwent much refinement between 1988 and the early 1990s. Though they continued to follow the framework presented in the

---

MATLAB file pathfollowTest.m: Code to call `pathfollow()` with random data

---

```
% generate a sparse random matrix of given density
m=10; n=100; density=0.2; A = sprandn(m,n,density);

% choose feasible x, y, s at random, with x and s each about
% half-full
xfeas = [rand(n/2,1); zeros(n-(n/2),1)];
sfeas = [zeros(n/2,1); rand(n-(n/2),1)];
xfeas = xfeas(randperm(n)); sfeas = sfeas(randperm(n));
yfeas = (rand(m,1)-0.5)*4;

% choose b and p to make this (x,y,s) feasible
b = A*xfeas; p=A'*yfeas+sfeas;

% call the solver
[x,y,s,f] = pathfollow(A,b,p);

fprintf(1,' final primal value: %12.6e \n', p'*x);
fprintf(1,' final dual   value: %12.6e \n', b'*y);
fprintf(1,' primal infeas: %12.6e \n', norm(A*x-b));
fprintf(1,' dual   infeas: %12.6e \n', norm(A'*y+s-p));
```

Path-Following Algorithm presented above, they introduced several small but significant variations to the basic approach and devised effective heuristics for choosing the initial point (x^0, y^0, s^0) and parameters such as σ_k and η_k. These heuristics decreased both the failure rate and the number of interior-point iterations required to find an approximate solution. We list here the most important modifications and enhancements.

1. An adaptive choice of the centering parameter σ_k at each iteration. The method first calculates the pure Newton (affine-scaling) step, obtained by replacing σ_k by 0 in (8.31). If it is possible to take a steplength of nearly 1 in this direction before violating the nonnegativity conditions on x and s, the method concludes that relatively little centering is needed, and so it sets σ_k close to zero. If it is possible to go only a short distance along the affine-scaling direction before violating the nonnegativity conditions, the method chooses a larger value of σ_k, placing more emphasis on centering the iterates (to balance the pairwise products $x_i s_i$, $i = 1, 2, \ldots, n$) rather than on making rapid progress toward the solution.

2. Adding a "corrector" component to the search direction. The method obtains its search direction by solving the following system:

$$\begin{bmatrix} A & 0 & 0 \\ 0 & A' & I \\ S^k & 0 & X^k \end{bmatrix} \begin{bmatrix} \Delta x^k \\ \Delta y^k \\ \Delta s^k \end{bmatrix} = - \begin{bmatrix} A x^k - b \\ A' y^k + s^k - p \\ X^k S^k e - \sigma_k \mu_k e + \Delta X_{\text{aff}}^k \Delta S_{\text{aff}}^k e \end{bmatrix}, \quad (8.38)$$

where ΔX_{aff}^k and ΔS_{aff}^k are diagonal matrices constructed from the x and s components of the affine-scaling direction computed in Step 1 above. The motivation for this additional component on the right-hand side comes from higher-order variants of Newton's method, in which we replace the usual linear (first-order) model that is the basis of Newton's method with an approximate second-order model.

3. A heuristic for choosing the starting point (x^0, y^0, s^0), based on a least-squares fit to the feasibility conditions $Ax = b$, $A'y + s = c$ (see Section 9.2.4), with adjustments to ensure that x^0 and s^0 are sufficiently positive. A starting point of the form $(x^0, y^0, s^0) = (\beta e, 0, \beta e)$ (for some large positive value of β) often suffices, and in fact we use a starting point of this form in `pathfollow()`. However, the least-squares heuristic leads to convergence on a wider range of problems.

4. A heuristic to determine η_k. Better efficiency can often be achieved by allowing this fraction to approach 1 as the iterates approach the solution. In `pathfollow()`, we use the simple assignment

$$\eta_k = \max(.9995, 1 - \mu_k).$$

Many more complex variants have been tried.

5. The use of different steplengths in the primal variable x and the dual variables (y, s) speeds the convergence and improves robustness in some cases. We can compute the maximum primal step from the formula (8.26) (alternatively, (8.27)) and compute the maximum dual step $\alpha_{s,\max}$ similarly. We then set

$$\alpha_x^k = \min(1, \eta_k \alpha_{x,\max}), \qquad \alpha_s^k = \min(1, \eta_k \alpha_{s,\max})$$

for the parameter $\eta_k \in (0, 1)$ mentioned above, and define the new iterates as follows:

$$x^{k+1} = x^k + \alpha_x^k \Delta x^k,$$
$$(y^{k+1}, s^{k+1}) = (y^k, s^k) + \alpha_s^k (\Delta y^k, \Delta s^k).$$

The pure Newton (affine-scaling) direction calculated in Step 1 is sometimes referred to as a "predictor" direction, and the whole technique is sometimes called "predictor-corrector" or "Mehrotra predictor-corrector" (honoring the paper by Mehrotra (1992), in which many of the techniques above were described). Unlike predictor-corrector techniques in other areas of computational mathematics, however, we do not actually take a step along the "predictor" direction but use it only as a basis for choosing the centering parameter σ_k. It would perhaps be more accurate to call it a "probing" direction.

The main cost of implementing an algorithm that combines all these heuristics is from the solution of two linear systems: A system of the form (8.31) with σ_k replaced by 0 (to obtain the affine-scaling step), and the system (8.38) (to obtain the search direction). Both linear systems have the same coefficient matrix, and so just a single matrix factorization is required. The technique described in the previous subsection can be used for this operation.

8.4   Interior-Point vs. Simplex

Are interior-point methods faster than simplex methods? This question has no simple answer. The practical efficiency of both methods depends strongly on how they are implemented. The number of iterations required by an interior-point method is typically between 10 and 100—almost always fewer than the number of simplex iterations, which is typically 2 to 3 times the number of primal and dual variables. However, each interior-point iteration is considerably more expensive, computationally speaking, than a simplex iteration.

After being relatively stable throughout the 1970s and 1980s, simplex codes became dramatically more efficient after 1990, in part because of the competition they faced from the new interior-point codes. New pricing strategies (for selecting variables to enter the basis) played a large role in this improvement, along with improvements in sparse linear algebra. Improvements in preprocessing (which involves reducing of the size of the linear program prior to solving it by inspecting the constraints of the primal and dual problems closely) also played a role in improving both simplex and interior-point codes.

In interior-point methods, the critical operation is the sparse matrix factorization operation (8.37). Different row/column ordering strategies (that is, different choices of the permutation matrix P) can lead to dramatically different factorization times. Moreover, the approach based on (8.36) needs to be modified when A contains dense columns, since these produce unacceptable "fill-in" in the product $A(S^{-1}X)A'$. Different implementations handle this issue in different ways.

As a general rule, interior-point methods are competitive with simplex (sometimes much faster) on large linear programs. On smaller problems, simplex is usually faster. Simplex methods also have the advantage that they are more easily warm-started. They can take advantage of prior knowledge about a solution (for example, a good initial estimate of the optimal basis) to greatly reduce the number of iterations. Warm-start information for interior-point methods leads to more modest improvements in run time. Production software for both simplex and interior-point methods has reached a fairly high degree of sophistication and can be used with confidence.

8.5   Extension to Quadratic Programming

The algorithms of the previous sections can be extended to convex quadratic programming problems. The underlying concepts are identical, though certain aspects of the implementation change. We show in this section how the path-following approach can be extended to quadratic programs expressed in the form (7.7). (The algorithms could just as easily be extended to the general form (7.12), but the notation would be more complicated.)

The KKT conditions for (7.7) are given as (7.8); we restate them here:

$$0 \leq x \;\perp\; Qx - A'u + p \geq 0,$$
$$0 \leq u \;\perp\; Ax - b \geq 0.$$

Here Q is a symmetric positive semidefinite $n \times n$ matrix, while A is an $m \times n$ matrix. By introducing slack variables v and s, we can rewrite these conditions as follows:

$$Qx - A'u - s = -p, \tag{8.39a}$$
$$Ax - v = b, \tag{8.39b}$$

$$0 \le x \perp s \ge 0, \tag{8.39c}$$

$$0 \le u \perp v \ge 0. \tag{8.39d}$$

By defining diagonal matrices X, S, U, V from the components of x, s, u, v, as in (8.6), we can rewrite the KKT conditions in a form similar to (8.7):

$$Qx - A'u - s = -p, \tag{8.40a}$$

$$Ax - v = b, \tag{8.40b}$$

$$XSe = 0, \tag{8.40c}$$

$$UVe = 0, \tag{8.40d}$$

$$x \ge 0, \quad s \ge 0, \quad u \ge 0, \quad v \ge 0. \tag{8.40e}$$

(Note that x and s are vectors in $\mathbf{R}^n$ while u and v are vectors in $\mathbf{R}^m$. The vector $e = (1, 1, \ldots, 1)'$ thus has m elements in (8.40d) and n elements in (8.40c), but we tolerate this minor abuse of notation.)

As for the linear programming algorithms developed above, we obtain path-following methods by applying modified Newton methods to the linear system formed by (8.40a), (8.40b), (8.40c), and (8.40d). As above, we require the primal and dual variables with nonnegativity constraints to stay strictly positive at all iterations, that is,

$$(x^k, s^k, u^k, v^k) > 0, \qquad k = 0, 1, 2, \ldots.$$

Following (8.8), we define the duality measure by summing together the violations of the complementarity conditions:

$$\mu = \frac{1}{m+n} \left(x's + u'v \right) = \frac{1}{m+n} \left(\sum_{i=1}^{n} x_i s_i + \sum_{j=1}^{m} u_j v_j \right).$$

(We denote μ_k to be the value of μ when evaluated at the point $(x, s, u, v) = (x^k, s^k, u^k, v^k)$.)

Given the elements above, we modify the Path-Following Algorithm of Section 8.3.2 to solve (7.7) (and (7.8)) as follows.

Path-Following Algorithm (Quadratic Programming)
Choose $\sigma_{\min}$ and $\sigma_{\max}$ such that $0 < \sigma_{\min} < \sigma_{\max} < 1$;
Choose initial point $(x^0, s^0, u^0, v^0) > 0$;
for $k = 0, 1, 2, \ldots$
 Choose $\sigma_k \in [\sigma_{\min}, \sigma_{\max}]$;
 Solve for $(\Delta x^k, \Delta s^k, \Delta u^k, \Delta v^k)$:

$$
\begin{bmatrix} Q & -I & -A' & 0 \\ A & 0 & 0 & -I \\ S^k & X^k & 0 & 0 \\ 0 & 0 & V^k & U^k \end{bmatrix}
\begin{bmatrix} \Delta x^k \\ \Delta s^k \\ \Delta u^k \\ \Delta v^k \end{bmatrix}
= -
\begin{bmatrix} Qx^k - A'u^k - s^k + p \\ Ax^k - v^k - b \\ X^k S^k e - \sigma_k \mu_k e \\ U^k V^k e - \sigma_k \mu_k e \end{bmatrix}, \tag{8.41}
$$

 where μ_k, X^k, S^k, etc. are defined above;
 Choose $\alpha_{\max}$ to be the largest positive value of α such that

$$(x^k, s^k, u^k, v^k) + \alpha(\Delta x^k, \Delta s^k, \Delta u^k, \Delta v^k) \geq 0;$$

Set $\alpha_k = \min(1, \eta_k \alpha_{\max})$ for some $\eta_k \in (0, 1)$;
Set $(x^{k+1}, s^{k+1}, u^{k+1}, v^{k+1}) = (x^k, s^k, u^k, v^k) + \alpha_k(\Delta x^k, \Delta s^k, \Delta u^k, \Delta v^k)$;
end (for).

Most of the modifications of Section 8.3.4, which accelerate the convergence of the path-following approach on practical problems, can also be applied to quadratic programs. In particular, the adaptive choice of the centering parameter σ_k, the predictor-corrector scheme, and the heuristic for determining the steplength scaling parameter η_k can be defined in an analogous fashion to the linear programming case. Because the Hessian Q causes a tighter coupling between primal and dual variables than for linear programming, however, most algorithms for quadratic programming *do not* take different primal and dual steplengths.

We discuss briefly solution of the linear systems at each path-following iteration, which have the following form:

$$\begin{bmatrix} Q & -I & -A' & 0 \\ A & 0 & 0 & -I \\ S & X & 0 & 0 \\ 0 & 0 & V & U \end{bmatrix} \begin{bmatrix} \Delta x \\ \Delta s \\ \Delta u \\ \Delta v \end{bmatrix} = \begin{bmatrix} r_p \\ r_b \\ r_{xs} \\ r_{uv} \end{bmatrix}.$$

By eliminating Δu and Δv, we obtain the following reduced system in $(\Delta x, \Delta s)$:

$$\begin{bmatrix} Q + X^{-1}S & -A' \\ A & U^{-1}V \end{bmatrix} \begin{bmatrix} \Delta x \\ \Delta s \end{bmatrix} = \begin{bmatrix} r_p + X^{-1}r_{xs} \\ r_b + U^{-1}r_{uv} \end{bmatrix}. \tag{8.42}$$

The coefficient matrix of this system is positive semidefinite but not symmetric. (Prove it!) If Q is diagonal, or if it has some other simple structure, we can eliminate Δx to obtain a system like the normal-equations form of (8.36). Generally, however, such a reduction is not possible. We can convert the matrix in (8.42) to symmetric indefinite form (a form to which some standard linear algebra software can be applied) by simply multiplying the second block row by -1 to obtain

$$\begin{bmatrix} Q + X^{-1}S & -A' \\ -A & -U^{-1}V \end{bmatrix} \begin{bmatrix} \Delta x \\ \Delta s \end{bmatrix} = \begin{bmatrix} r_p + X^{-1}r_{xs} \\ -r_b - U^{-1}r_{uv} \end{bmatrix}.$$

Further Reading

For more information on interior-point methods for linear programming, see Wright (1997). Karmarkar (1984) started the modern era of interior-point development. (Karmarkar's method is somewhat different from those discussed in this chapter, but it is responsible for motivating much of the subsequent research in the area.) An important exposé of the central path is provided by Megiddo (1989).

Some interior-point codes for linear programming are freely available on the Internet. One widely used such code is PCx (Czyzyk et al. (1999)). A number of high-quality commercial implementations are available. The interior-point code OOQP (Gertz & Wright (2003)), also freely available, solves convex quadratic programming problems. It is coded in object-oriented form, allowing users to supply linear algebra solver for the linear equations at each iteration that take advantage of the structure of the particular problem at hand.

Finally, we mention Khachiyan (1979), whose ellipsoid method was the first algorithm for linear programming with polynomial complexity but is not efficient in practice.

Chapter 9

Approximation and Classification

In this chapter, we examine the use of linear and quadratic programming techniques to solve applications in approximation/regression and for classification problems of machine learning. In approximation problems, our aim is to find a vector x that solves a system of equalities and/or inequalities "as nearly as possible" in some sense. Various "loss functions" for measuring the discrepancies in each equality and inequality give rise to different regression techniques. We also examine classification problems in machine learning, in which the aim is to find a function that distinguishes between two sets of labeled points in $\mathbf{R}^n$. Throughout this chapter, we use the concepts discussed in Appendix A, and so a review of that material may be appropriate.

9.1 Minimax Problems

In this section, we consider the solution of a modification of linear programming in which the linear objective function is replaced by a convex piecewise-linear function. Such a function can be represented as the pointwise maximum of a set of linear functions that we can reduce to a linear programming problem and solve using the techniques of this book. Recall that we have already seen an example of piecewise-linear convex functions during our study of parametric linear programming—the optimal objective value is piecewise linear when considered as a function of linear variations in the right-hand side.

Consider first the function f defined as follows:

$$f(x) := \max_{i=1,2,\dots,m} (c^i)'x + d_i, \tag{9.1}$$

where $c^i \in \mathbf{R}^n$ and $d_i \in \mathbf{R}$, $i = 1, 2, \dots, m$. We can prove that this function is convex by showing that its epigraph (defined in (A.3)) is a convex set. Writing

$$
\begin{aligned}
\text{epi}(f) &= \{(x, \mu) \mid f(x) \leq \mu\} \\
&= \left\{(x, \mu) \mid \max_{i=1,2,\dots,m} (c^i)'x + d_i \leq \mu\right\} \\
&= \{(x, \mu) \mid (c^i)'x + d_i \leq \mu \quad \forall i = 1, 2, \dots, m\} \\
&= \bigcap_{i=1,2,\dots,m} \{(x, \mu) \mid (c^i)'x + d_i \leq \mu\},
\end{aligned}
$$

217

we see from Lemmas A.2.1 and A.2.2 that this set is indeed convex.

For f defined in (9.1), consider the problem

$$\min_x \ f(x), \qquad Ax \le b, \qquad x \ge 0. \tag{9.2}$$

By introducing an artificial variable μ to play a similar role as in the definition of epi f, we can reformulate (9.2) as the following linear program:

$$
\begin{aligned}
\min_{(x,\mu)} \quad & \mu \\
\text{subject to} \quad & \mu \ge (c^i)'x + d_i \qquad \forall i = 1, 2, \ldots, m, \\
& Ax \le b, \\
& x \ge 0.
\end{aligned}
$$

Note that the constraints themselves do not guarantee that μ equals $f(x)$; they ensure only that μ is *greater than or* equal to $f(x)$. However, the fact that we are *minimizing* ensures that μ takes on the smallest value consistent with these constraints, and so at the optimum it is indeed equal to $f(x)$.

Exercise 9-1-1. Reformulate the following problem as a linear program and use MATLAB to solve it:

$$
\begin{aligned}
\min \quad & \max\{x_1 + x_2 - 1, x_1 - x_2 + 1\} \\
\text{subject to} \quad & x_1 + 4x_2 \le 2, \\
& 3x_1 + x_2 \le 4, \\
& x_1, x_2 \ge 0.
\end{aligned}
$$

Exercise 9-1-2. Use the simplex method to solve the following problem:

$$
\begin{aligned}
\min \quad & 7|x| - y + 3z \\
\text{subject to} \quad & y - 2z \le 4, \\
& -x - y + 3z \le -2, \\
& 3x - y + 3z \ge 0, \\
& y \ge 3, \ z \le 1.
\end{aligned}
$$

(Note that $|x| = \max\{x, -x\}$.)

Note that since f is defined as the pointwise maximum of a finite number of linear functions, this problem is sometimes called the discrete minimax problem. An elegant saddle-point theory is available for problems involving the supremum of a function with continuous variables; see, for example, Mangasarian (1969).

9.2 Approximation

In Section 1.3.2, we described the problem of fitting a linear surface to a set of observations. This problem is a good example of an approximation problem. Such problems seek a vector x that satisfies a set of (possibly inconsistent) linear equalities and inequalities as closely as possible. We define "closeness" in terms of norms, which are defined in Appendix A.

Consider first the following set of linear equations:

$$Ax = b,$$

where $A \in \mathbf{R}^{m \times n}$ that may or may not be consistent. We replace this problem with the following:

$$\min_{x \in \mathbf{R}^n} \quad \|Ax - b\|_p, \tag{9.3}$$

where p is usually 1, 2, or ∞. Because $\|r\|_p = 0$ for $r \in \mathbf{R}^m$ if and only if $r = 0$, any vector x that solves $Ax = b$ also is a minimizer of $\|Ax - b\|_p$, and conversely. If no solution to $Ax = b$ exists, the minimization problem will yield the x that minimizes the norm of the "error" or "residual" vector $r = Ax - b$ in the ℓ_p-norm sense.

We consider three norm choices $p = \infty$, $p = 1$, and $p = 2$ in turn, showing how the techniques of the earlier chapters can be used to solve the resulting minimization problems.

9.2.1 Chebyshev Approximation

When the ℓ_∞-norm is used to measure the residual norm, we obtain the following problem:

$$\min_{x \in \mathbf{R}^n} \|Ax - b\|_\infty = \min_{x \in \mathbf{R}^n} \max_{1 \leq i \leq m} |A_i.x - b_i|. \tag{9.4}$$

This problem is often referred to as *Chebyshev approximation*, after the famous Russian mathematician. Since $|A_i.x - b_i| = \max\{A_i.x - b_i, -(A_i.x - b_i)\}$, we can formulate this problem as a linear program by introducing a single variable ϵ to represent the objective function as follows:

$$\min_{\epsilon, x} \epsilon \quad \text{subject to} \quad -e\epsilon \leq Ax - b \leq e\epsilon, \tag{9.5}$$

where e represents a vector of ones with m elements. The constraints ensure that ϵ is no smaller than $|A_i.x - b_i|$, $i = 1, 2, \ldots, m$, while the fact that we are minimizing ensures that ϵ is set to the largest of these values (and no larger).

Lemma 9.2.1. *The problems (9.4) and (9.5) are equivalent.*

Proof. If $(\bar{\epsilon}, \bar{x})$ solves (9.5), then

$$\bar{\epsilon} = |A_j.\bar{x} - b_j|$$

for some j; otherwise the objective function can be improved by decreasing $\bar{\epsilon}$. Furthermore,

$$\bar{\epsilon} \geq |A_i.\bar{x} - b_i|$$

for all $i \neq j$ by feasibility. Hence $\bar{\epsilon} = \|A\bar{x} - b\|_\infty$. Now suppose that x is any point in $\mathbf{R}^n$. Then for $\epsilon := \|Ax - b\|_\infty$, (ϵ, x) is feasible for the linear program (9.5) and

$$\|Ax - b\|_\infty = \epsilon \geq \bar{\epsilon} = \|A\bar{x} - b\|_\infty,$$

where the inequality follows from the fact that $\bar{\epsilon}$ is the minimum value of (9.5).

Conversely, if $\bar{x}$ is the solution of (9.4), let $\bar{\epsilon} = \|A\bar{x} - b\|_\infty$. Then for any feasible point (ϵ, x) of (9.5), we have

$$\epsilon \geq \|Ax - b\|_\infty,$$

and so

$$\bar{\epsilon} = \|A\bar{x} - b\|_\infty \leq \|Ax - b\|_\infty \leq \epsilon.$$

Hence, $(\bar{\epsilon}, \bar{x})$ is a solution of (9.5), as required. $\quad\square$

By making use of the equivalence between (9.4) and (9.5), we can prove that (9.4) always exists.

Proposition 9.2.2. *A solution of* (9.4) *always exists.*

Proof. The equivalent linear program is always feasible ($\epsilon = \|b\|_\infty$, $x = 0$), and its objective is bounded below by zero. Hence, by strong duality, it has a solution. □

Example 9-2-1. Consider the following overdetermined system:

$$\begin{array}{rcrcl}
x_1 & + & x_2 & = & 1, \\
x_1 & - & x_2 & = & 1, \\
x_1 & & & = & 0.
\end{array} \qquad (9.6)$$

We solve the Chebyshev approximation problem for this system by means of the formulation (9.5). As we show, it is convenient to use the dual simplex method to solve it. Scheme II of Chapter 3 can be used to handle the components of x, which are free variables in (9.5).

The following MATLAB code constructs the formulation (9.5) from the data for this problem. We set up (9.6) as follows:

```
>> load ex9-2-1
>> [m,n] = size(A);
>> p = [zeros(n,1);1];
>> augA = [A ones(m,1);-A
   ones(m,1)];
>> T = totbl(augA,[b;-b],p);
>> T = relabel(T,'x3','eps');
```

	x_1	x_2	ϵ	1
$x_4 =$	1	1	1	−1
$x_5 =$	1	−1	1	−1
$x_6 =$	1	0	1	0
$x_7 =$	−1	−1	1	1
$x_8 =$	−1	1	1	1
$x_9 =$	−1	0	1	0
$z =$	0	0	1	0

In this implementation, x_1 and x_2 are the original variables and we relabel x_3 as ϵ. Following Scheme II of Section 3.6.3, we pivot x_1 and x_2 to the side of the tableau and then permute the rows to place them at the bottom. (Note that we can treat ϵ as a nonnegative variable, since the constraints in (9.5) cannot be satisfied if it has a negative value.)

```
>> T = ljx(T,1,1);
>> T = ljx(T,2,2);
>> T = permrows(T,[3:7 1 2]);
```

	x_4	x_5	ϵ	1
$x_6 =$	0.5	0.5	0	1
$x_7 =$	−1	0	2	0
$x_8 =$	0	−1	2	0
$x_9 =$	−0.5	−0.5	2	−1
$z =$	0	0	1	0
$x_1 =$	0.5	0.5	1	1
$x_2 =$	0.5	−0.5	0	0

Note that this tableau is dual feasible because the bottom row is nonnegative, and so we can apply the dual simplex method. (In fact, it is true for all problems (9.5) that we can apply dual simplex after using Scheme II, because the initial cost vector has only nonnegative entries, and this row of the tableau is not affected by the Scheme II pivots.) For this particular problem, a single dual simplex pivot leads to the optimal tableau.

$\gg$ `T = ljx(T,4,3);`

		x_4	x_5	x_9	1
x_6	$=$	0.5	0.5	0	1
x_7	$=$	-0.5	0.5	1	1
x_8	$=$	0.5	-0.5	1	1
ϵ	$=$	0.25	0.25	0.5	0.5
z	$=$	0	0	1	0.5
x_1	$=$	0.25	0.25	-0.5	0.5
x_2	$=$	0.5	-0.5	0	0

Reading the solution from this tableau, we see that

$$x_1 = 0.5, \qquad x_2 = 0, \qquad z = \epsilon = 0.5.$$

In the case in which A has linearly dependent columns, we will not be able to pivot all the original components of x to the side of the tableau. However, since the reduced cost for these components remains at zero throughout the Scheme II procedure, the column corresponding to the unpivoted variable will be entirely zero, and so the variable can be fixed at zero and removed from the problem. ∎

Exercise 9-2-2. Explain, by referring to the formulae for the Jordan exchange, why the reduced costs in the tableau are not affected by the Scheme II procedure for pivoting the original x components to the side of the tableau.

Exercise 9-2-3. Solve the Chebyshev approximation problem associated with the following overdetermined linear system:

$$\begin{array}{rcrcr} 5x_1 & - & x_2 & = & 3, \\ -x_1 & + & 2x_2 & = & -2, \\ x_1 & - & 3x_2 & = & 1. \end{array}$$

9.2.2 L_1 Approximation

We now consider the case in which the ℓ_1-norm is used to measure the size of the residual vector. In this case, (9.3) becomes

$$\min_{x \in \mathbf{R}^n} \|Ax - b\|_1 = \min_{x \in \mathbf{R}^n} \sum_{i=1}^m |A_i.x - b_i|. \tag{9.7}$$

We formulate this problem as a linear program by introducing a vector $y \in \mathbf{R}^m$ such that $y_i = |A_i.x - b_i|$ as follows:

$$\min_{(x,y)} e'y \quad \text{subject to} \quad -y \le Ax - b \le y. \tag{9.8}$$

Lemma 9.2.3. *The problems* (9.7) *and* (9.8) *are equivalent.*

Proof. The proof is similar to that of Lemma 9.2.1. □

Proposition 9.2.4. *A solution of* (9.7) *always exists.*

Proof. The equivalent linear program is always feasible (set $x = 0$, $y_i = |b_i|$, $i = 1, 2, \ldots, m$), and its objective is bounded below by zero. ☐

Example 9-2-4. We now form the simplex tableau corresponding to (9.8) for Example 9-2-1:

```
>> load ex9-2-1
>> [m,n] = size(A);
>> p = [zeros(n,1); ones(m,1)];
>> augA = [A eye(m);-A eye(m)];
>> T = totbl(augA, [b;-b],p);
>> T = relabel(T,'x3','y1');
>> T = relabel(T,'x4','y2',
   'x5','y3');
```

	x_1	x_2	y_1	y_2	y_3	1
$x_6 =$	1	1	1	0	0	−1
$x_7 =$	1	−1	0	1	0	−1
$x_8 =$	1	0	0	0	1	0
$x_9 =$	−1	−1	1	0	0	1
$x_{10} =$	−1	1	0	1	0	1
$x_{11} =$	−1	0	0	0	1	0
$z =$	0	0	1	1	1	0

As in the previous section, we pivot the free variables x_1 and x_2 to the side of the tableau and then move them to the bottom. Once again, these pivots do not affect the objective row, and thus the resulting tableau is dual feasible:

```
>> T = ljx(T,1,1);
>> T = ljx(T,2,2);
>> T = permrows(T,
   [3:7 1 2]);
```

	x_6	x_7	y_1	y_2	y_3	1
$x_8 =$	0.5	0.5	−0.5	−0.5	1	1
$x_9 =$	−1	0	2	0	0	0
$x_{10} =$	0	−1	0	2	0	0
$x_{11} =$	−0.5	−0.5	0.5	0.5	1	−1
$z =$	0	0	1	1	1	0
$x_1 =$	0.5	0.5	−0.5	−0.5	0	1
$x_2 =$	0.5	−0.5	−0.5	0.5	0	0

A single dual simplex pivot results in the optimal tableau given below:

```
>> T = ljx(T,4,5);
```

	x_6	x_7	y_1	y_2	x_{11}	1
$x_8 =$	1	1	−1	−1	1	2
$x_9 =$	−1	0	2	0	0	0
$x_{10} =$	0	−1	0	2	0	0
$y_3 =$	0.5	0.5	−0.5	−0.5	1	1
$z =$	0.5	0.5	0.5	0.5	1	1
$x_1 =$	0.5	0.5	−0.5	−0.5	0	1
$x_2 =$	0.5	−0.5	−0.5	0.5	0	0

A solution to the original problem can be obtained from this optimal tableau using the relations

$$x_1 = 1, \qquad x_2 = 0, \qquad z = y_1 + y_2 + y_3 = 1.$$

The resulting solution is $x_1 = 1$, $x_2 = 0$ with objective value 1. Note also that this solution is different from the one obtained as the Chebyshev solution in (9.6).

Exercise 9-2-5. Solve the ℓ_1-norm approximation problem for the overdetermined system in Exercise 9-2-3.

9.2.3 Approximate Solutions to Systems with Inequality Constraints

Sometimes the approximation problem also contains "hard" constraints—constraints that the solution must satisfy exactly, not just in a "near as possible" sense. When such constraints are present, we simply add them to the linear programming formulation explicitly. Specifically, the constrained Chebyshev approximation problem

$$\min_{x \in \mathbf{R}^n} \|Ax - b\|_\infty \quad \text{subject to} \quad Cx \geq d$$

can be formulated as follows:

$$\min_{\epsilon, x} \epsilon \quad \text{subject to} \quad -e\epsilon \leq Ax - b \leq e\epsilon, \quad Cx \geq d, \tag{9.9}$$

while the constrained L_1 approximation problem

$$\min_{x \in \mathbf{R}^n} \|Ax - b\|_1 \quad \text{subject to} \quad Cx \geq d$$

can be formulated as follows:

$$\min_{(x,y)} e'y \quad \text{subject to} \quad -y \leq Ax - b \leq y, \quad Cx \geq d.$$

Exercise 9-2-6. Can the procedure outlined above for solving (9.5) (that is, Scheme II followed by dual simplex) also be applied to (9.9), or does the presence of the additional constraints $Cx \geq d$ sometimes make it impossible to apply dual simplex directly after the Scheme II pivots?

The formulation techniques used for overdetermined systems of equalities in (9.5) and (9.8) can be generalized to systems of inequalities. Our approach here is essentially the same as in Phase I of the two-phase simplex method, in which we formulate a linear program whose solution is zero when a feasible point for the original problem has been found. The difference here is that we are interested in the solution even if the algebraic system is infeasible, since our goal is to find the vector that satisfies the system "as closely as possible" in some sense. Note that in contrast to the formulations outlined earlier in this section, the constraints $Cx \geq d$ are now allowed to be violated at the solution.

Consider the following system:

$$Ax = b, \qquad Cx \geq d. \tag{9.10}$$

(Note that any less-than inequalities can be converted to greater-than by simply multiplying both sides by -1.) The generalization of the Chebyshev approximation linear program (9.5) is as follows:

$$\min_{\epsilon, x} \epsilon \quad \text{subject to} \quad -e\epsilon \leq Ax - b \leq e\epsilon, \quad Cx + \epsilon e \geq d, \quad \epsilon \geq 0. \tag{9.11}$$

As before, ϵ represents the maximum violation of the constraints; the problem (9.11) seeks to minimize this violation. (We include the constraint $\epsilon \geq 0$ explicitly in the formulation in case A and b are null, since in this case, nonnegativity of ϵ is not guaranteed by the constraints.)

The generalization of the ℓ_1-norm approximation problem (9.8) is as follows:

$$\min_{(x,y,z)} \; e'y + e'z \quad \text{subject to} \quad -y \leq Ax - b \leq y, \quad Cx + z \geq d, \quad z \geq 0. \tag{9.12}$$

(Note the slight abuse of notation; the two e vectors in the objective, which both contain all 1's, may have different numbers of components.)

Exercise 9-2-7. Consider the following system of inequalities. We wish to find a vector x that (i) satisfies all these constraints *and* (ii) comes as close as possible to satisfying the four general inequalities as *equalities*, as closely as possible in the ℓ_1-norm sense. Formulate this problem as a linear program and solve using MATLAB.

$$
\begin{array}{rcrcl}
-4x_1 & + & 3x_2 & \leq & 12, \\
2x_1 & + & x_2 & \leq & 6, \\
x_1 & + & x_2 & \geq & 3, \\
5x_1 & + & x_2 & \geq & 4, \\
& & x_1, x_2 & \geq & 0.
\end{array}
$$

Does your solution change if the second constraint is replaced by

$$4x_1 + 2x_2 \leq 12 \; ?$$

9.2.4   Least-Squares Problems

We now consider using the ℓ_2-norm in the approximation problem. For the overdetermined system $Ax = b$ we are led to the following minimization problem:

$$\min_{x \in \mathbf{R}^n} \|Ax - b\|_2.$$

Since this function is uniformly nonnegative, the minimizer is unchanged if we replace it by its square: $\|Ax - b\|_2^2 = (Ax - b)'(Ax - b)$. Moreover, the squared function is quadratic, while the original function $\|Ax - b\|_2$ is nonsmooth (its derivative fails to exist when $A_i.x = b_i$ for some i). Hence, the problem we actually solve is

$$\min_{x \in \mathbf{R}^n} f(x) := (Ax - b)'(Ax - b). \tag{9.13}$$

This is known as a *least-squares* problem.

Note that f is a convex function; its Hessian $A'A$ is positive semidefinite, because $z'A'Az = \|Az\|_2^2 \geq 0$ for all z. Hence, by Corollary 7.1.2, a vector $\bar{x}$ is a minimizer of f if and only if $\nabla f(\bar{x}) = 0$; that is,

$$\nabla f(\bar{x}) = 2A'(A\bar{x} - b) = 0 \quad \Leftrightarrow \quad A'A\bar{x} = A'b. \tag{9.14}$$

We show now that an $\bar{x}$ satisfying this equation exists.

Theorem 9.2.5. *The system* (9.14) *always has a solution (and therefore the minimization problem* (9.13) *always has a solution) regardless of whether* $Ax = b$ *is solvable or not.*

Proof. If (9.14) has no solution, then the following linear program is infeasible:

$$\min_{x \in \mathbf{R}^n} \quad \left\{ 0'x \mid A'Ax = A'b \right\}.$$

The dual problem, namely,

$$\max_{u \in \mathbf{R}^n} \quad \left\{ b'Au \mid A'Au = 0 \right\},$$

is obviously feasible ($u = 0$ is a feasible point), and so by strong duality it is unbounded above. Therefore, there is some vector u with

$$A'Au = 0 \quad \text{and} \quad b'Au > 0.$$

For this u, we have

$$A'Au = 0 \implies u'A'Au = 0 \implies \|Au\|_2^2 = 0 \implies Au = 0 \implies b'Au = 0,$$

which contradicts $b'Au > 0$. Hence, (9.14) always has a solution. $\qquad \square$

The system of equations in (9.14) is called the *normal equations*. Solving (9.13) is therefore equivalent to solving the square system of linear equations (9.14). Computationally, this task is easier in general than solving a linear program, as is required for the ℓ_1-norm and the ℓ_∞-norm. This fact accounts in part for the popularity of the ℓ_2-norm objective.

Theorem 9.2.5 can be used to produce a simple proof of Theorem 4.9.2, the fundamental theorem of linear algebra, which states that $\ker A \oplus \operatorname{im} A' = \mathbf{R}^n$ for any $A \in \mathbf{R}^{m \times n}$. Given any vector $c \in \mathbf{R}^n$, we form the least-squares problem

$$\min_{v \in \mathbf{R}^m} (A'v - c)'(A'v - c).$$

By Theorem 9.2.5, this problem has a solution $\bar{v}$ satisfying $A(c - A'\bar{v}) = 0$. Hence, we can decompose c as follows:

$$c = (c - A'\bar{v}) + A'\bar{v}. \tag{9.15}$$

Since $A(c - A'\bar{v}) = 0$, the first term in (9.15) is in $\ker A$, while the second is obviously in $\operatorname{im} A'$, as required.

The following result shows that the solution to (9.14) is unique when rank $A = n$.

Lemma 9.2.6. *Let* $A \in \mathbf{R}^{m \times n}$. *If* rank $A = n$, *then* $(A'A)^{-1}$ *exists and*

$$\bar{x} = (A'A)^{-1}A'b$$

is the unique solution of (9.13).

Proof. Let x be a vector such that $A'Ax = 0$. Then $x'A'Ax = 0$, which means $\|Ax\|_2^2 = 0$. It follows that $Ax = 0$, and so by the full column rank assumption we must have $x = 0$. We have shown that $A'Ax = 0 \implies x = 0$, and therefore that $A'A$ is invertible. Hence, $\bar{x}$ is well defined, and $\bar{x}$ is the unique solution of (9.14) and therefore of (9.13). $\qquad \square$

In the alternative case of rank $A = k < n$, the matrix $A'A$ is singular. We know from Theorem 9.2.5, however, that (9.14) has a solution, which we can always find by using the techniques of Chapter 2. The tableau for (9.14) is

$$y \quad = \quad \begin{array}{c} x \qquad 1 \\ \boxed{\begin{array}{c|c} A'A & -A'b \end{array}} \end{array}$$

which after k Jordan exchanges has the following form (possibly after some reordering of rows and columns):

$$\begin{array}{rl} & \quad y_{I_1} \qquad x_{J_2} \qquad 1 \\ x_{J_1} = & \boxed{\begin{array}{c|c|c} H_{I_1 J_1} & H_{I_1 J_2} & h_{I_1} \\ \hline H_{I_2 J_1} & 0 & 0 \end{array}} \\ y_{I_2} = & \end{array}$$

(Consistency of the system (9.14) dictates that the bottom right partition of the tableau is zero.) From this tableau, we conclude that solutions of (9.14) have the following general form:

$$x_{J_1} = H_{I_1 J_2} x_{J_2} + h_{I_1}, \qquad x_{J_2} \quad \text{arbitrary.}$$

(When rank $A = n$, we have $J_2 = \emptyset$, and so the solution is unique.)

Example 9-2-8. We construct a least-squares solution to the overdetermined system (9.6) by minimizing $\|Ax - b\|_2$. To do this, we form the normal equations (9.14) and solve using Jordan exchange.

```
>> load ex9-2-1
>> T = totbl(A'*A,A'*b);
>> T = ljx(T,1,1);
>> T = ljx(T,2,2);
```

$$\begin{array}{rl} & \quad y_1 \qquad y_2 \qquad 1 \\ x_1 = & \boxed{\begin{array}{c|c|c} 0.33 & 0 & 0.66 \\ \hline 0 & 0.5 & 0 \end{array}} \\ x_2 = & \end{array}$$

The resulting least-squares solution is $x_1 = 2/3$, $x_2 = 0$ and is uniquely determined. The corresponding residual norm $\|Ax - b\|_2$ has the value $\sqrt{2/3}$ at this solution.

Note that this solution is different from that of the Chebyshev and L_1 approximation problems; each of three measures of "closeness" gives a different answer. ∎

If A is poorly conditioned (that is, the ratio of its largest to its smallest singular value is large), the solution $\bar{x}$ obtained by forming and solving the system (9.14) computationally may contain a large numerical error. An alternative approach is to use a QR factorization of A, which avoids the squaring of the condition number associated with the use of the normal equations. In fact, the QR technique is the one used by MATLAB. Given an overdetermined system $Ax = b$, we can find the least-squares solution in MATLAB by simply typing x = A\b. For details of the QR factorization technique, see Golub & Van Loan (1996, Chapter 5).

Exercise 9-2-9. Use the three approaches (namely tableau for normal equations, $(A'A)\backslash(A'b)$, and the QR factorization approach) to construct least-squares solutions for

$$A = \begin{bmatrix} 1 & -1 \\ 1 & -1 \\ 1 & -1 \end{bmatrix}, \qquad b = \begin{bmatrix} 1 \\ 2 \\ 3 \end{bmatrix}.$$

Exercise 9-2-10. Find an approximate solution using MATLAB to the following overde-termined system:

$$
\begin{aligned}
x_1 - x_2 &= 0, \\
2x_1 + x_2 &= 0, \\
x_2 &= 2
\end{aligned}
$$

by minimizing $\|Ax - b\|_p$ for $p = 1, 2, \infty$. Write down both the approximate solution and the resulting value of the $\|Ax - b\|_p$. By hand, plot the solution points in $\mathbf{R}^2$ along with lines representing the three equations.

When hard constraints of the form $Cx \geq d$ are also present, the formulation (9.13) is generalized as follows:

$$
\min_{x \in \mathbf{R}^n} (Ax - b)'(Ax - b) \quad \text{subject to} \quad Cx \geq d. \tag{9.16}
$$

This is a *quadratic program* and can be solved using the techniques for quadratic program-ming problems given in Chapter 7.

Exercise 9-2-11. Find values for x_1 and x_2 that make their sum as close to 1 as possible in the ℓ_2 norm, subject to the constraints that

$$
\begin{aligned}
2x_1 + 3x_2 &\leq 1, \\
x_1, x_2 &\geq 0.
\end{aligned}
$$

9.3   Huber Estimation

We now discuss an approximation scheme due to Huber (1981) in which the function to be minimized is in effect a hybrid of the L_1 function (9.7) and the least-squares function (9.13). Given the (possibly overdetermined and inconsistent) system $Ax = b$ and a fixed parameter $\gamma > 0$, Huber's estimation scheme minimizes the following function:

$$
\min_x \sum_{i=1}^m \rho(A_i.x - b_i), \tag{9.17}
$$

where the *loss function* $\rho : \mathbf{R} \to \mathbf{R}$ is defined as follows:

$$
\rho(t) = \begin{cases}
\frac{1}{2}t^2, & |t| \leq \gamma, \\
\gamma t - \frac{1}{2}\gamma^2, & t > \gamma, \\
-\gamma t - \frac{1}{2}\gamma^2, & t < -\gamma.
\end{cases} \tag{9.18}
$$

This function is plotted in Figure 9.1. Note that for $|t| \leq \gamma$, $\rho(t)$ is simply the standard least-squares function (9.13), while for $|t| > \gamma$, it behaves like the L_1 function; that is, the graph becomes a straight line but with a slope chosen to maintain continuity of the first derivative of ρ across the threshold value $t = \gamma$.

A primary motivation of the Huber approach is to overcome a disadvantage of the least-squares approach that "outliers" (that is, residuals $A_i.x - b_i$ that are large in magnitude) are squared in the least-squares objective (9.13) and therefore contribute heavily to the objective

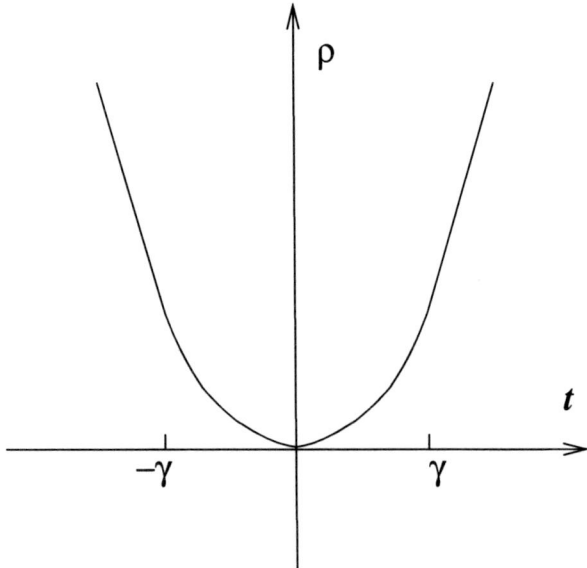

Figure 9.1. *Huber loss function ρ (9.18).*

value. Hence, these outliers may have an undue influence over the minimizing value x^* of (9.13). Since such outliers may arise from errors in gathering the data rather than from any intrinsic property of the underlying model, however, it may be more appropriate to de-emphasize these outliers rather than give them a strong influence over the solution. Huber's approach penalizes these outliers less heavily (it is easy to show that $\frac{1}{2}t^2 > \rho(t)$ when $|t| > \gamma$). For smaller residuals (whose magnitude is less than γ) it continues to use the least-squares objective, thereby retaining the nice statistical properties of this estimator.

Since the objective function in (9.17) is a continuously differentiable convex function, elementary theory for nonlinear optimization (see Corollary 7.1.2) tells us that its minimizer occurs where its derivative is zero, that is,

$$\sum_{i=1}^{m} A'_{i.}\rho'(A_{i.}x - b_i) = 0, \tag{9.19}$$

where $\rho'(t)$ denotes the derivative of ρ, which is

$$\rho'(t) = \begin{cases} t, & |t| \leq \gamma, \\ \gamma, & t > \gamma, \\ -\gamma, & t < -\gamma. \end{cases} \tag{9.20}$$

By defining the vector $w \in \mathbf{R}^n$ by

$$w_i = \rho'(A_{i.}x - b_i) = \begin{cases} A_{i.}x - b_i, & |A_{i.}x - b_i| \leq \gamma, \\ \gamma, & A_{i.}x - b_i > \gamma, \\ -\gamma, & A_{i.}x - b_i < -\gamma, \end{cases} \tag{9.21}$$

we can express the optimality condition (9.19) succinctly as

$$A'w = 0. \tag{9.22}$$

Perhaps surprisingly, we can find the solution of the Huber approximation problem (9.17) by solving a quadratic program. One such quadratic programming formulation is as follows:

$$\min \tfrac{1}{2}w'w + b'w \quad \text{subject to} \quad A'w = 0, \quad w \geq -\gamma e, \quad -w \geq -\gamma e, \tag{9.23}$$

where, as before, $e = (1, 1, \ldots, 1)' \in \mathbf{R}^m$.

We verify the equivalence of (9.23) and (9.17) by using the KKT conditions to show that the w that solves (9.23) satisfies the properties (9.21) and (9.22), where x is the Lagrange multiplier for the constraint $A'w = 0$ in (9.23). Using $\lambda \in \mathbf{R}^m$ and $\pi \in \mathbf{R}^m$ to denote the Lagrange multipliers for the constraints $w \geq -\gamma e$ and $-w \leq -\gamma e$, respectively, we can follow (7.13) to write the KKT conditions for (9.23) as follows:

$$w - Ax + b - \lambda + \pi = 0, \tag{9.24a}$$

$$A'w = 0, \tag{9.24b}$$

$$0 \leq w + \gamma e \perp \lambda \geq 0, \tag{9.24c}$$

$$0 \leq -w + \gamma e \perp \pi \geq 0. \tag{9.24d}$$

If we substitute from (9.24a) into (9.24c) and (9.24d), we get

$$0 \leq Ax - b + \lambda - \pi + \gamma e \perp \lambda \geq 0,$$
$$0 \leq -Ax + b - \lambda + \pi + \gamma e \perp \pi \geq 0.$$

It follows that if $A_i.x - b_i < -\gamma$, then $\lambda_i > 0$, and hence from (9.24c) it follows that $w_i = -\gamma$. Similarly, if $A_i.x - b_i > \gamma$, then $\pi_i > 0$ and $w_i = \gamma$. Otherwise, if $-\gamma \leq A_i.x - b_i \leq \gamma$, straightforward arguments guarantee that $\lambda_i = \pi_i = 0$ and hence that $w_i = A_i.x - b_i$. In all cases, w satisfies (9.21).

Example 9-3-1. The Huber estimation problem with $\gamma = 0.01$ and the data from (9.6) can be solved using Lemke's method for (9.24), following the method used in Example 7-4-7. For this, we need to pivot the w and x variables to the side of the tableau and eliminate the slack variables on (9.24a) and (9.24b). The resulting solution is $x_1 = 0.995$, $x_2 = 0$, which is different from the least-squares, Chebyshev, and L_1 approximate solutions. ∎

If $\bar{x}$ is the least-squares solution of $Ax = b$ (that is, $\bar{x}$ satisfies (9.14)), then provided that

$$\gamma \geq \|A\bar{x} - b\|_\infty,$$

$\bar{x}$ will also be a minimizer of the Huber function (9.17). We can verify this claim by setting

$$x = \bar{x}, \qquad w = A\bar{x} - b, \qquad \lambda = 0, \qquad \pi = 0$$

and noting that these values satisfy the KKT conditions (9.24).

Exercise 9-3-2. Write down the formulation (9.23) for the overdetermined linear system of Exercise 9-2-10. Find the range of values of γ for which the least-squares solution (computed in Exercise 9-2-10) coincides with the minimizer of the Huber function.

Exercise 9-3-3. Another quadratic programming formulation of the Huber estimation problem is given in Mangasarian & Musicant (2000) as

$$\min_{x,w,t} \tfrac{1}{2} w'w + \gamma e't \quad \text{subject to} \quad -t \le Ax - b - w \le t.$$

By forming the KKT conditions of this quadratic program, show (using a similar argument to that given above) that the solution vector w also satisfies (9.21).

9.4 Classification Problems

Consider now the situation in which we have two sets of points in $\mathbf{R}^n$ which are labeled as P_+ and P_-. Denoting any one of these points by x, we would like to construct a function f so that $f(x) > 0$ if $x \in P_+$ and $f(x) < 0$ if $x \in P_-$. The function f is known as a *classifier*. Given a new point x, we can use f to classify x as belonging to either P_+ (if $f(x) > 0$) or P_- (if $f(x) < 0$). We saw an example of such a problem in Section 1.3.5 (see Figure 1.4), where we constructed a linear function f (in a somewhat ad hoc fashion) to classify fine needle aspirates of tumors as either malignant or benign. We give a brief description of *support vector machines*, a modern tool for classification, in this section.

We start by describing the construction of a linear classifier, which has the form $f(x) = w'x - \gamma$, where $w \in \mathbf{R}^n$ and $\gamma \in \mathbf{R}$. Ideally, the hyperplane defined by $f(x) = 0$ should completely separate the two sets P_+ and P_-, so that $f(x) > 0$ for $x \in P_+$ and $f(x) < 0$ for $x \in P_-$. If such (w, γ) exist, then by redefining (w, γ) as

$$\frac{(w, \gamma)}{\min_{x \in P_+ \cup P_-} |w'x - \gamma|},$$

we have that

$$\begin{aligned} x \in P_+ &\quad \Rightarrow \quad f(x) = w'x - \gamma \ge 1, \\ x \in P_- &\quad \Rightarrow \quad f(x) = w'x - \gamma \le -1. \end{aligned} \tag{9.25}$$

To express these conditions as a system of linear inequalities, we assemble all points x row-wise into an $m \times n$ matrix A, which has $m = |P_+| + |P_-|$ rows, where $|P_+|$ and $|P_-|$ denote the number of points in P_+ and P_-, respectively. We then define a diagonal matrix D that labels each row of A as follows:

$$D_{ii} = \begin{cases} 1 & \text{if the point represented by row } i \text{ of } A \text{ belongs to } P_+; \\ -1 & \text{if the point represented by row } i \text{ of } A \text{ belongs to } P_-. \end{cases}$$

The conditions (9.25) can thus be written succinctly as follows:

$$D(Aw - e\gamma) \ge e, \tag{9.26}$$

where $e = (1, 1, \ldots, 1)'$ as usual. Figure 9.2 shows a separating hyperplane $w'x = \gamma$ for two point sets, obtained by finding a pair (w, γ) that is feasible for (9.26) as well as the bounding hyperplanes $w'x = \gamma \pm 1$.

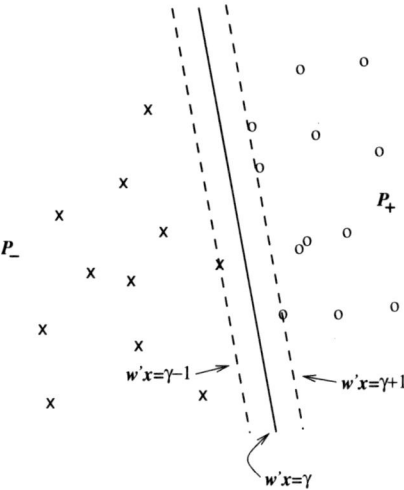

Figure 9.2. *Separable classes, a separating hyperplane $w'x = \gamma$, and the bounding hyperplanes $w'x = \gamma \pm 1$.*

If it is possible to separate the two sets of points, then it is desirable to maximize the distance (margin) between the bounding hyperplanes, which in Figure 9.2 is depicted by the distance between the two dotted lines. It can be shown (see Mangasarian (1999a)) that the separation margin measured by any norm $\| \cdot \|$ is

$$\frac{2}{\|w\|'},$$

where $\|w\|'$ denotes the dual norm, defined in Appendix A. If we take the norm to be the Euclidean (ℓ_2-) norm, which is self-dual, then maximization of $2/\|w\|'$ can be achieved by minimization of $\|w\|$ or $\|w\|^2 = w'w$. Hence, we can solve the following quadratic program to find the separating hyperplane with maximum (Euclidean) margin:

$$\min_{w,\gamma} \quad \tfrac{1}{2}w'w \tag{9.27}$$
$$\text{subject to} \quad D(Aw - e\gamma) \geq e.$$

The *support vectors* are the points x that lie on the bounding hyperplanes and such that the corresponding Lagrange multipliers of the constraints of (9.27) are positive. These correspond to the active constraints in (9.27).

In practice, it is usually not possible to find a hyperplane that separates the two sets because no such hyperplane exists. In such cases, the quadratic program (9.27) is infeasible, but we can define other problems that identify separating hyperplanes "as nearly as practicable" in some sense. As in Section 9.2.3, we can define a vector y whose components indicate the amount by which the constraints (9.26) are violated as follows:

$$D(Aw - e\gamma) + y \geq e, \qquad y \geq 0. \tag{9.28}$$

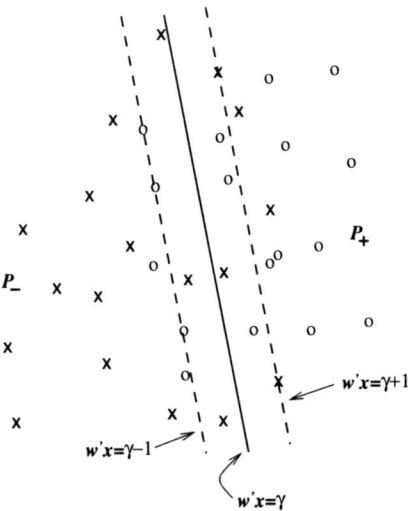

Figure 9.3. *Nonseparable classes and the hyperplane $w'x = \gamma$ obtained from (9.29).*

We could measure the total violation by summing the components of y, and add some multiple of this quantity to the objective of (9.27), to obtain

$$\min_{w,\gamma,y} \quad \tfrac{1}{2}w'w + \nu e'y$$
$$\text{subject to} \quad D(Aw - e\gamma) + y \geq e, \quad y \geq 0, \tag{9.29}$$

where ν is some positive parameter. This problem (9.29) is referred to as a (linear) *support vector machine* (see Vapnik (1995), Burges (1998), Mangasarian (2000), Schölkopf & Smola (2002)).

Figure 9.3 shows two linearly inseparable point sets and the hyperplane obtained by solving a problem of the form (9.29). The bounding hyperplanes are also shown. In this formulation, the support vectors are the points from each set P_- and P_+ that lie on the wrong side of their respective bounding hyperplanes (or are on their respective bounding planes and their corresponding Lagrange multipliers are positive). Again, these points correspond to the active constraints from the constraint set $D(Aw - e\gamma) + y \geq e$.

Using u to denote Lagrange multipliers for the constraints $D(Aw - e\gamma) + y \geq 0$ in (9.29), the KKT conditions (see (7.13)) for this problem can be written as follows:

$$0 = w - A'Du, \tag{9.30a}$$
$$0 = e'Du, \tag{9.30b}$$
$$0 \leq \nu e - u \perp y \geq 0, \tag{9.30c}$$
$$0 \leq D(Aw - e\gamma) + y - e \perp u \geq 0. \tag{9.30d}$$

It is not difficult to see that these are also the KKT conditions for the following problem, which is the dual of (9.29):

$$\min_{u} \quad \tfrac{1}{2}u'DAA'Du - e'u$$
$$\text{subject to} \quad 0 \leq u \leq ve, \quad e'Du = 0. \tag{9.31}$$

Indeed, it is often more convenient to solve this form of the problem for u and then recover w by setting $w = A'Du$ (from the first KKT condition (9.30a)) and recovering γ as the Lagrange multiplier for the constraint $e'Du = 0$ in (9.31).

We can obtain a linear programming alternative to (9.29) by replacing the quadratic term $\tfrac{1}{2}w'w$ by the ℓ_1-norm $\|w\|_1$ which corresponds to measuring the margin using the ℓ_∞-norm. By introducing a vector s whose elements are the absolute values of the corresponding elements of w, we obtain the following formulation:

$$\min_{w,\gamma,s,y} \quad e's + ve'y$$
$$\text{subject to} \quad D(Aw - e\gamma) + y \geq e, \quad s \geq w \geq -s, \quad y \geq 0. \tag{9.32}$$

(Since we are minimizing $e's$, the components of s will take on their smallest values compatible with the constraints, and so we have $\|w\|_1 = e's$ at the solution of (9.32).) This ℓ_1 formulation of the problem has been shown to set many components of w to zero and thereby select a small set of features that generate the best linear classification. An alternative formulation of (9.32) that appears to be more effective computationally is the following:

$$\min_{w^+,w^-,\gamma,y} \quad e'(w^+ + w^-) + ve'y$$
$$\text{subject to} \quad D(Aw - e\gamma) + y \geq e,$$
$$w = w^+ - w^-,$$
$$w^+, w^-, y \geq 0. \tag{9.33}$$

Note that $w_i^+ + w_i^- \geq |w_i|$, and at the solution we have that at least one of w_i^+ and w_i^- is zero for all i.

We refer to the formulations above as *linear* support vector machines, as they attempt to separate the data points with a (linear) hyperplane. A more general approach involves mapping each data point into a higher-dimensional space via a transformation Φ and then finding a separating hyperplane in this higher-dimensional space. When this hyperplane is mapped back into the original space $\mathbf{R}^n$ it describes a nonlinear surface, yielding greater flexibility and a potentially more powerful means of classification.

Suppose that the transformation Φ maps $\mathbf{R}^n$ to $\mathbf{R}^{n'}$. We seek a classifier of the form

$$f(x) = w'\Phi(x) - \gamma, \tag{9.34}$$

where w and $\Phi(x)$ now belong to the (typically larger) space $\mathbf{R}^{n'}$, which is sometimes called the *feature space*. Note that f is generally a nonlinear function, since Φ is nonlinear. Using $A_{i\cdot}$ to denote row i of the matrix A (which represents a point from one of the sets P_+ and P_-), we can reformulate the problem (9.29) in feature space as follows:

$$\min_{w,\gamma,y} \quad \tfrac{1}{2}w'w + ve'y$$
$$\text{subject to} \quad D_{ii}(w'\Phi(A_{i\cdot}') - \gamma) + y_i \geq 1, \quad y_i \geq 0, \quad i = 1, 2, \ldots, m. \tag{9.35}$$

The number of constraints is the same as in the original formulation, and so the dimension of the dual problem (9.31) does not change when we reset the problem in feature space. We have

$$\min_{u} \quad \tfrac{1}{2}u'DKDu - e'u$$
$$\text{subject to} \quad 0 \le u \le ve, \quad e'Du = 0, \tag{9.36}$$

where K is an $m \times m$ matrix whose elements are

$$K_{ij} = \Phi(A'_{i.})'\Phi(A'_{j.}) = k(A'_{i.}, A'_{j.}),$$

and the *kernel function* $k(\cdot, \cdot)$ is defined by

$$k(s, t) = \Phi(s)'\Phi(t). \tag{9.37}$$

Having solved (9.36) for u, we can recover w from the appropriate modification of (9.30a), that is,

$$w = \sum_{i=1}^{m} D_{ii}u_i\Phi(A'_{i.}).$$

The classifier (9.34) becomes

$$f(x) = w'\Phi(x) - \gamma = \sum_{i=1}^{m} D_{ii}u_i\Phi(A'_{i.})'\Phi(x) - \gamma = \sum_{i=1}^{m} D_{ii}u_i k(A'_{i.}, x) - \gamma,$$

where the kernel function k is defined in (9.37).

Note that to solve the problem (9.36) and to compute the classifier $f(x)$, we never need to know Φ explicitly; it is enough to know the kernel function k. In practice, we can forget about Φ altogether, which in general is hard to calculate. Instead we devise kernel functions that produce classifiers with powerful separation properties. Such kernels may not even have a product form as used in (9.37). The essential property of a kernel function is that it measures the proximity of its two vector arguments s and t. A widely used kernel is the *Gaussian* or *radial basis* kernel defined by

$$k(s, t) = \exp\{-\mu\|s - t\|_2^2\}, \tag{9.38}$$

where μ is a positive parameter. Other kernels include the linear kernel $k(s, t) = s't$ (which yields the linear support vector machine with which we started the discussion of this section) and the polynomial kernel

$$k(s, t) = (s't + 1)^d,$$

where d is a positive integer.

To demonstrate the effectiveness of the Gaussian kernel in generating a highly non-linear classifier, we depict in Figure 9.4 a classifier generated by a nonlinear support vector machine using a Gaussian kernel for the checkerboard data set (see Ho & Kleinberg (1996), Lee & Mangasarian (2001)). This data set consists of 1000 black and white points in $\mathbf{R}^2$ taken from 16 black and white squares of a checkerboard. Figure 9.4 displays the grid lines which constitute the nonlinear classifier generated by a Gaussian support vector machine. Note that these lines closely follow the actual topography of the checkerboard.

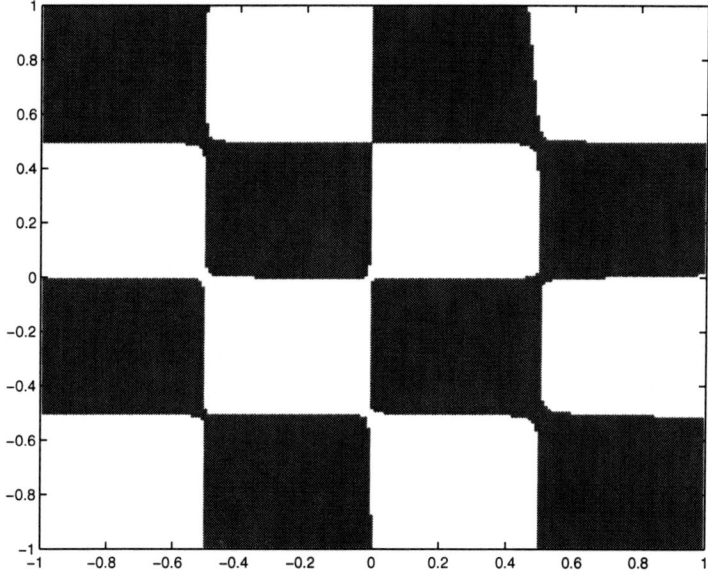

Figure 9.4. *Classifier using a Gaussian kernel to separate* 1000 *points selected randomly from a* 4 × 4 *checkerboard.*

How do we determine if the classifier is a good one, and how do we choose appropriate values for the parameter v in (9.29), (9.36), etc. and the parameter μ in the Gaussian kernel (9.38)? Typically, we use a *testing set* which consists of a subset of the data points that are removed from the full set of points before solving the support vector machines above. Once the classifiers are determined from this remaining data (known as the *training set*), we compare the predicted class for each data point in the testing set with its actual class and tabulate the number that are correctly classified. Various values of v can be tried; we usually choose the one for which the proportion of correctly classified testing points is maximized.

This process can be generalized further to a process called *cross-validation*, in which the entire data is split into a number of pieces (typically 10) and each of these pieces is used in turn as a testing set while the remaining points are used as the training set. The average testing set error is a better statistical measure of the likely performance of the classifier on unseen data (under the assumption it is drawn from the same distribution).

The notion of a testing set is further refined as a *tuning set*. This is a set of samples (distinct from the training set) that are left out of the training process. The tuning set is used in a similar manner to the testing set, except that the values of the parameter v (for example) can be chosen based on the tuning set error. Typically, many replications of the training process are carried out to tune parameters such as v and the Gaussian kernel parameter μ.

Appendix A

Linear Algebra, Convexity, and Nonlinear Functions

This appendix gives a brief introduction to convex sets and functions, vector norms, and quadratic functions. These ideas are used in Chapters 6, 7, and 9 and form part of the basic underpinning of the subject of optimization. We begin with some linear algebra.

A.1 Linear Algebra

We describe here some of the concepts and terms that are basic to our discussion of the linear algebra that underlies linear programming. We consider vectors, usually denoted by lower-case roman letters, to consist of a column of real numbers, whose individual components are indicated by a subscript. For instance, a vector $x \in \mathbf{R}^n$ has the form

$$x = \begin{bmatrix} x_1 \\ x_2 \\ \vdots \\ x_n \end{bmatrix}.$$

(Less frequently, we make use of row vectors, denoted by x', in which the components are arranged horizontally rather than vertically.) We write $x \geq 0$ to indicate that all components of x are nonnegative and $x \geq y$ (where both x and y are in $\mathbf{R}^n$) to indicate that $x_i \geq y_i$ for all $i = 1, 2, \ldots, n$.

In various places in the text (see especially Chapters 8 and 7), we use the notion of *complementarity* of two vectors. Given $u \in \mathbf{R}^n$ and $v \in \mathbf{R}^n$, the notation $\perp$ is defined as follows:

$$u \perp v \quad \Leftrightarrow \quad u'v = 0. \tag{A.1}$$

In describing optimality conditions, we frequently use notation such as

$$0 \leq u \perp v \geq 0,$$

which indicates that, in addition to satisfying $u'v = 0$, the vectors u and v have all nonnegative components.

Matrices are made up of rectangular arrangements of real numbers, whose individual components are indicated by double subscripts, the first subscript indicating row number and the second indicating column number. Thus the matrix $A \in \mathbf{R}^{m \times n}$ is

$$A = \begin{bmatrix} A_{11} & A_{12} & \dots & A_{1n} \\ A_{21} & A_{22} & \dots & A_{2n} \\ \vdots & \vdots & & \vdots \\ A_{m1} & A_{m2} & \dots & A_{mn} \end{bmatrix}.$$

Matrices are usually denoted by uppercase Roman letters. We use $A_{\cdot j}$ to denote the vector consisting of the jth column of A, while $A_{i \cdot}$ is the row vector consisting of row i from A.

The transpose of a matrix, indicated by a prime "$'$", is obtained by interchanging row and column subscripts. The transpose of the $m \times n$ matrix above is the following $n \times m$ matrix:

$$A' = \begin{bmatrix} A_{11} & A_{21} & \dots & A_{m1} \\ A_{12} & A_{22} & \dots & A_{m2} \\ \vdots & \vdots & & \vdots \\ A_{1n} & A_{2n} & \dots & A_{mn} \end{bmatrix}.$$

We can also take the transpose of a vector by arranging the elements in a row rather than a column. The transpose of the vector x above is

$$x' = \begin{bmatrix} x_1 & x_2 & \dots & x_n \end{bmatrix}.$$

A matrix is said to be *sparse* if the vast majority of its elements are zero. It is possible to store sparse matrices efficiently on a computer (by not storing the zero entries but rather storing only the nonzeros together with an indication of where they lie in the matrix). Moreover, it is possible to perform multiplications, factorizations, and other linear algebra operations efficiently with them.

A matrix is *diagonal* if its only nonzeros appear on its diagonal, that is, the positions in which the row index equals the column index. Formally, we say that A is diagonal if $A_{ij} = 0$ whenever $i \neq j$.

An important matrix is the *identity*, denoted by I. This is the square matrix of a given dimension that has the number 1 appearing on its diagonal, and zeros elsewhere. For example, the 3×3 identity is

$$\begin{bmatrix} 1 & 0 & 0 \\ 0 & 1 & 0 \\ 0 & 0 & 1 \end{bmatrix}.$$

Given a square matrix A, we define the *inverse* A^{-1} to be the matrix with the property that $AA^{-1} = I$. When it exists, the inverse is unique. Square matrices for which an inverse does not exist are said to be *singular*.

A matrix P is a *permutation matrix* if it is an identity matrix whose rows have been reordered. An example of a 3×3 permutation matrix is

$$\begin{bmatrix} 1 & 0 & 0 \\ 0 & 0 & 1 \\ 0 & 1 & 0 \end{bmatrix},$$

which is obtained by swapping the second and third rows of the identity matrix described above.

A matrix L is *lower triangular* if all entries above its diagonal are zero. Formally, the matrix L is lower triangular if $L_{ij} = 0$ whenever $j > i$; that is, its column index is greater than its row index. The following is a 4×3 lower triangular matrix:

$$\begin{bmatrix} 2 & 0 & 0 \\ 1 & -1 & 0 \\ 2 & 1.5 & -2 \\ 1 & 1 & 3 \end{bmatrix}.$$

Matrices and vectors can be added together if they have the same dimensions. Specifically, if $x \in \mathbf{R}^n$, $y \in \mathbf{R}^n$, then $x + y$ is the vector in $\mathbf{R}^n$ whose ith entry is $x_i + y_i$. We can also multiply a vector or matrix by a scalar α by multiplying each individual element by α. That is, given the matrix $A \in \mathbf{R}^{m \times n}$, the matrix αA is the matrix in $\mathbf{R}^{m \times n}$ whose (i, j) element is αA_{ij}.

The product AB of two matrices A and B can be formed, provided that the number of columns of A equals the number of rows of B. If $A \in \mathbf{R}^{m \times n}$ and $B \in \mathbf{R}^{n \times q}$, then the product AB is in $\mathbf{R}^{m \times q}$ and its (i, j) component is given by the formula

$$(AB)_{ij} = \sum_{k=1}^{n} A_{ik} * B_{kj}, \qquad i = 1, 2, \ldots, m, \qquad j = 1, 2, \ldots, q.$$

Given a matrix $A \in \mathbf{R}^{m \times n}$ and a vector $x \in \mathbf{R}^n$, the product Ax is a vector in $\mathbf{R}^m$ whose ith element is $A_i.x$.

A.2 Convex Sets

Given any two points $x, y \in \mathbf{R}^n$, the *line segment* joining x and y, written (x, y), is defined by

$$(x, y) = \{(1 - \lambda)x + \lambda y \mid 0 < \lambda < 1\}.$$

A set $C \subset \mathbf{R}^n$ is a *convex set* if the line segment between any two points $x, y \in C$ lies in C, that is,

$$x, y \in C \implies (x, y) \subset C.$$

The point $z = (1 - \lambda)x + \lambda y$ for some $0 \leq \lambda \leq 1$ is said to be a *convex combination* of x and y. Figure A.1 shows two sets in $\mathbf{R}^2$; the left one is convex and the right one is not.

The proof of the following result exhibits a standard technique of showing the convexity of a set.

Lemma A.2.1. *Let $A \in \mathbf{R}^{m \times n}$, $b \in \mathbf{R}^m$. The following sets are convex:*

$$C_1 = \{x \mid Ax \geq b\} \quad and \quad C_2 = \{x \mid x \geq 0\}.$$

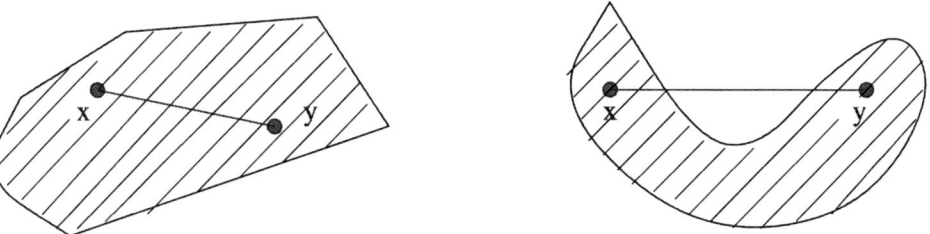

Figure A.1. *Example of a convex set and a nonconvex set.*

Proof. Let x, y be any two points in C_1, and consider $z = (1-\lambda)x + \lambda y$ for some $\lambda \in (0, 1)$. Then

$$Az = A((1-\lambda)x + \lambda y) = (1-\lambda)Ax + \lambda Ay \geq (1-\lambda)b + \lambda b = b.$$

The inequality in this relationship follows from the facts that $Ax \geq b$ and $Ay \geq b$ (since both x and y are in C_1) and also $\lambda \geq 0$ and $(1-\lambda) \geq 0$. We conclude that $z \in C_1$ for any $\lambda \in [0, 1]$, and so C_1 is convex.

C_2 is a special case of C_1 for which $A = I$ and $b = 0$, and so its convexity follows from the first part of the proof. ☐

We can generalize the definition of a convex combination of points beyond just two endpoints. Formally, a point $x \in \mathbf{R}^n$ is a *convex combination* of the points $\{x^1, x^2, \ldots, x^r\}$ in $\mathbf{R}^n$ if for some real numbers $\lambda_1, \lambda_2, \ldots \lambda_r$ which satisfy

$$\sum_{i=1}^{r} \lambda_i = 1 \quad \text{and} \quad \lambda_i \geq 0, \qquad 1 \leq i \leq r,$$

we have

$$x = \sum_{i=1}^{r} \lambda_i x^i. \tag{A.2}$$

(A *linear combination* of the points $\{x^1, x^2, \ldots, x^r\}$ is a vector of the form (A.2) in which there are no restrictions on the scalars $\lambda_1, \lambda_2, \ldots \lambda_r$.)

Exercise A-2-1. Show that a set S in $\mathbf{R}^n$ is convex if and only if every convex combination of a finite number of points of S is in S.

We now demonstrate some ways of generating new convex sets from given convex sets.

Lemma A.2.2. *If C_1 and C_2 are convex sets, then so is their intersection $C_1 \cap C_2$.*

Proof. If $C_1 \cap C_2$ is empty, there is nothing to prove. Otherwise, let x, $y \in C_1 \cap C_2$, and consider $z = (1-\lambda)x + \lambda y$ for some $\lambda \in [0, 1]$. Since C_1 is convex, it follows from $x \in C_1$ and $y \in C_1$ that $z \in C_1$. By similar reasoning, $z \in C_2$. Hence $z \in C_1 \cap C_2$, and so we conclude that $C_1 \cap C_2$ is convex. ☐

Note that $C_1 \cup C_2$ is not convex in general for convex C_1 and C_2.

Exercise A-2-2. Give algebraic definitions of two convex sets C_1 and C_2 in $\mathbf{R}^2$ whose union is not convex. Draw your sets.

It follows from Lemma A.2.2 that the feasible region of a linear program in standard form, that is,

$$\{x \mid Ax \geq b\} \cap \{x \mid x \geq 0\},$$

is convex.

For any set $S \subset \mathbf{R}^n$, we define the *convex hull* conv(S) to be the intersection of all convex sets that contain S.

Exercise A-2-3. If S is already a convex set, show that conv(S) = S.

Lemma A.2.3. *Consider the matrix $A \in \mathbf{R}^{m \times n}$ and the convex set $C \subset \mathbf{R}^n$. Then the set $AC := \{Ax \mid x \in C\} \subset \mathbf{R}^m$ is convex.*

Proof. Let $x, y \in AC$, and consider $z = (1 - \lambda)x + \lambda y$ for some $\lambda \in (0, 1)$. Then $x = A\hat{x}$ and $y = A\hat{y}$ for some $\hat{x}, \hat{y} \in C$. Furthermore,

$$z = (1 - \lambda)x + \lambda y = (1 - \lambda)A\hat{x} + \lambda A\hat{y} = A((1 - \lambda)\hat{x} + \lambda\hat{y}).$$

Since C is convex, it follows that $(1 - \lambda)\hat{x} + \lambda\hat{y} \in C$ and thus $z \in AC$. Hence AC is convex, as claimed. $\square$

We leave the proof of the following result as an exercise.

Lemma A.2.4. *If $C_1 \subset \mathbf{R}^m$ and $C_2 \subset \mathbf{R}^n$ are convex sets, then so is*

$$C_1 \times C_2 := \left\{ \begin{bmatrix} x \\ y \end{bmatrix}; \ x \in C_1, \ y \in C_2 \right\}.$$

Corollary A.2.5. *If $C_1 \subset \mathbf{R}^n$ and $C_2 \subset \mathbf{R}^n$ are convex sets, then so are $C_1 + C_2 := \{x + y \mid x \in C_1, y \in C_2\}$ and $\gamma C_1 := \{\gamma x \mid x \in C_1\}$ for any $\gamma \in \mathbf{R}$.*

Proof. The result can be proved directly using the standard technique for convexity proofs. It also follows immediately from Lemmas A.2.3 and A.2.4 if we note that $C_1 + C_2 = [I \ \ I](C_1 \times C_2)$ and $\gamma C_1 = (\gamma I)C_1$. $\square$

Exercise A-2-4. Show that if the optimal value of the objective function of a linear programming problem is attained at k points, $x^1, x^2, \ldots, x^k$, then it is also attained at any convex combination $\bar{x}$ of these k points (convex hull):

$$\bar{x} = \sum_{i=1}^{k} \lambda_i x^i, \qquad \lambda_i \geq 0, \qquad \sum_{i=1}^{k} \lambda_i = 1.$$

(Hint: show that $\bar{x}$ is feasible and that it has the same objective function value as each $x^1, x^2, \ldots, x^k$.)

A.3 Smooth Functions

Suppose we have a function $f: D \to \mathbf{R}$ defined on a set $D \subset \mathbf{R}^n$, and let x be a point in the interior of D. We define the *gradient* of f at x to be the vector of first partial derivatives of f at x, and denote it by $\nabla f(x)$. That is, we have

$$\nabla f(x) = \begin{bmatrix} \frac{\partial f(x)}{\partial x_1} \\ \frac{\partial f(x)}{\partial x_2} \\ \vdots \\ \frac{\partial f(x)}{\partial x_n} \end{bmatrix}.$$

(Naturally, this definition makes sense only if the derivatives exist. If they do not, then f is nonsmooth or discontinuous, and more general concepts from analysis are required.) If the second partial derivatives exist, we can assemble them into an $n \times n$ *Hessian* matrix, which is denoted by $\nabla^2 f(x)$ and defined as follows:

$$\nabla^2 f(x) = \begin{bmatrix} \frac{\partial^2 f(x)}{\partial x_1^2} & \frac{\partial^2 f(x)}{\partial x_1 \partial x_2} & \cdots & \frac{\partial^2 f(x)}{\partial x_1 \partial x_n} \\ \frac{\partial^2 f(x)}{\partial x_2 \partial x_1} & \frac{\partial^2 f(x)}{\partial x_2^2} & \cdots & \frac{\partial^2 f(x)}{\partial x_2 \partial x_n} \\ \vdots & \vdots & \ddots & \vdots \\ \frac{\partial^2 f(x)}{\partial x_n \partial x_1} & \frac{\partial^2 f(x)}{\partial x_n \partial x_2} & \cdots & \frac{\partial^2 f(x)}{\partial x_n^2} \end{bmatrix}.$$

A.4 Convex Functions

Consider a function $f: D \to \mathbf{R}$ as in Section A.3. The domain of f can be extended to the whole set $\mathbf{R}^n$ by defining $f(x) = \infty$ if $x \notin D$. We indicate this extension by writing

$$f: \mathbf{R}^n \to \mathbf{R} \cup \{\infty\}.$$

The original domain D, which satisfies $D = \{x \mid f(x) < \infty\}$, is called the *effective domain* of f.

Introduction of ∞ into the range of f raises some issues regarding how ∞ should be treated arithmetically. The following conventions are used:

$$\begin{aligned} \lambda\infty &= \infty && \text{if } \lambda > 0, \\ \lambda\infty &= -\infty && \text{if } \lambda < 0, \\ 0\infty &= 0, \\ -\infty \leq x \leq \infty && \forall x \in \mathbf{R} \cup \{-\infty, \infty\}, \\ \infty + \infty &= \infty, \\ \infty + (-\infty) && \text{not defined.} \end{aligned}$$

The *epigraph* of f is defined by

$$\operatorname{epi} f := \{(x, \mu) \mid f(x) \leq \mu\} \subset \mathbf{R}^{n+1}. \tag{A.3}$$

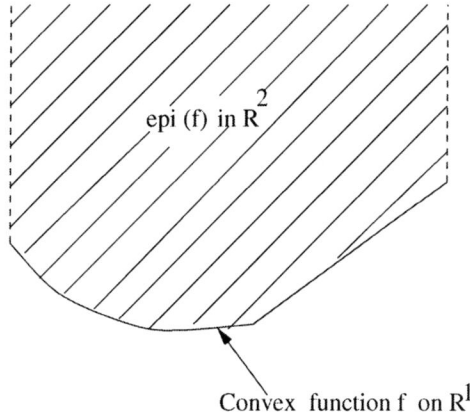

epi (f) in R^2

Convex function f on R^1

Figure A.2. *A convex function and its epigraph.*

We say that f is a *convex function* if its epigraph is a convex set in $\mathbf{R}^{n+1}$. Figure A.2 illustrates the epigraph of a convex function.

The following theorem gives a useful characterization of convexity.

Theorem A.4.1. *The function f is convex if and only if*

$$f((1 - \lambda)x + \lambda y) \le (1 - \lambda)f(x) + \lambda f(y) \tag{A.4}$$

for all $\lambda \in (0, 1)$ and all $x, y \in \mathbf{R}^n$. In other words, linear interpolation never underestimates f.

Proof. Let $x, y \in \mathbf{R}^n$, and let $\lambda \in (0, 1)$. If $f(x) = \infty$ or if $f(y) = \infty$, there is nothing to prove, since the right-hand side of (A.4) is ∞. Otherwise, by the convexity definition for f, we have that $(x, f(x)), (y, f(y)) \in$ epi f, and so it follows from the convexity of epi f that $((1 - \lambda)x + \lambda y, (1 - \lambda)f(x) + \lambda f(y)) \in$ epi f and hence $f((1 - \lambda)x + \lambda y) \le (1 - \lambda)f(x) + \lambda f(y)$.

Conversely, let $(x, \mu), (y, \gamma) \in$ epi f and $\lambda \in (0, 1)$. Then

$$f((1 - \lambda)x + \lambda y) \le (1 - \lambda)f(x) + \lambda f(y) \le (1 - \lambda)\mu + \lambda \gamma,$$

so that $((1 - \lambda)x + \lambda y, (1 - \lambda)\mu + \lambda \gamma) \in$ epi f. We conclude that epi f is convex, as claimed. $\square$

The linear function $f(x) = p'x$ is therefore a convex function, with domain $D = \mathbf{R}^n$. If f is convex, then D must be convex by Lemma A.2.3, since

$$D = [I \quad 0]\,\text{epi}\, f.$$

The same idea can be used to prove the following result.

Corollary A.4.2. *If f is a convex function, then each* level set

$$L(\alpha) := \left\{ x \in \mathbf{R}^n \mid f(x) \le \alpha \right\}, \qquad \alpha \in \mathbf{R},$$

is a convex set but not conversely.

Proof. We can define $L(\alpha)$ as the following projected intersection:

$$L(\alpha) = P \cdot \text{epi } f \bigcap \{(x, \mu) \mid \mu \le \alpha\},$$

where P is the $n \times (m+1)$ projection matrix $[I\ 0]$. By applying Lemmas A.2.2 and A.2.3, we obtain that $L(\alpha)$ is convex.

If $f(x) = x^3$, $x \in \mathbf{R}$, then $L(\alpha)$ is a convex set for every α, but f is not a convex function. □

Define $(f + g)(x) = f(x) + g(x)$ and $(\gamma f)(x) = \gamma f(x)$ for all x. The following result is established using Theorem A.4.1.

Corollary A.4.3. *If f and g are convex functions on $\mathbf{R}^n$, then so are $f + g$ and γf for $\gamma \ge 0$.*

Proof.

$$
\begin{aligned}
(f + g)((1 - \lambda)x + \lambda y) &= f((1 - \lambda)x + \lambda y) + g((1 - \lambda)x + \lambda y) \\
&\le (1 - \lambda)f(x) + \lambda f(y) + (1 - \lambda)g(x) + \lambda g(y) \\
&= (1 - \lambda)(f + g)(x) + \lambda(f + g)(y).
\end{aligned}
$$

A similar proof holds for γf. □

A function f is a *concave function* if $-f$ is a convex function.

A.5 Quadratic Functions

Consider the quadratic function $f(x) := \frac{1}{2}x'Qx$. We may assume that Q is symmetric, since we can replace it by its symmetric part $(Q + Q')/2$ without changing the value of f, as the following relations show:

$$x'Qx = 1/2x'Qx + 1/2(x'Qx)' = x'\left(\frac{Q + Q'}{2}\right)x.$$

Given that Q symmetric, the gradient and Hessian of f can be written as

$$\nabla f(x) = Qx, \qquad \nabla^2 f(x) = Q,$$

respectively.

The following proposition gives a characterization of convex quadratic functions.

Proposition A.5.1. *Let $f(x) := \frac{1}{2}x'Qx$, where Q is a symmetric $n \times n$ matrix. The following are equivalent:*

(a) f is convex.

(b) The linearization of f at any point x never overestimates f, that is,

$$f(y) \geq f(x) + \nabla f(x)'(y - x) \qquad \forall y. \tag{A.5}$$

(c) The function f has nonnegative curvature, that is,

$$y'\nabla^2 f(x)y = y'Qy \geq 0 \qquad \forall x, y \in \mathbf{R}^n.$$

Proof. The fact that (b) is equivalent to (c) follows from the following identity (easily established for quadratic functions):

$$f(u) = f(v) + \nabla f(v)'(u - v) + \frac{1}{2}(u - v)'\nabla^2 f(v)(u - v). \tag{A.6}$$

Theorem A.4.1 states that (A.4) holds for all $\lambda \in (0, 1)$ and $x, y \in \mathbf{R}^n$. By rearranging this inequality and using (A.6) with $u = (1 - \lambda)x + \lambda y$ and $v = x$, we have

$$f(y) - f(x) \geq \frac{f((1 - \lambda)x + \lambda y) - f(x)}{\lambda} = \nabla f(x)'(y - x) + \frac{\lambda}{2}(y - x)'Q(y - x).$$

Thus (a) implies (b) by taking $\lambda \to 0$.

For the reverse implication (b) implies (a), note that for $\lambda \in (0, 1)$, we have by redefining x and y appropriately in (A.5) that

$$f(x) - f((1 - \lambda)x + \lambda y) \geq \lambda \nabla f((1 - \lambda)x + \lambda y)(x - y)$$

and

$$f(y) - f((1 - \lambda)x + \lambda y) \geq -(1 - \lambda)\nabla f((1 - \lambda)x + \lambda y)(x - y).$$

Multiplying the first inequality by $(1 - \lambda)$, the second by λ, and adding gives

$$f((1 - \lambda)x + \lambda y) \leq (1 - \lambda)f(x) + \lambda f(y),$$

which indicates that f is convex, as required. $\quad\Box$

Part (b) of Proposition A.5.1, which says that a tangent plane to the quadratic function f never overestimates f anywhere, is in fact true for any differentiable convex function. Part (c) has a more general analogue for twice differentiable convex functions.

A matrix Q satisfying $y'Qy \geq 0$ for all $y \in \mathbf{R}^n$ is called *positive semidefinite*. It is *positive definite* if

$$x'Qx > 0 \qquad \forall x \neq 0.$$

Proposition A.5.1 shows that the quadratic function f is convex if and only if its Hessian matrix Q is positive semidefinite. A matrix that is not positive semidefinite or whose negative is not semidefinite is called *indefinite*.

The following result shows that Proposition A.5.1 continues to hold when a linear term $p'x + \alpha$ is added to the purely quadratic term $\frac{1}{2}x'Qx$.

Corollary A.5.2. *Let Q be a symmetric $n \times n$ matrix and $f(x) = \frac{1}{2}x'Qx + p'x + \alpha$. Then $\nabla f(x) = Qx + p$, $\nabla^2 f(x) = Q$, and statements (a), (b), and (c) of Proposition A.5.1 are equivalent.*

Proof. f is convex if and only if $\frac{1}{2}x'Qx$ is convex by Corollary A.4.3. The other two properties of Proposition A.5.1 hold if and only if they hold for f. □

It follows from the above discussion that $f(x) := \frac{1}{2}x'Qx + p'x + \alpha$ is a convex function if and only if Q is positive semidefinite.

The definitions of positive semidefinite and positive definite matrices do not require symmetry of the matrix Q. However, a matrix is positive (semi)definite if and only if its symmetric part is positive (semi)definite. It is clear that if Q is positive definite, then it is invertible and positive semidefinite. The converse is not true, however, as can be seen by the example

$$Q = \begin{bmatrix} 1 & -1 \\ 1 & 0 \end{bmatrix}.$$

Note that Q is positive semidefinite if and only if the eigenvalues of $Q + Q'$ are nonnegative (see Strang (1993) for a definition of matrix eigenvalues). The MATLAB command `eig(Q+Q')` will return the eigenvalues. In the simple case above, these eigenvalues would be 2 and 0, and hence Q is positive semidefinite. (Note that the eigenvalues of Q are generally not the same as those of $(1/2)(Q + Q')$ unless Q is symmetric.)

Another technique to check for convexity of a quadratic function is to use the notion of "completing the square." We now show how this process works.

Example A-5-1. Consider $x_1^2 + 3x_1x_2 + x_2^2$. Does this function have a nonnegative value for all choices of x_1 and x_2? Rewriting as

$$x_1^2 + 3x_1x_2 + x_2^2 = \left(x_1 + \frac{3}{2}x_2 \right)^2 - \frac{5}{4}x_2^2,$$

we see that the function is negative whenever we choose x_1 and x_2 in such a way that $x_2 \neq 0$ and $(x_1 + (3/2)x_2) = 0$ (for example, $x_2 = 1$ and $x_1 = -3/2$). We can write this function in the form $f(x) = (1/2)x'Qx$ by choosing Q to be the following symmetric matrix:

$$Q = \begin{bmatrix} 2 & 3 \\ 3 & 2 \end{bmatrix}.$$

The eigenvalues of Q are -1 and 5, confirming that Q is *not* positive semidefinite. Hence, this function does not satisfy the property in Proposition A.5.1(c) and is therefore not convex.

A second example is the following:

$$x_1^2 + x_1x_2 + x_2^2 = \left(x_1 + \frac{1}{2}x_2 \right)^2 + \frac{3}{4}x_2^2.$$

By completing the square, we have expressed this function as the sum of two nonnegative entities. Since it can also be written as $(1/2)x'Qx$ by setting

$$Q = \begin{bmatrix} 2 & 1 \\ 1 & 2 \end{bmatrix},$$

it follows that Q is positive semidefinite (in fact positive *definite*), and hence the function f is convex. (An alternative way to see this is to evaluate the eigenvalues of Q which are 1 and 3.) ∎

The final result of this section is a simple corollary of Proposition A.5.1. It holds not just for quadratics but also for general differentiable convex functions.

Corollary A.5.3. *Suppose* $f(x) = \frac{1}{2}x'Qx + p'x + \alpha$, *where* $Q \in \mathbf{R}^{n \times n}$ *is a symmetric matrix. Then* f *is convex if and only if for all* $x, y \in \mathbf{R}^n$, *we have*

$$(\nabla f(x) - \nabla f(y))(x - y) \geq 0, \tag{A.7}$$

that is,

$$(x - y)'Q(x - y) \geq 0,$$

for all $x, y \in \mathbf{R}^n$.

Proof. If f is convex, then $(x - y)'Q(x - y) \geq 0$ for all $x, y \in \mathbf{R}^n$ by Proposition A.5.1(c). Hence, we have

$$0 \leq (x - y)'Q(x - y) = (Qx - Qy)'(x - y)$$
$$= [(Qx + p) - (Qy + p)]'(x - y) = (\nabla f(x) - \nabla f(y))'(x - y),$$

and so (A.7) holds. The converse is immediate by taking $y = 0$ in (A.7) and Proposition A.5.1(c). □

If $F : \mathbf{R}^n \to \mathbf{R}^n$ satisfies $(F(x) - F(y))'(x - y) \geq 0$, then F is called a *monotone map*. The above result states that the gradient of a convex (quadratic) function is a monotone map.

Further generalizations of all the results of this section can be found in the following texts on convex analysis and nonlinear programming: Rockafellar (1970), Mangasarian (1969), Nocedal & Wright (2006).

A.6   Norms and Order Notation

We introduce the concept of a "norm" of a vector. A norm is a special type of convex function that measures the length of a vector x.

The *norm* of a vector $y \in \mathbf{R}^m$, denoted by $\|y\|$, is any function from $\mathbf{R}^m$ into $\mathbf{R}$ with the following properties:

1. $\|y\| \geq 0$, and $\|y\| = 0$ if and only if $y = 0$;

2. $\|\alpha y\| = |\alpha|\|y\|$ for $\alpha \in \mathbf{R}$;

3. $\|y + z\| \leq \|y\| + \|z\|$ for all y and z (triangle inequality).

The classical norms in $\mathbf{R}^m$ are differentiated by a subscript p in the range $[1, \infty]$:

$$\|y\|_p := \left(|y_1|^p + \cdots + |y_m|^p\right)^{\frac{1}{p}}, \qquad 1 \leq p < \infty,$$
$$\|y\|_\infty := \max_{1 \leq i \leq m} |y_i|.$$

The most widely used values are $p = 1, 2$, and ∞. For $p = 1$ and $p = 2$ the definition above specializes to the following:

$$\|y\|_1 = \sum_{i=1}^{m} |y_i|,$$

$$\|y\|_2 = \sqrt{\sum_{i=1}^{m} |y_i|^2} = \sqrt{y'y}.$$

For example, if $y = (1, 2, -3)'$, we have

$$\|y\|_1 = 6, \qquad \|y\|_2 = \sqrt{14}, \qquad \|y\|_\infty = 3.$$

The triangle inequality for $\|y\|_p$ is known as the Minkowski inequality, and its proof can be found, for example, in Rudin (1974).

Note that all norms (in $\mathbf{R}^m$) are *equivalent* in the sense that

$$\alpha_{pq}\|y\|_q \leq \|y\|_p \leq \beta_{pq}\|y\|_q,$$

where α_{pq} and β_{pq} are positive numbers.
In particular, for $y \in \mathbf{R}^m$, we have

$$\|y\|_\infty \leq \|y\|_2 \leq \|y\|_1 \leq \sqrt{m}\|y\|_2 \leq m\|y\|_\infty.$$

Theorem A.6.1. *The norm function $f(y) := \|y\|$ is a convex function on $\mathbf{R}^n$.*

Proof. Using the triangle inequality, we have that for $0 \leq \lambda \leq 1$

$$\begin{aligned}
f((1-\lambda)x + \lambda y) &= \|(1-\lambda)x + \lambda y\| \\
&\leq \|(1-\lambda)x\| + \|\lambda y\| \\
&= (1-\lambda)\|x\| + \lambda\|y\| = (1-\lambda)f(x) + \lambda f(y).
\end{aligned}$$

Hence, by Theorem A.4.1, f is convex. $\square$

Since each norm is a convex function, the unit ball defined by $\{x \mid \|x\| \leq 1\}$ defines a convex set, because of Corollary A.4.2.

For a given norm $\|\cdot\|$ the dual norm $\|\cdot\|'$ is defined by

$$\|x\|' := \sup_{y:\|y\|\leq 1} y'x.$$

The Euclidean norm $\|\cdot\|_2$ is self-dual, while $\|\cdot\|'_\infty = \|\cdot\|_1$ and $\|\cdot\|'_1 = \|\cdot\|_\infty$.

In analyzing nonlinear problems, we frequently need to work with rough estimates of certain quantities, rather than precise measures. *Order notation* is particularly useful in this context. We define a certain usage of this notation here.

Suppose that we have two quantities η and ν such that η is a function of ν—we write $\eta(\nu)$ to emphasize the dependence. (η and ν can be scalars, vectors, or matrices.) We write

$$\eta = O(\|\nu\|)$$

if $\eta(v) \to 0$ whenever $v \to 0$. (Verbally, we say that "η is of the order of v" or "η is of the order of the norm of v.") This property indicates that η goes to zero at least as fast as v.

We write

$$\eta = o(\|v\|)$$

if the ratio $\|\eta(v)\| / \|v\| \to 0$ whenever $v \to 0$. (Verbally, and somewhat colloquially, we say that "η is little-o of v.") This is a stronger property, indicating that η is approaching zero at a faster rate than v.

A.7 Taylor's Theorem

Taylor's theorem is an important result from calculus that is used to form simplified approximations to nonlinear functions in the vicinity of a point at which information about the function and its derivatives is known. We state here a variant that applies to functions that map $\mathbf{R}^N$ to $\mathbf{R}^N$ of the type discussed in Chapter 8.

Theorem A.7.1. *Suppose that $F : \mathbf{R}^N \to \mathbf{R}^N$ is continuously differentiable in some convex open set $\mathcal{D}$ and that z and $z + p$ are vectors in $\mathcal{D}$. Then*

$$F(z + p) = F(z) + \int_0^1 J(z + tp)p\, dt,$$

where J is the $N \times N$ Jacobian matrix whose (i, j) element is $\partial F_i / \partial z_j$.

It follows immediately from this result that

$$F(z + p) = F(z) + J(z)p + \int_0^1 [J(z + tp) - J(z)]p\, dt. \tag{A.8}$$

Because J is a continuous function, we can use the order notation introduced in Section A.6 to write

$$[J(z + tp) - J(z)]p = o(\|p\|).$$

Hence, we can estimate the integral term in (A.8) as follows:

$$\left| \int_0^1 [J(z + tp) - J(z)]p\, dt \right| \le \int_0^1 \|[J(z + tp) - J(z)]p\|\, dt = o(\|p\|). \tag{A.9}$$

When J is not close to being singular, the term $J(z)p$ in (A.8) dominates the final $o(\|p\|)$ term, when p is small, and so we can write

$$F(z + p) \approx F(z) + J(z)p.$$

This observation, which we note also in (8.14), is used to design the interior-point methods of Chapter 8.

Another useful corollary of Theorem A.7.1 occurs when we apply it to a function of the form

$$F(\lambda) := f(u + \lambda v),$$

where $f : \mathbf{R}^n \to \mathbf{R}$ is a continuously differentiable function of n variables, $u \in \mathbf{R}^n$ and $v \in \mathbf{R}^n$ are two given vectors, and λ is a scalar variable. Thus $F : \mathbf{R} \to \mathbf{R}$ is a continuously differentiable scalar function, and so we can apply Theorem A.7.1 with $N = 1$ to find an estimate of $F(\lambda)$ for λ near 0. From the chain rule of multidimensional calculus, we have

$$F'(0) = \nabla f(u)'v,$$

and so by applying Theorem A.7.1 with $z = 0$, $p = \lambda$ and using the estimate (A.9), we have

$$F(\lambda) = F(0) + \lambda F'(0) + o(|\lambda|) \quad \Leftrightarrow \quad f(u + \lambda v) = f(u) + \lambda \nabla f(u)'v + o(|\lambda|). \quad \text{(A.10)}$$

(Note that $\|\lambda\| = |\lambda|$ because λ is a scalar.)

We can use (A.10) to generalize a result proved earlier for convex quadratic functions— Proposition A.5.1(b)—to nonlinear convex functions.

Proposition A.7.2. *Let $f : \mathbf{R}^n \to \mathbf{R}$ be a continuously differentiable convex function. Then the linearization of f at any point x never overestimates f, that is,*

$$f(y) \geq f(x) + \nabla f(x)'(y - x) \quad \forall y. \quad \text{(A.11)}$$

Proof. Suppose for contradiction that there are vectors x and y for which the claim does not hold, that is,

$$f(y) < f(x) + \nabla f(x)'(y - x) - \epsilon, \quad \text{(A.12)}$$

where ϵ is a positive constant. By the convexity of f, we have from (A.4) that for $0 < \lambda < 1$

$$f((1 - \lambda)x + \lambda y) \leq (1 - \lambda)f(x) + \lambda f(y)$$
$$\Leftrightarrow \quad f(x + \lambda(y - x)) \leq f(x) + \lambda(f(y) - f(x)).$$

By rearranging this expression, we have that

$$f(y) \geq f(x) + \frac{f(x + \lambda(y - x)) - f(x)}{\lambda}.$$

We now apply (A.10) to the numerator of the last term, with $u = x$ and $v = (y - x)$, to obtain

$$f(y) \geq f(x) + \frac{\lambda \nabla f(x)'(y - x) + o(|\lambda|)}{\lambda} = f(x) + \nabla f(x)'(y - x) + o(|\lambda|)/\lambda.$$

By combining this inequality with (A.12), we obtain

$$0 < \epsilon \leq o(|\lambda|)/\lambda,$$

and taking limits as $\lambda \to 0$, we have by the definition of $o(\cdot)$ that $o(|\lambda|)/\lambda \to 0$ and therefore $\epsilon = 0$, a contradiction. $\square$

Appendix B

Summary of Available MATLAB Commands

B.1   Basic MATLAB Operations

In our MATLAB codes, vectors are usually denoted by lowercase letters. For example, if the following commands are given in MATLAB

```
≫ x = [1; 7; 4];
≫ y = [2 1];
```

then x is interpreted as a three-dimensional column vector and y as a two-dimensional row vector. The ith element of x is denoted by x_i. The MATLAB statement

```
≫ i=3; x(i)
```

prints out x_3, which in this case has the value 4.

The set of all n-dimensional real vectors is denoted by $\mathbf{R}^n$. For example, $\mathbf{R}$ or $\mathbf{R}^1$ is the real line and $\mathbf{R}^2$ is the real two-dimensional space. $\mathbf{R}^n_+$ represents the nonnegative orthant in $\mathbf{R}^n$, that is, the set of all the n-dimensional real vectors that satisfy $x_i \geq 0$, $i = 1, 2, \ldots, n$, also written as $x \geq 0$.

Matrices in MATLAB are usually denoted as uppercase letters. The following are examples of matrices specified in MATLAB of dimensions 2×3, 3×2, 1×1, and 1×2, respectively:

```
≫ A = [4 1 2; 1 4 1]
≫ B = [1 3; 1 0; 5 4]
≫ C = [4]
≫ D = [2 1]
```

Recall that the ith row of the matrix A can be represented in mathematical notation by $A_{i\cdot}$. The MATLAB notation is similar but uses ":" in place of "·". Thus the following MATLAB statements print out $A_{2\cdot}$, $A_{\cdot 1}$, and A_{21}, respectively:

```
≫ i=2; A(i,:)
```

```
>> j=1; A(:,j)
>> A(i,j)
```

As for the mathematical notation introduced in Section A.1, the MATLAB notation for transpose of a matrix A is the prime " $'$ "; that is,

```
>> A'
```

Addition of matrices and vectors whose dimensions are the same, and multiplication of matrices and vectors by scalars, are represented in MATLAB in the obvious way:

```
>> x = [1 2 3]; y = [2 -3 4]; x+y
>> alpha = 7; alpha*x
```

Matrix-matrix multiplication in MATLAB is also denoted by a "$*$"; for example,

```
>> A*B
```

MATLAB will generate an error message if A and B do not conform, that is, if the number of columns of A does not equal the number of rows in B. When A has a single row and B has a single column the result of the MATLAB operation $A * B$ is a scalar; we call this result the *scalar product* of the two arguments.

B.2   MATLAB Functions Defined in This Book

We list below the output of the standard `help` command of MATLAB, applied to some of the MATLAB functions defined in this book. Note the inclusion of the semicolon at the end of some of the commands to avoid undesirably verbose output.

```
>> help addcol

  syntax: H = addcol(H,x,lbl,s);
  add column x with label lbl as column s to the tableau H

>> help addrow

  syntax: H = addrow(H,x,lbl,r);
  add row x with label lbl as row r of the tableau H

>> help bjx

  syntax: B = bjx(A,R,S)
  input: tableau A, integer vectors R,S
  perform a block Jordan exchange with pivot A(R,S)

>> help delcol

  syntax: H = delcol(H,s);
```

```
   delete col numbered s (or labeled s) of the tableau H
```

```
>> help delrow
```

```
   syntax: H = delrow(H,r);
   delete row numbered r (or labeled r) of the tableau H
```

```
>> help dualbl
```

```
   syntax: H = dualbl(H);
   adds dual row and column labels in last two rows and
   columns
```

```
>> help jx
```

```
   syntax: B = jx(A,r,s)
   input: matrix A, integers r,s
   perform a Jordan exchange with pivot A(r,s)
```

```
>> help lemketbl
```

```
   syntax: H = lemketbl(M,q);
   generate labeled Lemke tableau H from square matrix M and
   vector q
```

```
>> help ljx
```

```
   syntax: B = ljx(A,r,s);
   input: labeled tableau A, integers r,s
   perform a labeled Jordan exchange with pivot A(r,s)
```

```
>> help newton
```

```
 syntax: z = newton(z0, @myF, @myJ, eps, itmax)
```

```
   performs Newton's method from starting point z0, terminating
   when 2-norm of step is shorter than eps or when at itmax steps
   have been taken, whichever comes first. Call as follows:
```

```
   where z0 is the starting point, myF and myJ are the actual
   names of the function and Jacobian evaluation routines;
   method terminates when length of Newton step drops below eps
   or after at most itmax iterations (whichever comes first).
```

```
>> help pathfollow
```

```
syntax: [x,y,s,f] = pdip(A,b,p)

path-following primal-dual interior-point method for
problem

PRIMAL: min p'x s.t. Ax=b, x>=0,
DUAL:   max b'y  s.t. A'y+s=p,   s>=0.

 input: A is an m x n SPARSE constraint matrix
        b is an m x 1 right-hand side vector
        p is an n x 1 cost vector

output: x is the  n x 1 solution of the primal problem
        y is the m x 1 dual solution
        s is the n x 1 vector of "dual slacks"
        f is the optimal objective value

internal parameters:
        itmax is the maximum number of iterations allowed
        tol is the convergence tolerance
        bigMfac is the factor used to define the starting
        point
        maxDiag is the element for the X^{-1}S matrix
        etaMin is the minimum value of the steplength scale
        parameter eta

>> help permrows

 syntax: H= permrows(H,p);
 permute rows of H according to the permutation in p

>> help relabel

 syntax: H = relabel(H,x,old1,new1,old2,new2,...);
 update label old to be label new

>> help rsm

 syntax: [x_B,B] = rsm(A,b,p,B)
 A revised simplex routine for min p'x s.t. Ax=b, x>=0.
 on input A is m x n, b is m x 1, p is n x 1
 B is 1xm index vector denoting the basic columns.

>> help rsmbdd
```

```
syntax: [x,B,N] = rsmbdd(A,b,p,lb,ub,B,N)
A revised simplex routine for min p'x s.t. Ax=b, lb<=x<=ub.
B is 1 x m index vector denoting the basic columns.
N is 1 x (1-m) index vector denoting the nonbasic columns
and bounds (+/-1).
```

```
>> help rsmupd
```

```
syntax: [x_B,B] = rsmupd(A,b,p,B)
A revised simplex routine for min p'x s.t. Ax=b, x>=0.
on input A is m x n, b is m x 1, p is n x 1
B is 1 x m index vector denoting the basic columns.
```

```
>> help simplex
```

```
syntax: x = simplex(A,b,p)
A two-phase simplex routine for min p'x s.t. Ax>=b, x>=0.
on input A is m x n, b is m x 1, p is n x 1
```

```
>> help steplength
```

```
syntax: [alpha, alphax, alphas] =
steplength(x, s, Dx, Ds, eta)

given current iterate (x,s) and steps (Dx,Ds), compute
steplengths that ensure that x + alphax*Dx>0 and
s + alphas*Ds>0, and alpha = min(alphax,alphas). eta
indicates the maximum fraction of step to the boundary
(typical value: eta=.999)
```

```
>> help tbl
```

```
syntax: tbl(H)
print out the tableau given in H
```

```
>> help totbl
```

```
syntax: H = totbl(A,b,p);
make the matrices A,b,p into tableau H
```

Bibliography

Ahuja, R. K., Magnanti, T. L. & Orlin, J. B. (1993), *Network Flows: Theory, Algorithms, and Applications*, Prentice–Hall, Englewood Cliffs, New Jersey.

Bartels, R. H. & Golub, G. H. (1969), 'The simplex method of linear programming using LU decomposition', *Communications of the Association for Computing Machinery* **12**, 266–268.

Beale, E. M. L. (1955), 'Cycling in the dual simplex algorithm', *Naval Research Logistics Quarterly* **2**, 269–275.

Bland, R. G. (1977), 'New finite pivoting rules for the simplex method', *Mathematics of Operations Research* **2**, 103–107.

Bosch, R. A. & Smith, J. A. (1998), 'Separating hyperplanes and the authorship of the disputed federalist papers', *American Mathematical Monthly* **105**, 601–608.

Burges, C. J. C. (1998), 'A tutorial on support vector machines for pattern recognition', *Data Mining and Knowledge Discovery* **2**, 121–167.

Chvátal, V. (1983), *Linear Programming*, W. H. Freeman and Company, New York.

Cottle, R. W. & Dantzig, G. B. (1968), 'Complementary pivot theory of mathematical programming', *Linear Algebra and Its Applications* **1**, 103–125.

Cottle, R. W., Pang, J. S. & Stone, R. E. (1992), *The Linear Complementarity Problem*, Academic Press, Boston.

Czyzyk, J., Mehrotra, S., Wagner, M. & Wright, S. J. (1999), 'PCx: An interior-point code for linear programming', *Optimization Methods and Software* **11/12**, 397–430.

Dantzig, G. B. (1963), *Linear Programming and Extensions*, Princeton University Press, Princeton, New Jersey.

Fletcher, R. & Matthews, S. P. J. (1984), 'Stable modifications of explicit *LU* factors for simplex updates', *Mathematical Programming* **30**, 267–284.

Forrest, J. J. & Goldfarb, D. (1992), 'Steepest-edge simplex algorithms for linear programming', *Mathematical Programming* **57**, 341–374.

257

Forrest, J. J. H. & Tomlin, J. A. (1972), 'Updating triangular factors of the basis to maintain sparsity in the product form simplex method', *Mathematical Programming* **2**, 263–278.

Frank, M. & Wolfe, P. (1956), 'An algorithm for quadratic programming', *Naval Research Logistics Quarterly* **3**, 95–110.

Gass, S. (1985), *Linear Programming Methods and Applications*, fifth ed., Boyd and Fraser, Danvers, Massachusetts.

Gay, D. M. (1978), On combining the schemes of Reid and Saunders for sparse LP bases, *in* I. S. Duff & G. W. Stewart, eds., 'Sparse Matrix Proceedings 1978', SIAM, Philadelphia, pp. 313–334.

Gertz, E. M. & Wright, S. J. (2003), 'Object-oriented software for quadratic programming', *ACM Transactions on Mathematical Software* **29**, 58–81.

Gill, P. E., Murray, W., Saunders, M. A. & Wright, M. H. (1989), 'A practical anti-cycling procedure for linearly constrained optimization', *Mathematical Programming* **45**, 437–474.

Goldfarb, D. & Reid, J. K. (1977), 'A practical steepest-edge simplex algorithm', *Mathematical Programming* **12**, 361–371.

Golub, G. H. & Van Loan, C. F. (1996), *Matrix Computations*, third ed., The Johns Hopkins University Press, Baltimore.

Harris, P. M. J. (1973), 'Pivot selection methods of the Devex LP code', *Mathematical Programming* **5**, 1–28.

Ho, T. K. & Kleinberg, E. M. (1996), Building projectable classifiers of arbitrary complexity, *in* 'Proceedings of the 13th International Conference on Pattern Recognition', Vienna, Austria, pp. 880–885, http://cm.bell-labs.com/who/tkh/pubs.html. Checker dataset at ftp://ftp.cs.wisc.edu/math-prog/cpo-dataset/machine-learn/checker.

Huber, P. J. (1981), *Robust Statistics*, John Wiley & Sons, New York.

Kantorovich, L. V. (1960), 'Mathematical methods in the organization and planning of production', *Management Science* **6**, 366–422. English translation, Russian version 1939.

Karmarkar, N. (1984), 'A new polynomial time algorithm for linear programming', *Combinatorica* **4**, 373–395.

Karush, W. (1939), Minima of functions of several variables with inequalities as side conditions, Master's thesis, Department of Mathematics, University of Chicago.

Khachiyan, L. G. (1979), 'A polynomial algorithm for linear programming', *Soviet Mathematics Doklady* **20**, 191–194.

Klee, V. & Minty, G. J. (1972), How good is the simplex algorithm?, *in* 'Inequalities, III', Academic Press, New York, pp. 159–175.

Kotiah, T. C. T. & Steinberg, D. I. (1977), 'Occurrences of cycling and other phenomena arising in a class of linear programming models', *Communications of the ACM* **20**, 107–112.

Kotiah, T. C. T. & Steinberg, D. I. (1978), 'On the possibility of cycling with the simplex method', *Operations Research* **26**, 374–376.

Kuhn, H. W. & Tucker, A. W. (1951), Nonlinear programming, *in* J. Neyman, ed., 'Proceedings of the Second Berkeley Symposium on Mathematical Statistics and Probability', University of California Press, Berkeley and Los Angeles, California, pp. 481–492.

Lee, Y.-J. & Mangasarian, O. L. (2001), RSVM: Reduced support vector machines, *in* 'Proceedings of the First SIAM International Conference on Data Mining', Chicago, 2001, CD-ROM, ftp://ftp.cs.wisc.edu/pub/dmi/tech-reports/00-07.pdf.

Lemke, C. E. (1965), 'Bimatrix equilibrium points and mathematical programming', *Management Science* **11**, 681–689.

Lemke, C. E. & Howson, J. T. (1964), 'Equilibrium points of bimatrix games', *SIAM Journal on Applied Mathematics* **12**, 413–423.

Mangasarian, O. L. (1969), *Nonlinear Programming*, McGraw–Hill, New York. (Corrected reprint: SIAM Classics in Applied Mathematics 10, SIAM, Philadelphia, 1994).

Mangasarian, O. L. (1999*a*), 'Arbitrary-norm separating plane', *Operations Research Letters* **24**, 15–23.

Mangasarian, O. L. (1999*b*), Regularized linear programs with equilibrium constraints, *in* M. Fukushima & L. Qi, eds., 'Reformulation: Nonsmooth, Piecewise Smooth, Semismooth and Smoothing Methods', Kluwer Academic Publishers, Dordrecht, The Netherlands, pp. 259–268.

Mangasarian, O. L. (2000), Generalized support vector machines, *in* A. Smola, P. Bartlett, B. Schölkopf & D. Schuurmans, eds., 'Advances in Large Margin Classifiers', MIT Press, Cambridge, Massachusetts, pp. 135–146.

Mangasarian, O. L. & Musicant, D. R. (2000), 'Robust linear and support vector regression', *IEEE Transactions on Pattern Analysis and Machine Intelligence* **22**, 950–955.

Mangasarian, O. L., Street, W. N. & Wolberg, W. H. (1995), 'Breast cancer diagnosis and prognosis via linear programming', *Operations Research* **43**, 570–577.

Megiddo, N. (1989), Pathways to the optimal set in linear programming, *in* N. Megiddo, ed., 'Progress in Mathematical Programming: Interior-Point and Related Methods', Springer-Verlag, New York, Chapter 8, pp. 131–158.

Mehrotra, S. (1992), 'On the implementation of a primal-dual interior point method', *SIAM Journal on Optimization* **2**, 575–601.

Murty, K. G. (1976), *Linear and Combinatorial Programming*, John Wiley & Sons, New York.

Nazareth, J. L. (1987), *Computer Solution of Linear Programs*, Oxford University Press, Oxford, UK.

Nemhauser, G. L. & Wolsey, L. A. (1988), *Integer and Combinatorial Optimization*, John Wiley & Sons, New York.

Nocedal, J. & Wright, S. J. (2006), *Numerical Optimization*, second ed., Springer-Verlag, New York.

Reid, J. K. (1982), 'A sparsity-exploiting variant of the Bartels-Golub decomposition for linear programming bases', *Mathematical Programming* **24**, 55–69.

Rockafellar, R. T. (1970), *Convex Analysis*, Princeton University Press, Princeton, New Jersey.

Rudin, W. (1974), *Real and Complex Analysis*, second ed., McGraw–Hill, Tokyo, Japan.

Schölkopf, B. & Smola, A. (2002), *Learning with Kernels*, MIT Press, Cambridge, Massachusetts.

Schrijver, A. (1986), *Theory of Linear and Integer Programming*, John Wiley & Sons, New York.

Sigmon, K. & Davis, T. A. (2004), *MATLAB Primer*, seventh ed., Chapman and Hall/CRC, Boca Raton, Florida.

Strang, G. (1993), *Introduction to Linear Algebra*, Wellesley–Cambridge Press, Wellesley, Massachusetts.

Vanderbei, R. J. (1997), *Linear Programming: Foundations and Extensions*, Kluwer Academic Publishers, Boston.

Vapnik, V. N. (1995), *The Nature of Statistical Learning Theory*, Springer-Verlag, New York.

Winston, W. L. & Venkataramanan, M. (2003), *Introduction to Mathematical Programming*, Vol. 4, Brooks/Cole–Thomson Learning, Pacific Grove, California.

Wolsey, L. A. (1998), *Integer Programming*, Series in Discrete Mathematics and Optimization, John Wiley & Sons, New York.

Wright, S. J. (1997), *Primal-Dual Interior-Point Methods*, SIAM, Philadelphia.

Index